Human Herpesvirus-6, Second edition

General Virology, Epidemiology and Clinical Pathology

PERSPECTIVES IN MEDICAL VIROLOGY

Volume 12

Series Editors

A.J. Zuckerman

Royal Free and University College Medical School
University College London
London, UK

I.K. Mushahwar

Abbott Laboratories
Viral Discovery Group
Abbott Park, IL, USA

Human Herpesvirus-6, Second edition
General Virology, Epidemiology and Clinical Pathology

Editors

Gerhard Krueger

The University of Texas Medical School at Houston
Department of Pathology
Laboratory Medicine
Houston, TX, USA

Dharam Ablashi

HHV-6 Foundation
Santa Barbara, CA, USA

ELSEVIER

Amsterdam – Boston – Heidelberg – London – New York – Oxford – Paris
San Diego – San Francisco – Singapore – Sydney – Tokyo

Elsevier
Radarweg 29, PO Box 211, 1000 AE Amsterdam, The Netherlands
The Boulevard, Langford Lane, Kidlington, Oxford OX5 1GB, UK

First edition 2006

Library of Congress Cataloguing-in-Publication Data
A catalog record for this book is available from the Library of Congress

British Library Cataloguing in Publication Data
A catalogue record for this book is available from the British Library

ISBN-13: 978-0-444-52063-0
ISBN-10: 0-444-52063-5
ISSN: 0168-7069

Printed and bound in Italy

06 07 08 09 10 10 9 8 7 6 5 4 3 2 1

Dedication to Kristin S. Loomis, Executive Director, HHV-6 Foundation

We want to pay tribute to Kristin for her hard and never-ending work on promoting HHV-6 research, educating practicing physicians and disseminating information about HHV-6 to patients and their family members.

In April 2004, Kristin and Annette Whittemore co-founded the HHV-6 Foundation, and Kristin became the Executive Director. Kristin's motivation, imaginative nature, hard labor and devotion to promote understanding of this pathogen and to look for a cure to help many patients is as strong now as when she began this journey. Her knowledge of HHV-6 and her drive to make things happen have impressed and intrigued many experts on HHV-6 and persuaded them to join in her efforts. She is confident that we will one day conquer HHV-6 infection and alleviate the suffering.

Kristin became interested in HHV-6 when her eldest child was diagnosed with HHV-6A CNS infection, with cognitive disorders. Although she had no medical background (Wellesley BA in economics and Harvard MBA) Kristin then taught herself about HHV-6 by reading scientific articles and by speaking with scientists and clinicians.

We, the scientists, are very grateful to Kristin for her encouragement, support and for making so many resources available through the HHV-6 Foundation. We sincerely hope that her dreams will become a reality and that would be the greatest reward of her life.

Dharam Ablashi
Santa Barbara, CA, USA

Contents

PART III: CLINICAL PATHOLOGY

Foreword

In some ways human herpesvirus-6 (HHV-6) is a real oddity. The way it was discovered and the extraordinary opposed feelings about its importance, ranging from belief in its role in a host of diseases (real ones and perhaps imaginary ones as well) to its total disregard are two good examples. I would like to comment on both.

In the middle of the 1980s, early 1985 to be precise, our group at National Cancer Institute (NCI) was rather overwhelmed by the HIV and still facing new concepts with HTLV-1 and HTLV-2. Chiefly, however, we focused on HIV pathogenesis, and as a part of these studies we decided to include studies of the mechanisms involved in HIV's role in carcinogenesis, i.e. its role as a powerful co-carcinogen. Because Kaposi sarcoma and B-cell lymphomas were far, far the most common neoplasias associated with HIV infection and virtually no group at NCI was studying them, I turned my attention to a study of their pathogenesis. It was the work with B-cell lymphomas that led to the discovery of HHV-6. In my thinking B-cell lymphomas were mainly caused by accidents in the Ig gene rearrangements that normally take place in B-cells, and this rare event would be more probable if there were a chronic antigenic drive such as malaria in Burkitt's lymphoma and presumably HIV antigens in HIV-infected persons. Also, those B-cells infected by Epstein–Barr virus (EBV) would more likely be selected because such cells can be immortalized by this herpesvirus. Further, I knew that only approximately one-third of all B-cell lymphomas associated with HIV infection were EBV positive. Consequently, I speculated that it was likely that another herpesvirus would be involved which, like EBV, could immortalize B-cells, and if we looked for them we would discover them. As is often the case, the idea was productive but its details were wrong. Soon patients with B-cell lymphoma became available. We obtained peripheral blood mononuclear cells (PBMCs) and splenic tissue, obtained preliminary data for a herpesvirus and soon isolated HHV-6 from it. We called the virus HBLV for human B-lymphotropic virus. Who were "we"?—A technical assistant, Mr. Z. Salahuddin, who was with me culturing hematopoietic and related cells since the 1970s and the newly acquired Dr. Dharam Ablashi. Fortunately, I had Dharam received into my group from Dr. Stuart Aaronson's department. Suffice it is to say Dharam was pivotal to all of this early work, but soon others in my laboratory joined in an effort to better define its properties and its prevalence. One such individual was my post-doctoral fellow, Paolo Lusso. He was important to our studies showing that this new herpesvirus was pan-T tropic (mainly mature T-cells) and selectively neurotropic. We renamed

HBLV as HHV-6, and certainly no one could debate the choice of this safe name. (See his chapter in this book for an update on this topic.)

At this point (1986–1987), my interest in HHV-6 became two-fold: (1) Did it have any role in cancer, i.e., in any B-cell lymphomas? We found no evidence for this. (2) Was it a co-factor in AIDS progression? This was a suspicion we harbored because HHV-6 killed many infected T-cells when actively replicating, and presumably an HIV infection would ultimately lead to the loss of control of HHV-6. I do not believe we ever proved its co-factor role in AIDS progression, but this point remains of interest and will surely be discussed in some of the chapters in this book. Fascinating and tantalizing results have also been reported of a linkage of HHV-6 to multiple sclerosis, especially by S. Jacobson and his colleagues, which are also reviewed here. I believe this will remain tantalizing but not demonstrated until a day arrives when we have a safe and specific drug that inhibits HHV-6, and is clinically shown to improve (or not) the waxing stages of MS.

The second oddity, namely, the passionate feelings pro and con of HHV-6's importance in human disease is more difficult to understand. Once in an introduction on me, Tony Epstein (of EBV) noted that HHV-6 was the first new human herpesvirus found in over 25 years. This drought may have led herpesvirologists and others moving into the field of HHV-6 to be more than average in their enthusiasm and passion for their work. More often than we would wish, this "passion" has spilled over into the popular and the not so popular press and has hurt the whole field. As a result, this book, composed of articles written by experts many of who have brought this subject into modern science, fulfills a real need. In addition to those topics I have already mentioned, the articles in this book are rather all encompassing and extend from classification (by Ablashi, a pioneer of HHV-6), HIV structure and components (see chapters by Zhou/Krueger, Gompels and Mori) to the more applied—diagnosis and epidemiology, serology, organ pathology (a host of authors), animal models (Genain) and theory (Brewer and DeClercq/Naesens). Without doubt, this book covers our current knowledge on this virus, lets the reader know how much we do not know, and how real is what we believe we know.

Robert C. Gallo
Baltimore, MD, USA
August 2005.

Foreword

It has been almost 20 years since human herpesvirus-6 (HHV-6) was isolated. The last comprehensive book concerning HHV-6 was published in 1992. Since its publication, there has been a rapid and continual increase in our understanding of the virology, as well as, the pathophysiology of HHV-6 infection. The genome of the virus was sequenced and two major strains of HHV-6 were identified. Extensive research has been performed in determining the regulation of HHV-6 gene expression and many of the proteins encoded by the virus have been characterized. While this suggests that the molecular characterization of HHV-6 and the proteins that it encodes is nearing completion, there is still much that we have to elucidate concerning how these proteins participate/contribute to the pathology and pathogenesis of HHV-6-associated disease.

HHV-6 has been shown to be a significant pathogen related to diseases in young children contributing to risks for encephalitis, severe convulsions and fever. There are also several reports that suggest an association of HHV-6 with multiple sclerosis. There is evidence that the virus can modulate the physiology of secondary lymphoid organs by direct infection of T-lymphocytes resulting in immune dysregulation. The broad range of illnesses associated with HHV-6 even include a role in the risk for basal cell carcinomas. Further studies are needed to understand these relationships. Additional studies are also needed to understand how HHV-6 interacts with other human herpesviruses in infected cells, as well as, with human immunodeficiency virus. This book will not only be an important reference source for HHV-6 studies in years to come, but it should stimulate increased interest in this unusual human herpesvirus.

Ronald Glaser
Marshall V. Williams

Preface

Human-herpesvirus-6 (HHV-6) was discovered over 20 years ago. The first comprehensive account of HHV-6 in book form was published in 1992. Since then nearly 1000 scientific articles have appeared in various journals, leading to a greater understanding of this pathogen. It has become apparent from many studies that HHV-6 is a serious pathogen in certain populations and under certain conditions. There is also growing evidence that low-grade chronic infections can cause CNS pathology and trigger autoimmune disease. HHV-6 is rarely considered in the clinical differential diagnoses. Clinicians typically do not search for HHV-6, and if they do, they will find only few laboratories providing the serum PCR tests that can differentiate between active and latent viruses. The PCR tests are reasonably sensitive for picking up acute reactivation and roseola, but cannot detect low-grade chronic disease. The test available at most hospitals can only indicate that an infection has happened in the past, which given the high occurrence of the virus, is an ineffective test, other than for the pediatrician in identifying primary roseola. Most commercial PCR DNA tests currently on the market for reactivated HHV-6 are of poor sensitivity and often cannot detect HHV-6 even in patients with HHV-6 encephalitis with seizures. While scientists may disagree about whether serological or molecular assays will eventually result in the best assay, there is no disagreement about the inadequacy of the existing assays.

Even in the age of molecular biology, the clinical diagnosis of a certain disease caused by a virus, remains a clinical diagnosis and is not determined by the sole evidence for the presence of a viral genome. Even the rise of certain antibody titers does not necessarily indicate that this virus is the cause of the disease, and it is difficult to apply Koch's classic criteria to many viral infections, especially to those of high lifetime prevalence like HHV-6. Thus, our knowledge of etiology and pathogenesis of HHV-6-associated diseases can only come from the combined efforts of clinicians, virologists, molecular biologists and pathologists.

It is the purpose of this book to provide an up-to-date and concise overview of what is currently known about HHV-6. We have attempted to cover the entire field of clinical, epidemiological, immunological and molecular biology of HHV-6.

We are grateful to have found the collaboration of leading experts contributing various topics of individual chapters. As their experience naturally varies according to their own activities, it cannot be avoided that certain differences in opinion may be voiced in the various chapters. We have not attempted to interfere with the author's choice of materials to be discussed and with their interpretation. We hope to provide with this book a stimulus for future investigations and—above all—for better diagnosis and treatment of our patients.

The present monograph is organized in four parts: Part I provides a brief overview of general virology; Part II reviews current diagnostic techniques and epidemiologic data; Part III focuses on HHV-6-associated diseases in individual organ systems with special chapters on AIDS, transplantation, chronic fatigue syndrome, multiple sclerosis, other CNS disorders and on treatment; and Part IV finally concludes the book by referring to some new avenues of research. The book should thus be a valuable reference for both clinical and basic scientists including epidemiologists, virologists, pathologists and essentially all scientists entering the field of herpes virus research.

The editors of this volume, who worked together in HHV-6 studies for all of the 20 years since the virus was detected, are especially grateful to Elsevier Science Publishers and their series editors, Professor Arie J. Zuckerman and Isa K. Mushahwar, who always provided us with an excellent opportunity to publicize the knowledge of this virus. It started with the special volume of the *Journal of Virological Methods* (Vol. 21: Diagnostic Techniques of Persistent Active Herpesvirus Infections) and continued with two volumes of *Perspectives in Medical Virology* (Vol.4 and the current issue). It was a great pleasure working with Elsevier's Paul Taylor (previous issues) and Lisa Tickner, Clare Rathbone and Joanna De Souza (current issue). Without their understanding support our work would probably not have been accomplished. Last but not least, we are gratefully mentioning again Drs. van Dommelen of Organon Teknika NV, The Netherlands, and Albert Ramon, associate professor of Cell Biology, University of Antwerp, Belgium, who constructively fostered some early herpesvirus studies and our previous publications.

Gerhard R.F. Krueger
Houston, Texas & Cologne, Germany

Dharam V. Ablashi
Santa Barbara, California & Washington, DC

PART I: GENERAL VIROLOGY

Perspectives in Medical Virology
Gerhard Krueger and Dharam Ablashi (Editors)

DOI 10.1016/S0168-7069(06)12001-7

Discovery and Classification of Human Herpesvirus-6 (HHV-6)

Dharam V. Ablashi[a,b]
[a]*HHV-6 Foundation, 285 San Ysidro Road, Santa Barbara, CA 93108, USA*
[b]*Department of Microbiology & Immunology, Georgetown University School of Medicine, Washington, DC, USA*

Discovery

The discovery of herpesvirus-6 (HHV-6) dates back to early 1985 when Zaki Salahuddin, in Dr. Robert Gallo's Laboratory of Tumor Cell Biology, was establishing long-term cultures from peripheral blood and splenic tissue of AIDS patients. He frequently found large syncytia that were distinct from HIV-1-induced syncytia. What he really saw in the peripheral blood mononuclear cells (PBMCs) of at least 6–8 patients with B-cell lymphoma were large, refractile cells (Fig. 1), always either single or, occasionally, two or more together. These cells began to disappear after a few days in culture, even in the presence of IL-2. The individuals with these cells were all AIDS patients with or without lymphoma. When these cells were stained with Giemsa, they were often multinucleated, or two large nuclei basically covered the entire cell (Fig. 2). The PBMCs of one particular lymphocytic leukemia patient, a 17-year-old boy, received in March 1985 from Dr. Gregory Halligan of Philadelphia, showed these bizarre-looking cells following mitogen stimulation. These PBMCs were sent to Dr. Matthew Gonda at the Frederick Cancer Research Center, Frederick, MD. Herpesvirus-like particles were observed in large numbers (Fig. 3), and a great majority of these particles were extra cellular, with an enveloped virion diameter of 160–200 nm. A repeat sample of PBMCs from this patient was obtained on April 17, 1985, and similar cells reappeared in the culture.

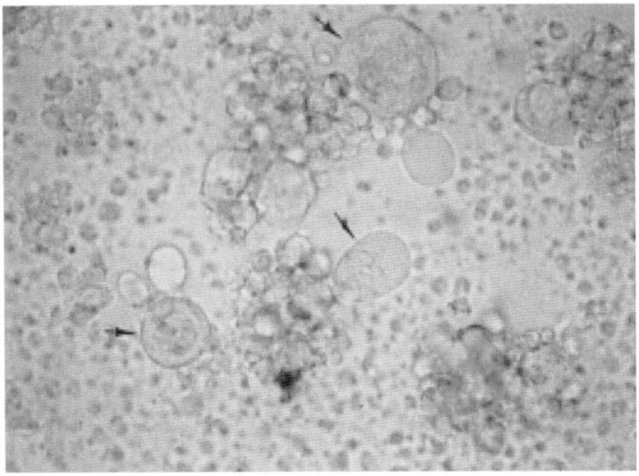

Fig. 1 Peripheral blood mononuclear cells (PHA stimulated), cultured from AIDS patient with B-cell lymphoma, showing large refractile cells. (for colour version: see colour section on page 345).

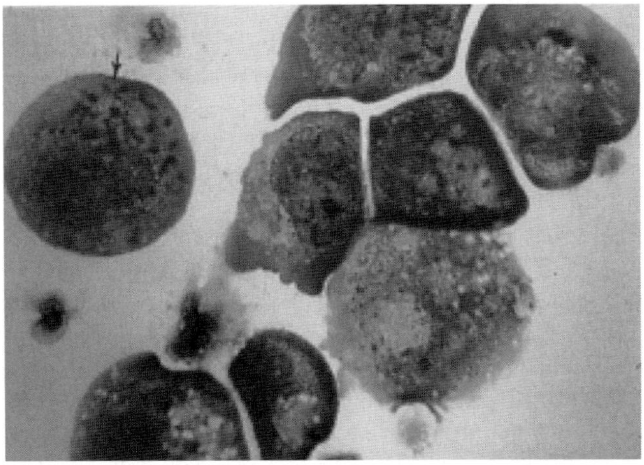

Fig. 2 Giemsa-stained peripheral blood mononuclear cells, containing refractile cells, showing multinucleated giant cells. (for colour version: see colour section on page 345).

After careful analysis for the presence of HIV-1, HTLV-I, and HTLV-II, the only virus particles evident were the herpesvirus-like particles. Since it was a herpesvirus and found in a PBMCs culture, not much was done except storing the virus in a –70 °C freezer, and cells were stored in liquid nitrogen.

Because of my interest in the role of EBV in AIDS B-cell lymphomas, I was invited to join Dr. Gallo's group in June 1985 and I began to look at the infection of EBV with HIV-1 (IIIB) (these data were published). During these investigations, it was evident that B-cells lacking EBV could not be infected with HIV-1. When,

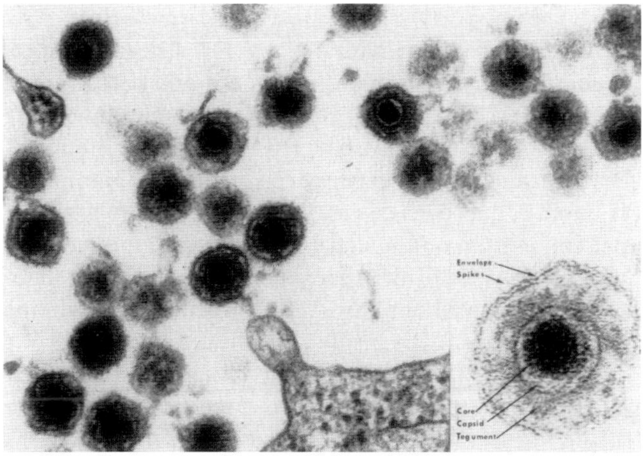

Fig. 3 Herpesvirus particles in various stages of maturation obtained from cultured PBMCs of a patient (GS) co-cultivated with human cordblood mononuclear cells.

however, B-cells were converted to EBV, positivity or cell lines carrying the EBV genome expressing $CD4^+$, receptor could be infected with HIV-1. One day, while this line of research was progressing, Zaki Salahuddin and Dr. Gallo asked me to look at another herpesvirus isolate, which they had frozen. They felt that since it was found in the PBMCs of their AIDS patient with B-cell lymphoma, it might be a variant of EBV that might be immortalizing cells. Zaki and I were able to transmit the cell-free supernatant obtained from cell cultures infected with this virus to fetal cord blood mononuclear cells. It was very clear that PHA must stimulate such cells; otherwise, the infection would not be effective. More than 50% of the inoculated cord blood mononuclear cells showed bizarre-looking, extremely large cells, which appeared between three and seven days, post infection. We called these cells as "juicy cells". We consulted Dr. Bernard Kramarsky (then from Electronucleonics, Inc., of Silver Spring, MD), who had knowledge and experience with the ultrastructure of viruses. He believed that these herpesvirus particles could not be EBV since this virus shows very few extracellular particles. He also noted that the tegument was much more pronounced than that of EBV and similar to CMV (Salahuddin et al., 1986; Biberfeld et al., 1987).

The task assigned to me by Dr. Gallo was to characterize the herpesvirus isolate and to rule out any possibility of contamination by other viruses. I spent about six months studying its biologic and immunologic characteristics. At this time, Dr. Gary Pearson of the Georgetown University School of Medicine was brought into the picture. He provided me with all the EBV and CMV monoclonal antibodies as well as some good suggestions. I tested all the available monoclonal antibodies of human herpesviruses and herpesvirus *Saimiri*, a primate herpesvirus that can grow in human cells. I also used sera from Rhesus monkeys, African green monkeys, chimpanzees, and baboons to see whether there was any cross-reactivity. The results were clearly negative. Dr. Steve Josephs, a molecular virologist in Dr. Gallo's laboratory, trained in retrovirology by Dr. Flossy Wong-Staal, worked with me to analyze the isolate by generating probes and performing hybridization with human and simian herpesviruses. Our first dot-blot hybridization showed slight reactivity to CMV, but not to EBV, HSV, or VZV (Josephs et al., 1986).

Since Dr. Pearson and I were experienced with EBV immunofluorescence, I used the indirect immunofluorescence assay (IFA) to test the patient's sera for IgG antibody. The patient's serum and infected cord blood mononuclear cells provided us with the way to screen sera from other patients (Fig. 4). Dr. Peter Biberfeld of the Karolinska Institute, Stockholm, Sweden, performed the immune electron microscopy using patients' serum. His results, which were published in the *Journal of the National Cancer Institute* (Biberfeld et al., 1987), showed very strong positivity. We also performed the adsorption studies using the patients' plasma and showed the specificity of the serum to this virus after adsorption with other herpesviruses (Buchbinder et al., 1989). We were then convinced that we had a new herpes agent, which we called *GS* isolate. Since we had found this agent in AIDS patients more frequently, we thought it might be the causative agent of AIDS-associated lymphoma. Since our original patients were either AIDS or other lymphoproliferative disorder or lymphoma patients, we decided to call this virus *HBLV*. When we presented our data to Dr. Gallo, he was of the opinion that we should be certain that this was not a contamination by another herpesvirus. He also said that since he was not a herpes virologist, we should seek the opinion of an established herpes virologist. Dr. Gallo, at my suggestion, called Dr. Bernard Roizman of the University of Chicago, to discuss our HBLV data with him. Dr. Roizman suggested that since Dr. Elliot Kieff, his former student and an expert on EBV, was coming to Washington, we should show our data to him, which we did. After looking through

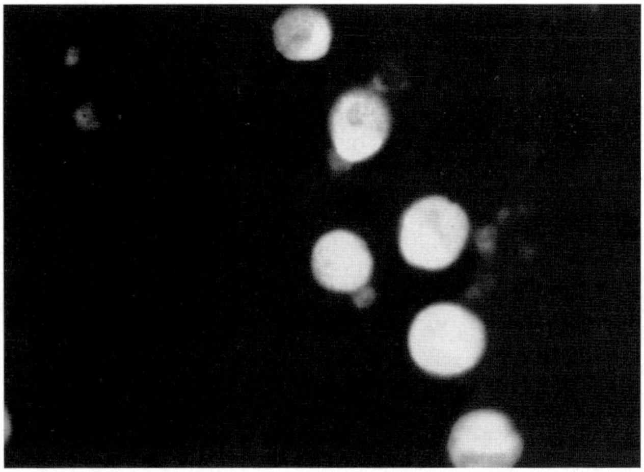

Fig. 4 Immunofluorescent staining of HHV-6-infected human cordblood mononuclear cells with patient GS serum. (for colour version: see colour section on page 346).

our data books, Dr. Kieff asked whether we could give him some viral DNA so that he could check it in his own laboratory. After about two weeks, Dr. Kieff called us to say that we could publish our data on this virus, as a new herpesvirus. Dr. Roizman was also comfortable with these data. The last human herpesvirus reported prior to this was EBV in 1966 by Sir M.A. Epstein. Because of such a long gap in the discovery of a new herpesvirus, most people we talked to were unwilling to believe that we had found a new herpesvirus. In fact, those who reviewed our manuscript for *Science* were very critical and believed that we had found a CMV variant. Before our two papers on HBLV appeared in *Science* (Josephs et al., 1986; Salahuddin et al., 1986), Dr. Debra Barrens from *Science* visited our lab and spent two hours with us looking through the HBLV cultures, IFA slides, and other data. She was very excited when she realized that this was something new and could be associated with AIDS. After the papers on these studies were published in *Science*, people called and wrote to us saying that they had also seen these strange-looking cells in the culture of PBMCs from AIDS patients.

The name HBLV—was it a mistake? I do not say that it was. At present, all lymphomas found with HHV-6 DNA are of B-cell origin (Josephs et al., 1988), with the exception of one disseminated T-cell lymphoma (Jarrett et al., 1988). Later, I was unable to infect B-cells with HHV-6 *in vitro* unless EBV DNA was present. In fact, the laboratories of Dr. José Menezes, University of Montreal, Canada, and Dr. Jonas Luka, Eastern Virginia Medical School in Norfolk, VA, showed that not only are the EBV genome-positive B-cells infectable with HHV-6, but that HHV-6 can also activate EBV antigens such as EA, VCA, and Zebra protein (Flamand et al., 1993). After Dr. Paolo Lusso, then a post-doc in Dr. Gallo's laboratory, characterized the infected cells as T-cells (Lusso et al., 1987, 1988), we were the first group to change the name from HBLV to HHV-6 in a brief report (Ablashi et al., 1987). Later Lusso et al. (1988) conducted more detailed studies of the T-cells infectable with HHV-6. So far, it is evident that HHV-6 has a somewhat wide host range (Ablashi et al., 1988).

To summarize, was the discovery of HHV-6 a lucky chance, or was it keen observation on the part of Zaki Salahuddin and Dharam Ablashi? Once this occurred, Zaki and I, along with Drs. Joseph, Kramarsky, and Lusso, helped to characterize HBLV. We would have never found this virus if our clinical collaborators, Drs. Halligan and Mark Kaplan of the North Shore University Hospital, Long Island, NY, had not provided us with the specimens. Drs. Robert Gallo and Flossy Wong-Staal not only gave us moral support and encouragement, but also made necessary resources available to us for the discovery of HHV-6. We also acknowledge the critical help of Drs. Roizman and Kieff. The rest is history, because we can now look back and say that we made a valuable contribution to the advancement of science.

Nomenclature and classification

HHV-6 is a double-stranded DNA virus belonging to the human herpesviridae family (Braun et al., 1997). Herpesviruses are generally highly disseminated in nature. To date, herpesviruses examined from animals and humans are able to remain latent in their natural host. The cell harboring latent virus genomes take the form of closed circular molecules and only a small subset of viral genes are expressed. Latent genes retain the capacity to replicate and cause disease upon reactivation from the latent state, with HHV-6 not being an exception to this process. This may differ from one virus to another. The International Committee on the Taxonomy of Viruses (ICTV) endorsed nomenclature consists of the designation of herpesviruses by serial Arabic number and the family or sub-family of the natural host of the virus (e.g. HHV-6, HHV-7, etc.). The ICTV classified human herpesviruses into three sub-families, i.e. alpha, beta, and gamma, and eight human herpesviruses, i.e. HSV-1, HSV-2, VZV, CMV, HHV-6, HHV-7, EBV, and HHV-8 were put into these sub-families (Fig. 5) on the basis of their biological properties, before DNA sequences of the individual members of the family were known. ICTV also classified a small number of herpesviruses as to genera, based on DNA sequence homology and similarities in genomic sequence demonstrated by immunologic methods.

HHV-6 is ubiquitous, with 90% seropositivity in adults. HHV-6 variants HHV-6A and HHV-6B are classified as members of the beta-herpesvirus sub-family (Braun et al., 1997; Campadelli-Fiume et al., 1999; Krueger and Ablashi, 2003; DeBolle et al., 2005). The other two members are human cytomegalovirus (HCMV) and HHV-7. A non-exclusive characteristic of the members of beta-herpesvirinae is a restricted host range. The reproductive cycle is long and the infection progresses

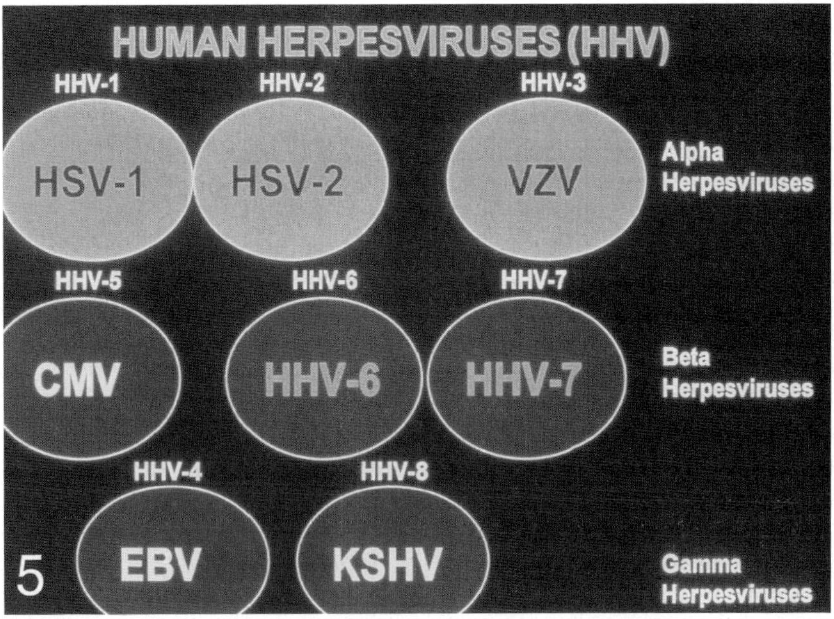

Fig. 5 Sub-families of Human Herpesviruses. (for colour version: see colour section on page 346).

slowly in culture. Infected cells more frequently become enlarged, as seen with HHV-6 (Fig. 1) and carrier cultures are established readily. The virus can be maintained in the latent state in secretory glands, lymphoreticular cells, and other tissues. This sub-family containing the genera cytomegalovirus and Roseolavirus (HHV-6A and HHV-6B, HHV-7) are characterized by growth in T-cells, although they infect other cells. The HHV-6 genome consists of a large (160–170 kb) unique (u) region flanked by sorter (8–9 kb) direct repeats (DR). Some reading frames (ORFs) are found in all herpesviruses, some are in Roseolaviruses only and are unique to HHV-6. The members of the Roseola genus share the standard feature of herpes virion structure; an icosahedral capsid 90–110 nm in diameter, containing 145–170 kb double-stranded DNA genome, a tegument surrounding the capsid and lipid-layer envelope. This is studded with viral-specified integral membrane protein and glycoproteins-enveloped extracellular virion (160–200 nm in diameter). Compared with CMV and HHV-7, HHV-6 teguments are smooth and fill the spaces between the capsid and envelope (Biberfeld et al., 1987; Braun et al., 1997).

The two HHV-6 variants HHV-6A and HHV-6B (Ablashi et al., 1993) are quite distinct molecularly and biologically, although they are closely related and share 90% of nucleotide sequences. This makes them different viruses, since regions of their genome differ by as much as 31%. Cell tropism in HHV-6 is notably lymphotropic and neurotropic, but it infects and replicates in a wide range of human cells, both *in vitro* and *in vivo*, probably due to the ubiquity of the major HHV-6 receptor CD46 (Gompels et al., 1995; Krueger and Ablashi, 2003; Braun et al., 1997; DeBolle et al., 2005). The most efficient replication is in $CD4^+$ T-cells, but it also grows readily in activated PBMCs. HHV-6A and HHV-6B are distinguished by the cultured lymphocyte cell lines in which they replicate. Both HHV-6A and HHV-6B can be detected in the lymphocytes, monocytes/macrophages, PBMCs, salivary glands, and the central nervous system (oligodendrocytes) (Inoue et al., 1994; Braun et al., 1997; Clark, 2000; Krueger and Ablashi, 2003; DeBolle et al., 2005).

The route of transmission of HHV-6A is unclear, but the transmission of HHV-6B is through saliva (Braun et al., 1997; Campadelli-Fiume et al., 1999; Clark, 2000). The HHV-6 genome is transcribed in three distinct phases: (i) immediate early (IE), (ii) early, and (iii) late. The highest degree of sequence divergence is found in IE region and may be target in developing assays that would differentiate variants HHV-6A and HHV-6B infection.

There has been a lot of discussion and argument about what to call the HHV-6 variants. Drs. S. Roizman and Philip Pellet mentioned in the chapter on "The family of herpesviridae: A brief introduction" (Roizman et al., 2001) that thus far, nine human herpesviruses have been isolated from humans, i.e. HSV-1, HSV-2, VZV, HCMV, EBV, HHV-6A, HHV-6B, HHV-7, and HHV-8. The reasoning behind reclassifying or renaming these two HHV-6 variants according to herpesvirus nomenclature is based on the facts that related viruses should be classified as distinct species and that their genomes differ in a readily assayable and distinct manner, or across the entire genome and that the virus can be shown to have distinct epidemiological and biological characteristics. To support epidemiological studies of variants HHV-6A and HHV-6B, assays that can differentiate its antibodies directed to variant HHV-6A or HHV-6B are needed. Most of the immunologic assays to detect antigens or antibodies are crucial to the pathobiology of the virus and epidemiology. The detailed characteristics of HHV-6 properties, replication, reactivation, latency, persistent, and lytic infection are covered in various chapters in this book.

References

Ablashi DV, Agut H, Berneman Z, Campadelli-Fiume G, Carrigan D, Ceccerini-Nelli L, Chandran B, Chou S, Collandre H, Cone R, Dambaugh T, Dewhurst S, DiLuca D, Foa-Tomasi L, Fleckenstein B, Gallo R, Gompels U, Hall C, Jones M, Lawrence G, Martin M, Montagnier L, Neipel F, Nicholas J, Pellett P, Razzaque A, Torrelli G, Thomson B, Salahuddin S, Wyatt L, Yamanishi K. Arch Virol 1993; 129: 363–366.

Ablashi DV, Lusso P, Hung CL, Salahuddin SZ, Josephs SF, Llana T, Kramarsky B, Biberfeld P, Markham PD, Gallo RC. Int J Cancer 1988; 42: 787–791.

Ablashi DV, Salahuddin SZ, Josephs SF, Iman F, Lusso P, Gallo RC, Hung CL, Lemp J, Markham PD. Nature 1987; 329: 207.

Biberfeld P, Kramarsky B, Salahuddin SZ, Gallo RC. J Natl Cancer Inst 1987; 79: 933–942.

Braun DK, Dominguez G, Pellet PE. Clin Microbiol Rev 1997; 10: 521–561.

Buchbinder A, Ablashi DV, Saxinger C, Josephs SF, Salahuddin SZ. Lancet 1989; 1: 217.

Campadelli-Fiume G, Mirandola P, Menott L. Emerg Inf Dis 1999; 5: 353–366.

Clark DA. Rev Med Virol 2000; 10: 155–178.

DeBolle L, Naesens L, DeClercq E. Clin Microbiol Rev 2005; 18: 217–245.

Flamand L, Stefanscu I, Ablashi DV, Menezes J. J Virol 1993; 67: 6768–6777.

Gompels UA, Nicholas J, Lawrence G, Jones M, Thomson BJ, Martin MED, Efstathiou S, Craxton M, Macauley HA. Virology 1995; 209: 29–51.

Inoue N, Dambaugh TR, Pellet PE. Infec Agents Dis 1994; 184: 343–360.

Jarrett RF, Gledhill S, Qureshi F, Crae SH, Madhok I, Brown I, Evans A, Kraiewski A, O'Brien CJ, Cartwright RA, Venables P, Onion DE. Leukemia 1988; 2: 496–502.

Josephs SF, Buchbinder A, Streicher HZ, Ablashi DV, Salahuddin SZ, Guo H, Wong-Staal F, Cossman J, Raffield M, Sundeen J, Levine PH, Biggar R, Krueger GRF, Fox RI, Gallo RC. Leukemia 1988; 2: 132–135.

Josephs SF, Salahuddin SZ, Ablashi DV, Schachler F, Wong-Staal F, Gallo RC. Science 1986; 234: 601–603.

Krueger GRF, Ablashi DV. Intervirol 2003; 46: 257–269.

Lusso P, Markham PD, Schachler F, Veronese FD, Salahuddin SZ, Ablashi DV, Pahwa S, Krohn K, Gallo RC. J Exp Med 1988; 167: 1659–1670.

Lusso P, Salahuddin SZ, Ablashi DV, Gallo RC, Veronese FD, Markham PD. Lancet 1987; 11: 743–744.

Roizman B, Pellett PE. The family of herpesviridae: a brief introduction. In: Fields Virology (Knipe DM, Howley PM, editors). vol. 2. Lippincott, Williams & Wilkins; 2001; p. 2381.

Salahuddin SZ, Ablashi DV, Markham PD, Josephs SF, Sturzenegger S, Kaplan M, Halligan G, Biberfeld P, Wong-Staal F, Kramarsky B, Gallo RC. Science 1986; 234: 596–601.

Human Herpesvirus-6: General Virology, Epidemiology and Clinical Pathology
Perspectives in Medical Virology
Gerhard Krueger and Dharam Ablashi (Editors)

DOI 10.1016/S0168-7069(06)12002-9

Ultrastructure and Assembly of Human Herpesvirus-6 (HHV-6)

Z. Hong Zhou[a], James K. Stoops[a], Gerhard R.F. Krueger[a,b]
[a]*Department of Pathology and Laboratory Medicine, University of Texas Health Science Center at Houston Medical School, Houston, TX 77030, USA*
[b]*Department of Anatomy II, The University of Cologne, Cologne 50924, Germany*

Introduction

Human herpesvirus-6 (HHV-6) is a ubiquitous member of the betaherpesvirus subfamily of the *Herpesviridae* family. The HHV-6 genome is arranged colinearly and codes for approximately 67% of proteins in common with human cytomegalovirus (HCMV), and 21% with all other herpesviruses. Sequence comparison shows that it is closely related to HHV-7 and HCMV.

HHV-6 is a lymphotropic herpesvirus infecting up to 90%of the population and establishing latent or persistent infections for a lifetime (Salahuddin et al., 1986; Josephs et al., 1988; Levine et al., 1992; Krueger et al., 1998a). Clinical features of HHV-6 infection are described in Part II of this book.

There are two variants of HHV-6, HHV-6A and HHV-6B with obvious genomic polymorphism within a variant (Ablashi et al., 1991, 1993; Schirmer et al., 1991). Both also vary in their tissue distribution in human and in their tissue culture cells for propagation (Lusso et al., 1988; Black et al., 1989; Ablashi et al., 1991; Di Luca et al., 1994). Some virologists recently regard HHV-6A and HHV-6B as separate herpesviruses (see Chapters 1 and 6 in this book). All ultrastructural features described in the following are derived from HHV-6A grown in HSB2 cells.

Despite variations in host range, genome size, and composition, HHV-6 shares common virion structures with other herpesviruses (Kramarsky and Sander, 1992) showing four basic elements: the core, the capsid, the tegument, and the envelope (Fig. 1). The core of the mature virion consists of double-stranded DNA (dsDNA) closely packed in a spherical capsid (Zhou et al., 1999). The capsid is a rigid icosahedral protein shell, 1200–1300 Å in diameter, that encloses and protects the dsDNA core. The tegument is a poorly defined, asymmetric layer of host and viral proteins between the capsid and the envelope. It varies in thickness and distribution around the capsid with some of its proteins in close proximity and anchored to the capsid. The envelope is a host-derived lipid bilayer containing spikes of viral glycoproteins. The entire virion varies in diameter from 1400 to 3000 Å, depending on the thickness of the tegument and the integrity of the envelope.

The eight known human herpesviruses are classified into three subfamilies (alpha-, beta- and gammaherpesviruses) based on shared biological properties (Table 1) (Roizman and Pellett, 2001). Alphaherpesviruses have a variable host range, short reproductive cycle, and rapid spread in culture. They establish latent infection primarily in neurons. This subfamily includes the human pathogens herpes simplex virus types 1 and 2 (HSV-1 and -2 or HHV-1 and -2) and varicella-Zoster virus (VZV or HHV-3). Betaherpesviruses have a more restricted host range, longer reproductive cycle, and slower growth in culture. The virus can remain latent in salivary glands, neurons, lymphocytes, and possibly other tissues. HCMV (i.e. HHV-5) and human herpesvirus types 6 and 7 (HHV-6, HHV-7) are members of this subfamily. Gammaherpesviruses include Epstein–Barr virus (EBV or HHV-4) and Kaposi's sarcoma-associated herpesvirus (KSHV) or HHV-8, both associated with certain lymphomas and other cancers. HHV-8 appears to have a more

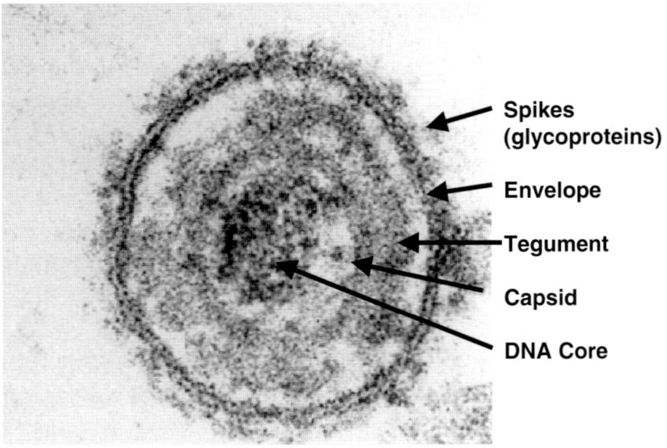

Fig. 1 Basic architecture of the herpesvirus virion. The ds DNA genome is coiled in the protein capsid. The capsid is surrounded by various tegument proteins, some of which are anchored to the capsid, while others are free-floating. The envelope is a host-derived lipid bilayer containing spikes of viral glycoprotein. (Adapted from Kramarsky and Sander (1992) with permission from the publisher).

Table 1

The known human herpesviruses

Virus	Subfamily	Genome length (kb)	Preferred host cells	Selective-associated diseases[a]
HHV-1 (HSV-1)	α	152	Neuroectodermal	Cold sores
HHV-2 (HSV-2)	α	152	Neuroectodermal	Genital sores
HHV-3 (VZV)	α	125	Neuroectodermal	Chicken pox; shingles
HHV-4 (EBV)	γ	172	Lymphohematopoietic; ectodermal	Infectious mononucleosis; Burkitt's lymphoma; nasopharyngeal carcinoma
HHV-5 (HCMV)	β	230	Mesodermal, incl. lymphohematopoietic	Infectious mononucleosis; sialoadenitis
HHV-6	β	160	Lymphohematopoietic	Exanthem subitum; infectious mononucleosis; Kikuchi's lymphadenitis; infantile febrile seizures
HHV-7	β	145	Lymphohematopoietic	Nonspecific lymphadenitis
HHV-8 (KSHV)	γ	165	Lymphocytic; fibrohistiocytic, incl. endothelial cells	Kaposi's sarcoma; serosa-associated lymphoma

[a]Additional diseases will occur in immunodeficient patients.

restricted host range than the other herpesviruses. In the following, structural features are reviewed of viruses representative for each of the three subfamilies.

Herpesvirus-6 assembly and maturation

Our current understanding of HHV-6 life cycle is supported by electron microscopic observations of viral morphology in thin-sectioned infected cells (Fig. 2). Electron microscopic data of HHV-6 infection compare well with other herpesviruses (Dalton and Manaker, 1967; Biberfeld et al., 1987; Kramarsky and Sander, 1992; Klussmann et al., 1997). The prominent tegument resembles closely HCMV (Smith and De Harven, 1973; Heine and Cottler-Fox, 1975). Progeny virus of HHV-6 assembles like cytomegalovirus. On day 3 after infection of HSB2 cells with HHV-6 (strain GS or Co6), uncoated capsids appear in the nuclei of about 1% of

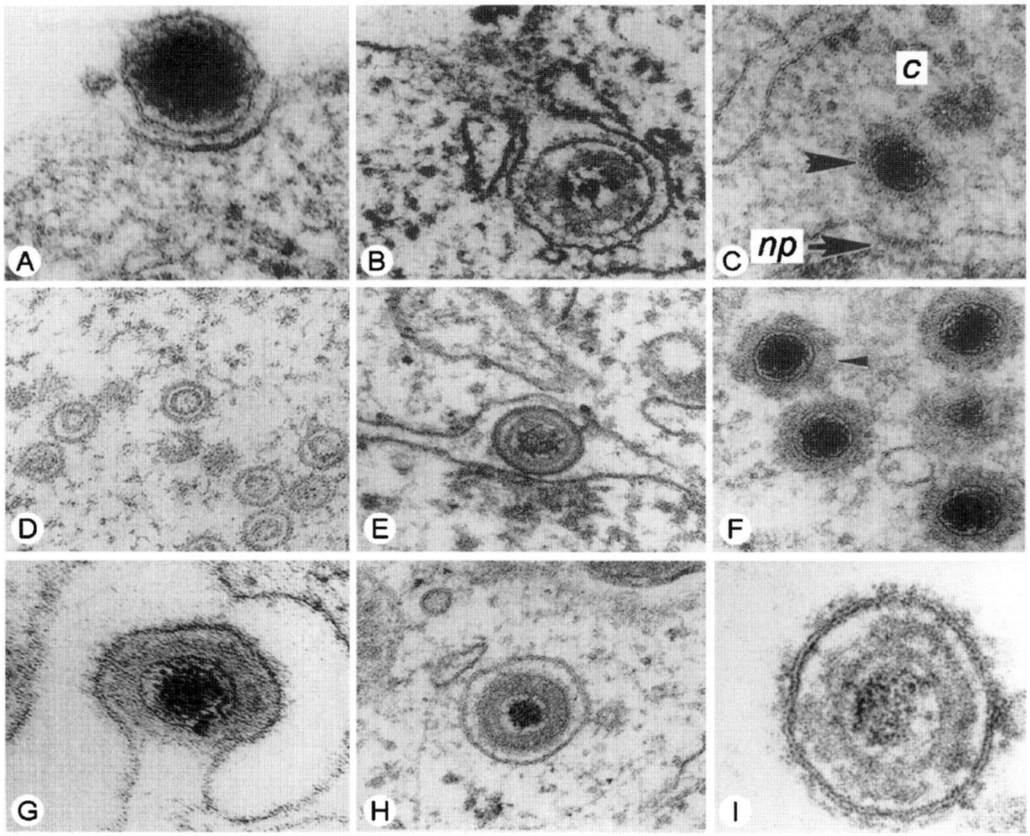

Fig. 2 Thin sections showing stages of HHV-6 infection of HSB2 cells in tissue culture. (A) HHV-6 particle attaches to cell membrane, which forms endocytotic pit (15–30 min). (B) Virion is endocytosed in about 6 h. (C) Tegumented nucleocapsid in cytosol after completion of endocytosis is transported to the nucleus; c, cytosol; np, nuclear pore. (D) Immature capsids in the nucleus (about 3 days). (E) Immature virion in perinuclear cisterna (4–6 days). (F) Viral DNA core, capsid, and tegument—tegumented capsid—in cytosol (5–6 days). (G, H) Viral budding into Golgi vesicle acquires final mature envelope (about 6 + days). (I) Mature virion with glycoprotein spikes on envelope in extracellular space (6 + days). Reproduced from Kramarsky and Sander (1992).

the cells. On day 6, they are observed in nearly all of the cells, and most capsids contain a nucleic acid core. Subsequently, virus particles with immature tegument are budding into the perinuclear space where they may acquire an initial smooth envelope. Unenveloped virus particles with teguments then appear in the cytosol, from where they bud into Golgi vesicles to acquire their final structure. They are transported via smooth endoplasmic reticulum to the cell surface and subsequently released. Budding of mature virus from the cytoplasmic membrane is not observed. The replication cycle of HHV-6B appears to be slower.

Herpesvirus infection begins with the attachment of virus particles to glycoprotein receptors on the surface of the host cell (Fig. 2A). Cellular receptor for HHV-6 attachment have been identified as the complement receptor CD46 (Lusso et al., 1994; Santoro et al., 1999). Upon binding the receptor, the viral envelope fuses with the membrane of an endocytic vesicle, leading to receptor-mediated endocytosis of the capsid. The cytoskeletal network, including the dynein and dynactin components, is used to transport capsids to the nuclear pores (Dohner et al., 2002). The DNA is injected into the nucleus, while the empty capsids remain outside.

In the nucleus the DNA circularizes and one of two pathways is initiated. The viral genome can be maintained as a circular episome and undergoes latent replication without producing infectious progeny. It is during this pathway that the genes for cellular transformation are expressed. A latent virus can be reactivated by various factors, such as excessive exposure to ultraviolet light (Krueger et al., 1998b), endotoxins, plant agglutinins, steroids, chemical cocarcinogens (i.e. tumor-promoting agents), other infections, and increased cytokine production. Once reactivated, persistent viral activity can be supported by decreased immune responsiveness resulting from stress or other factors (Krueger et al., 1998b).

During lytic replication, viral DNA is synthesized as a concatemer using the "rolling circle" method. Meanwhile, capsid assembly begins with the formation of a spherical, scaffold-containing procapsid that is similar to the prohead of bacteriophages (Dalton and Hageneau, 1973; Heine, 1974; Newcomb et al., 1996; Rixon and McNab, 1999; Newcomb et al., 2000; Yu et al., 2005).

The mechanism by which the capsids mature into infectious virions remains controversial. The generally accepted mechanism involves the acquisition by the capsid of their envelope from the inner nuclear membrane, and budding from the cell surface via the secretory pathway. However, new evidence also suggests that HSV-1 capsids are enveloped through the inner nuclear membrane, which then fuses with the outer nuclear membrane to release naked capsids into the cytoplasm. The capsids acquire additional tegument proteins and envelope by budding into the *trans*-Golgi network (Skepper et al., 2001). Viral nucleoid synthesis and envelope production may proceed independently under certain conditions, leading to over-production and release of empty capsids (A capsids) and capsid-devoid, tegument-containing enveloped particles (called L particles in HSV-1 and dense bodies in HCMV). While these particles are noninfectious, it remains to be determined whether these may still interfere with cell function and/or autoimmunity (Klussmann et al., 1997).

Three-dimensional structure of HHV-6 capsid

Cell lines. For the HHV-6 capsid structure presented here, we used an immature T-lymphocyte cell line, HSB2 (ATCC CLL 120.1 CCRF HSB2), which was originally obtained from Robert C. Gallo's laboratory at NCI, NIH, USA, and kept in the Immunopathology Laboratory, University of Cologne (GRFK), Germany, since 1986. The cells were grown in Hepes (20 mM)-buffered RPMI 1640 (Sigma–Aldrich) supplemented with 2 mM glutamine, 10% fetal calf serum (Biochrome KG), 100 U of penicillin, and 100 μg streptomycin (Sigma). Cells were stimulated before viral infection with 5 μg/ml polybrene (Merck).

Propagation of virus and purification of capsids. Polybrene-stimulated HSB2 cells were infected with HHV-6A (strain Co6: Ablashi et al., 1991; Krueger et al., 1991) by growing them in pooled tissue culture media from previous HHV-6 replicating cultures, containing approximately 10^6 infectious particles per milliliter. Fresh uninfected medium was added whenever necessary. Cultures of infected cells were terminated when, according to previous electron microscopic studies, a maximum of capsids and nucleocapsids were observed, i.e. about 8 days after infection. Cells were harvested by centrifugation, pooled, and stored for further use in 10 mM Tris-HCl buffer, pH 7.5, at −70°C.

CryoEM and three-dimensional (3D) *reconstruction.* Electron cryomicroscopy (cryoEM) was performed for purified HHV-6 capsids in a 100 kV-JEOL1200 microscope as previously described (Zhou et al., 2001) (Fig. 3a). All data processing steps were carried out in SGI Octane dual processor workstations (Silicon Graphics, Inc.) using our 3D reconstruction software package, called IMIRS (Zhou et al., 1998; Liang et al., 2002). Briefly, individual HHV-6 capsid particle images were first boxed out manually. The orientation determination and 3D reconstruction were carried out using programs for refinement (Zhou et al., 1998) and reconstruction (Johnson et al., 1997), which are based on Fourier common-lines (Crowther, 1971; Fuller et al., 1996) and Fourier–Bessel synthesis (Crowther et al., 1970), respectively. The final reconstruction was calculated by merging 30 capsid particles at ~30 Å resolution. The 3D visualization was carried out using the Iris Explorer (NAG, Inc., Downers Grove, IL) with custom-designed modules. The maps were displayed at a threshold level of approximately 1σ (standard deviation) above the mean density of the map.

The HHV-6 capsid structure has the 5:3:2 symmetry that is characteristic of an icosahedron, which is composed of 20 triangular faces and 12 vertices. The capsid has 6 fivefold ('5') symmetry axes passing through the vertices (pentons), 10 threefold (passing through triplex Tf) axes passing through the faces, and 15 twofold ('2') axes passing through the edges (Fig. 3b). The major capsid protein (MCP) forms the basic structural building block of the herpesvirus capsid. MCP proteins are arranged into groups of six ("hexons") or 5 ("pentons"), with the pentons forming the vertices of the icosahedron and the hexons filling out the faces. The hexons have slightly different geometries depending on their location, and are thus designated as P (penton-adjacent), E (edge), or C (center-face). Connecting the

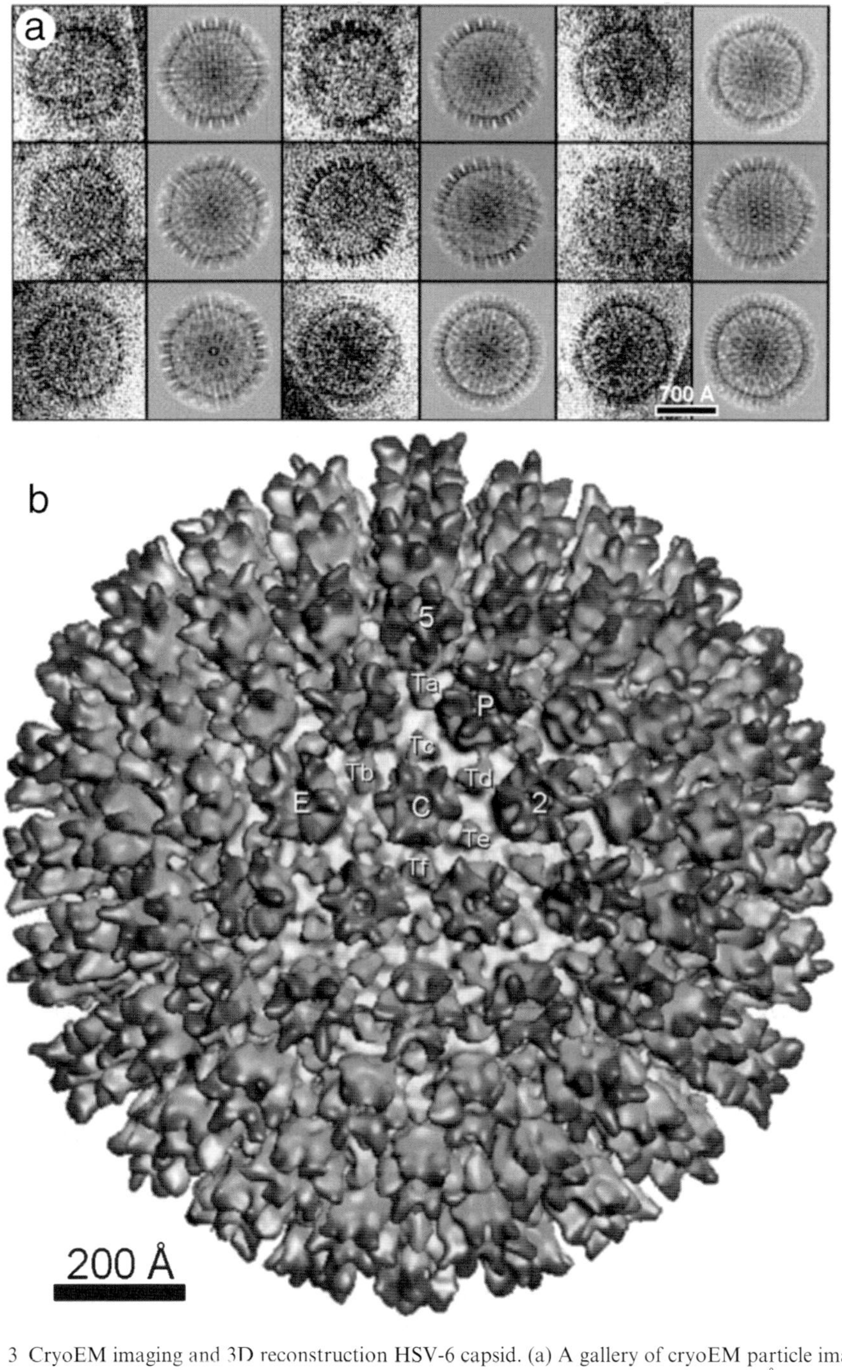

Fig. 3 CryoEM imaging and 3D reconstruction HSV-6 capsid. (a) A gallery of cryoEM particle images of HHV-6 capsids. (b) Shaded surface representation of HHV-6 capsid reconstruction at 30 Å resolution. The structure is color coded according to capsid radius so that the capsid shell is in yellow, the triplexes are in green, and the upper domains of the pentons and hexons are in purple. (for colour version: see colour section on page 347).

pentons and hexons are triplexes, which, by analogy to other human herpesvirus capsids, consist of one triplex monomer protein (TMP) and two triplex dimer proteins (TDP). There are 6 types of triplexes, designated Ta–Tf depending on their location in the asymmetric unit. Each icosahedral capsid consists of 12 pentons, 150 hexons, and 320 triplexes and they are arranged in a triangulation number T = 16 icosahedral lattice. Other components of the capsid include the small capsid protein (SCP), the viral protease, and the scaffold proteins.

Comparision of the HHV-6 cappsid structure with those of other human herpesviruses

As noted above, the gross morphology of the capsid is highly conserved among herpesviruses. Although the pentons, hexons, and triplexes diverge slightly in shape (Fig. 4) due to differences in amino acid sequences of the capsid proteins (see Wu et al., 2000; Trus et al., 2001, for detailed comparison of capsomers), and the HSV-1 capsid is slightly more angular than that of HHV-8, the stoichiometry and overall organization of the MCP and two triplex proteins are the same.

More detailed studies are available from the prototypical herpesvirus, HSV-1 (Zhou et al., 2000) using a 400 kV electron microscope to image 5860 HSV-1 capsid

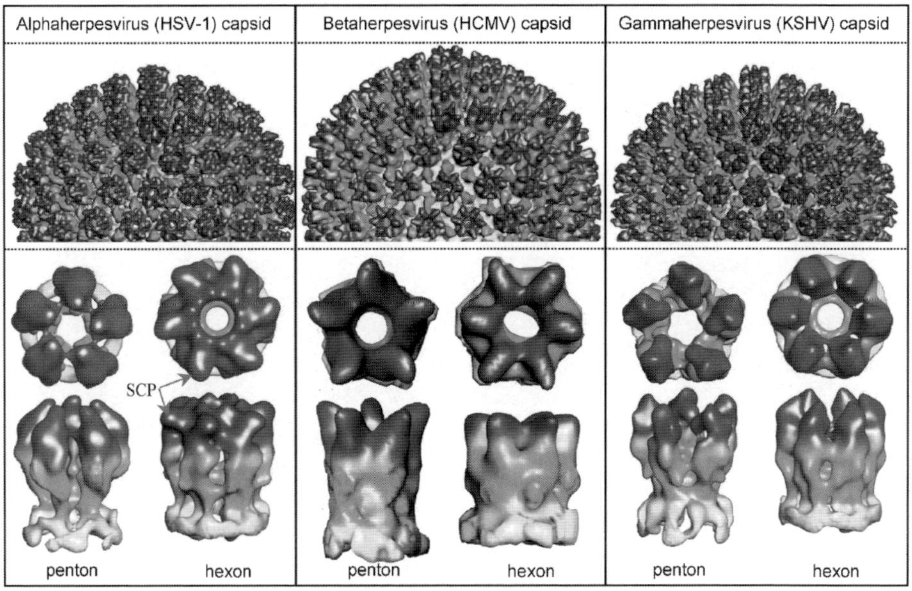

Fig. 4 Comparison of HSV-1, HCMV, and KSHV capsids. A penton and a hexon are extracted from the cryoEM structures for detail comparison. Overall structure is similar; minor differences in hexon morphology may be a result of different structures of the distally located SCPs. (Modified from a figure by P. Lo). (for colour version: see colour section on page 348).

particles that were merged to obtain an 8.5 Å structure of the HSV-1 capsid. The capsid shell has a total mass of about 200 MDa, with 960 copies of MCP VP5, 320 copies of TMP VP19c, 640 copies of TDP VP23, and 900 copies of small capsid protein VP26. As the main building block of the HSV-1 capsid, VP5 was subjected to further structural characterization. At 8.5 Å resolution, it is possible to visualize secondary structural elements that are not visible at lower resolutions. For example, α-helices appear as extended, cylindrical rods 5–7 Å in diameter. Thirty-nine α-helices and four β-sheets were identified from the VP5 subunits of the 8.5 Å cryoEM structure and found a striking resemblance with a domain in the annexin protein family (Baker et al., 2003).

The 8.5 Å structure also gave deeper insight into the interesting quaternary structure of the triplexes that are composed of two molecules of VP23 and one molecule of VP19c. The lower portion of the triplexes, which interacts with the floor domains is nearly 3-fold of the pentons and hexons, is nearly 3-fold symmetric with three subunits that are roughly equivalent. This arrangement alters through the middle of the triplex such that the upper portion is composed mostly of VP23 in a dimeric configuration. All three subunits of the triplex are required for the correct tertiary structure to fold properly. VP23 in isolation exists only as a molten globule with no distinct tertiary structure (Kirkitadze et al., 1998).

The asymmetric tegument and envelope components of HSV-1 were also recently characterized using cryoET (Grunewald et al., 2003). The virions in that study were shaped like slightly irregular ovals, with the capsid eccentrically located near one pole. The particulate tegument contained short, actin-like filaments, and was asymmetrically distributed around the capsid. The viral membranes were smooth and contained varying numbers of heterogeneous glycoprotein spikes that tended to be more densely packed around the pole distal from the capsid. This distribution indicates some sort of functional clustering that may help the virion engage cell receptors during infection. Similar studies still need to be done for such a detailed structural characterization for HHV-6.

Conclusion

Structural studies have revealed both similarities and some differences in the assembly and 3D structures between HHV-6 and other human herpesviruses. This information will help us understand the unique characteristics of their assembly, infection, replication, and pathogenesis, and may ultimately aid in the design of specific antiviral therapies.

References

Ablashi DV, Agut H, Berneman Z, Campadelli-Fium G, Carrigan D, Ceccerini-Nelli L, Chandran B, Chou S, Collandre H, Cone R, Dambaught T, Dewhurst S, DiLucca D, Foa-Tomasi L, Fleckenstein B, Frenkel N, Gallo R, Gompels U, Hall C, Jones M,

Lawrence G, Martin M, Montagnier L, Neipel F, Nicholas J, Pellett P, Razzaque A, Torrelli G, Thomson B, Salahuddin S, Wyatt L, Yamanishi K. Arch Virol 1993; 129: 363–366.
Ablashi DV, Balachandran N, Josephs SF, Hung CL, Krueger GR, Kramarsky B, Salahuddin SZ, Gallo RC. Virology 1991; 184: 545–552.
Baker ML, Jiang W, Bowman BR, Zhou ZH, Quiocho FA, Rixon FJ, Chiu W. J Mol Biol 2003; 331: 447–456.
Biberfeld P, Kramarsky B, Salahuddin SZ, Gallo RC. J Natl Cancer Inst 1987; 79: 933–941.
Black JB, Sanderlin KC, Goldsmith CS, Gary HE, Lopez C, Pellett P. J Virol Methods 1989; 26: 133–146.
Crowther RA. Philos Trans R Soc Lond B Biol Sci 1971; 261: 221–230.
Crowther RA, DeRosier DJ, Klug A. Proc Roy Soc London 1970; 317: 319–340.
Dalton AJ, Hageneau F. Ultrastructure of Animal Viruses and Bacteriophages. New York: Academic Press; 1973.
Dalton AJ, Manaker RA. Carcinogenesis: A Broad Critique. Baltimore: Williams & Wilkins; 1967. pp. 59–90.
Di Luca D, Dolcetti R, Mirandola P, De Re V, Secchiero P, Carbone A, Boiocchi M, Cassai E. J Infect Dis 1994; 1760: 211–215.
Dohner K, Wolfstein A, Prank U, Echeverri C, Dujardin D, Vallee R, Sodeik B. Mol Biol Cell 2002; 13: 2795–2809.
Fuller SD, Butcher SJ, Cheng RH, Baker TS. J Struct Biol 1996; 116: 48–55.
Grunewald K, Desai P, Winkler DC, Heymann JB, Belnap DM, Baumeister W, Steven AC. Science 2003; 302: 1396–1398.
Heine UI. The Cell Nucleus. vol. III. New York: Academic Press; 1974. pp. 489–536.
Heine U, Cottler-Fox M. Electron microscopic observations on the composition of herpes type virions. In: Oncogenesis and Herpesviruses II (de The G, Epstein MA, zur Hausen H, editors). Lyon: IARC Sci. Publ 1975; pp. 103–110.
Johnson O, Govindan V, Park Y, Zhou ZH. Proceedings of the 4th International Conference on High Performance Computing. Los Alamitos, CA: IEEE Computer Society Press; 1997. pp. 517–521.
Josephs SF, Ablashi DV, Salahuddin SZ, Kramarsky B, Franza BR, Pellett P, Buchbinder A, Memon S, Wong-Staal F, Gallo RC. J Virol Methods 1988; 21: 179–190.
Kirkitadze MD, Barlow PN, Price NC, Kelly SM, Boutell CJ, Rixon FJ, McClelland DA. J Virol 1998; 72: 10066–10072.
Klussmann JP, Krueger E, Sloots T, Berneman Z, Arnold G, Krueger GR. Virchows Arch 1997; 430: 417–426.
Kramarsky B, Sander C. Electron microscopy of human herpesvirus-6 (HHV-6). In: Human Herpesvirus-6 (Ablashi DV, Krueger GRF, Salahuddin SZ, editors). Amsterdam: Elsevier; 1992; pp. 59–68.
Krueger GRF, Ablashi DV, Whitman JI, Luka J, Rojo J. Rev Med Hosp Gen Mexico 1998b; 61: 226–240.
Krueger GRF, Koch B, Leyssens N, Berneman Z, Rojo J, Horwitz C, Sloots T, Margalith M, Conradie JD, Imai S, Urasinski I, de Bruyere M, Ferrer Argote V, Krueger J. Vox Sang 1998a; 75: 193–197.
Krueger GRF, Sander C, Hoffmann A, Barth A, Koch B, Braun M. In Vivo 1991; 5: 217–226.

Levine P, Jarrett R, Clark DA. The epidemiology of human herpesvirus-6. In: Human Herpesvirus-6 (Ablashi DV, Krueger GRF, Salahuddin SZ, editors). Amsterdam: Elsevier; 1992; pp. 9–23.
Liang Y, Ke EY, Zhou ZH. J Struct Biol 2002; 137: 292–304.
Lusso P, Markham PD, Tschachler E, Di Marzo Veronese F, Salahuddin SZ, Ablashi DV, Pahwa S, Krohn K, Gallo RC. J Exp Med 1988; 167: 1659–1670.
Lusso P, Secchiero P, Crowley RW, Garzino-Demo A, Berneman ZN, Gallo RC. Proc Natl Acad Sci USA 1994; 91: 3872–3876.
Newcomb WW, Homa FL, Thomsen DR, Booy FP, Trus BL, Steven AC, Spencer JV, Brown JC. J Mol Biol 1996; 263: 432–446.
Newcomb WW, Trus BL, Cheng N, Steven AC, Sheaffer AK, Tenney DJ, Weller SK, Brown JC. J Virol 2000; 74: 1663–1673.
Rixon FJ, McNab D. J Virol 1999; 73: 5714–5721.
Roizman B, Pellett PE. Herpesviridae: A Brief Introduction. In: Fields Virology (Fields BN, Knipe DM, Howley PM, Griffin DE, Lamb RA, Martin MA, Roizman B, Strauss SE, editors). vol. 2. Philadelphia: Lippincott, Williams & Wilkins; 2001; pp. 2381–2398.
Salahuddin SZ, Ablashi DV, Markham PD, Josephs SF, Sturzenegger S, Kaplan M, Halligan G, Biberfeld P, Wong-Staal F, Kramarsky B, Gallo RC. Science 1986; 234: 596–601.
Santoro F, Kennedy PE, Locatelli G, Malnati MS, Berger EA, Lusso P. Cell 1999; 99: 817–827.
Schirmer EC, Wyatt LS, Yamanishi K, Rodriguez WJ, Frenkel N. Proc Natl Acad Sci USA 1991; 88: 5922–5926.
Skepper JN, Whiteley A, Browne H, Minson A. J Virol 2001; 75: 5697–5702.
Smith JD, De Harven E. J Virol 1973; 12: 919–930.
Trus BL, Heymann JB, Nealon K, Cheng N, Newcomb WW, Brown JC, Kedes DH, Steven AC. J Virol 2001; 75: 2879–2890.
Wu L, Lo P, Yu X, Stoops JK, Forghani B, Zhou ZH. J Virol 2000; 74: 9646–9654.
Yu X, Trang P, Shah S, Atanasov I, Kim YH, Bai Y, Zhou ZH, Liu F. Proc Natl Acad Sci USA (Track II) 2005; 102: 7103–7108.
Zhou ZH, Chen DH, Jakana J, Rixon FJ, Chiu W. J Virol 1999; 73: 3210–3218.
Zhou ZH, Chiu W, Haskell K, Spears Jr H, Jakana J, Rixon FJ, Scott LR. Biophys J 1998; 74: 576–588.
Zhou ZH, Dougherty M, Jakana J, He J, Rixon FJ, Chiu W. Science 2000; 288: 877–880.
Zhou ZH, Liao W, Cheng RH, Lawson JE, McCarthy DB, Reed LJ, Stoops JK. J Biol Chem 2001; 276: 21704–21713.

Human Herpesvirus-6: General Virology, Epidemiology and Clinical Pathology
Perspectives in Medical Virology
Gerhard Krueger and Dharam Ablashi (Editors)

DOI 10.1016/S0168-7069(06)12003-0

HHV-6 Genome: Similar and Different

U.A. Gompels[a], F.C. Kasolo[a,b,c]
[a]*Department of Infectious and Tropical Diseases, London School of Hygiene & Tropical Medicine, University of London, UK*
[b]*Virology Unit, University Teaching Hospital, Lusaka, Zambia*
[c]*AFRO VPD, World Health Organisation, Harare, Zimbabwe*

Genome classification and biology

Roseoloviruses, human herpesviruses 6 and 7 (HHV-6, HHV-7) are widespread T lymphotropic and neurotropic viruses causing mostly benign infections. However, particularly for HHV-6, during some primary as well as secondary reactivated infections, which can follow immune aberrations or deficiencies, there can be severe complications which can lead to fatalities. Thus, this is of relevance for immunosuppressed HIV/AIDS or transplantation patients, as well as increasingly, for those with neurological disease, including encephalitis and a link with multiple sclerosis (primarily HHV-6A). Understanding when and how this virus does or does not cause disease is key to developing effective treatments plus evaluating the impact of HHV-6 infections on worldwide populations. Studies on the virus genome provide a foundation for this exploration and can guide the way towards development of new anti-virals as well as possible novel treatments for immune-related pathologies using this well-adapted virus as a guide.

The general properties of HHV-6 and the closely related HHV-7 and their genomes have been summarized in reviews and the original reports of their complete genomic sequences (Gompels et al., 1995; Nicholas, 1996; Megaw et al., 1998; Dominguez et al., 1999; Isegawa et al., 1999; Gompels, 2004). This chapter reviews overall properties of the genome of HHV-6 with some updates, while further details can be found in the first sequence papers as well as in Genbank nucleotide sequence and genome entries. There are two strain groups for HHV-6 (Ablashi et al., 1993), the prototypes are strain U1102 for HHV-6 variant A (HHV-6A) and strain Z29 for variant B (HHV-6B). HHV-6A strain U1102 was identified and first characterized in the UK (Downing et al., 1987) and HHV-6B strain Z29 in the USA (Lopez et al., 1988); both genomes are sequenced (Gompels et al., 1995; Dominguez et al., 1999). These laboratory strains are from adult HIV/AIDS patients from African countries: U1102 from Uganda and Z29 from Zaire, where the virus has reactivated from the immunosuppression giving a systemic infection. Other laboratory strains studied include HHV-6A strain GS, the first report of HHV-6 infection, from an adult HIV/AIDS patient in USA (Salahuddin et al., 1986), HHV-6A strain AJ, from an adult HIV/AIDS patient in UK (Tedder et al., 1987), and HHV-6B strain HST, from an exanthem subitum pediatric patient in Japan (Yamanishi et al., 1988). There are partial, or fragments of sequence available for strains GS and AJ. While for HHV-6B strain HST, the complete nucleotide sequence was also derived and represents the only primary childhood infection isolate of the genomes analyzed (Isegawa et al., 1999). These are closely related strains differing on average by 5% with increases in variation primarily at the ends of the genomes overlapping repetitive sequences as discussed further below. There is also one hypervariable gene at the centre of the genome which also encodes a variable glycoprotein, U47 or gO, and marks a region of genomic reorganization between herpesvirus subgroups as shown below (Gompels et al., 1995; Kasolo et al., 1997; Dominguez et al., 1999). The two strains of HHV-6B with genomic sequences available, not only show less variation than between the variants, HHV-6A and HHV-6B, strains, but also display increasing variation towards the ends of the genome and overlapping repetitive sequences (Dominguez et al., 1999; Isegawa et al., 1999). The genomes of two strains of HHV-7 have also been sequenced, JI and RK, which show less variation than between the HHV-6 strains (Nicholas, 1996; Megaw et al., 1998).

HHV-6B strains seem more prevalent in primary pediatric infections, where tested, primarily in European countries, USA and Japan, with occasional HHV-6A infection (Hall et al., 1994; van Loon et al., 1995; Ward et al., 2005; Zerr et al., 2005), and some evidence for increased congenital infection with HHV-6A (Hall et al., 1994, 2004; Ward et al., 2005). There also appears to be geographic variation, as in an African country, Zambia, childhood HHV-6A infections appear more frequent (Kasolo et al., 1997). These studies have been performed directly on blood samples taken during acute infections followed by DNA PCR and nucleotide sequencing. However, true distribution by strain-specific serology has yet to be performed, hindered by problems of the antibody cross-reactivity against lysates of whole virus antigen used in serological assays and the lack of defined single antigens with a combination of 100% specificity and immunodominance.

Genomes, cellular tropism and laboratory culture

Both HHV-6 and HHV-7 are T lymphotropic and neurotropic. Furthermore, it is likely that *in vivo*, the lytic and latent cell types that HHV-6 infects are highly specific subsets. While *in vitro*, virus is cultured in laboratory-adapted cell lines, and studies suggest that this may influence genomic composition. Such changes have been recently demonstrated for the related betaherpesvirus, human cytomegalovirus (HCMV), where serial passage in fibroblast rather than endothelial or leucocyte cell types has resulted in large genomic deletions (Cha et al., 1996; Murphy et al., 2003; Dolan et al., 2004). For HHV-6, some differences upon serial passage of strain Z29 in culture have been observed including expansion of repetitive sequences from the origin of lytic replication, terminal direct repeats, het region, and other rearrangements are possible (Gompels et al., 1995; Stamey et al., 1995; Dominguez et al., 1999; Gompels, unpublished). There may be less deletions observed for HHV-6 than in HCMV, since the reference strains Z29 and U1102 have both been isolated and propagated initially in primary cord blood cells, although passage in various leukaemic cell lines have also been reported. Thus, considerations of the genome of HHV-6 must also address issues of cellular tropism and possible changes arising from *in vitro* culture.

As described in other chapters, for routine culture both HHV-6 and HHV-7 have been adapted to grow in CD4+ T-leukaemic cell lines, for example, J-JHAN (Jurkat), HSB2, or Molt-3 for HHV-6, and SupT-1 for HHV-6 and HHV-7. In addition, there are some differences reported in growth of HHV-6 strains in leukaemic cell lines. However, preferential growth for both HHV-6 and HHV-7 are in activated cord blood lymphocytes or mononuclear cells (CBL, CBMC) or in peripheral blood lymphocytes or mononuclear cells (PBL, PBMC). Screened cord blood is preferred, as infection with laboratory strains can result in reactivation of resident latent virus from adult blood, although there is occasionally a similar risk from cord blood.

Both HHV-6 and HHV-7 have a cellular tropism for T lymphocytes as shown *in vivo* during viremia from acute infection as well as *in vitro*. Infection and lytic replication results in a characteristic cytopathic effect of large cells (cytomegalia) and ballooning cells. The cells are completely permissive for replication and virus production, with infection resulting in cell death by necrotic lysis. Although, there is also *in vitro* and some *in vivo* evidence that infection also causes cell death by apoptosis in uninfected or non-productively infected bystander cells (Inoue et al., 1997; Secchiero et al., 1997; Yasukawa et al., 1998). Studies show that CD4+, CD8+ and gamma/delta T lymphocytes can be infected, but overall data suggest that activated CD4+ T lymphocytes are the preferential target of fully permissive infection *in vivo* (Takahashi et al., 1989; Lusso et al., 1991). Antibody to the T-cell-specific marker and signal transduction molecule, CD3 (OKT3) has been shown to augment infection of HHV-6 in both primary (Roffman and Frenkel, 1991) and T-leukaemic cell lines (H.A. Macaulay and U.A. Gompels, unpublished results). This antibody is also often used in transplantation patients and may aid virus replication.

Latent infection has been demonstrated within monocytic/macrophage cells as well as bone marrow progenitor cells similar to that observed for HCMV, and may be a general property of betaherpesvirus infection (Kondo et al., 1991; Gompels et al., 1993, 1994; Kempf et al., 1997; Yasukawa et al., 1997). A strong interaction with monocytic/macrophage cells has been recorded during HHV-6 primary infection and may also include a form of latency with specific restricted transcripts as well as replication within some differentiated subsets (Kondo et al., 2002a,b, 2003). Similarly, latent infections of primary macrophages have also been observed for HHV-7 (Zhang et al., 2001). Higher levels of HHV-7 compared to HHV-6 can be detected by PCR in blood and saliva of asymptomatic adults (Di Luca et al., 1995; Kidd et al., 1996; Gautheret-Dejean et al., 1997). Given the similar prevalence, this suggests that HHV-6 has more stringent regulation of latency. Studies also show that HHV-7 can also act to reactivate latent HHV-6 infections (Katsafanas et al., 1996). Additional HHV-6-specific latency-associated transcripts have been identified from the U94/Rep gene; it is highly conserved between strains, but deleted in HHV-7 (Nicholas, 1996; Megaw et al., 1998; Rapp et al., 2000). This HHV-6 gene has roles in gene expression and replication modulation that may contribute to HHV-6 latency regulation (Rotola et al., 1998; Mori et al., 2000; Rapp et al., 2000; Dhepakson et al., 2002; Turner et al., 2002; Caselli et al., 2005). In rare cases, evidence for viral genome integration has been observed, which can result in high levels of persisting, circulating HHV-6 DNA, not always with concomitant reactivation and gene expression (Tanaka-Taya et al., 2004; Ward et al., 2005). This may be mediated by the numerous repetitive sequences in the genome with similarities to the host genome. The clinical significance, genomic structure and composition are under evaluation.

The genome detailed in this chapter is from the reference HHV-6A strain U1102 with comparisons made to HHV-6B strains HST and Z29 as well as comparisons to HHV-7 and other human herpesviruses. The genomic sequence was derived from plasmid clones from early passage virus propagated in cord blood (Martin et al., 1991; Gompels et al., 1995). Only one region was intractable for plasmid cloning, covering the R3 repeat, and this sequence was determined directly from PCR amplified products. Sequencing from uncultivated virus directly from tissue or blood samples has only been conducted on small fragments to compare strains.

Genome structure and repetitive sequence

HHV-6A strain U1102 has a genome of 159,321 bp, while HHV-6B strains HST and Z29 are slightly larger, the respective sizes are 161,573 bp and 162,114 bp (Gompels et al., 1995; Dominguez et al., 1999; Isegawa et al., 1999). Much of the differences in length accommodated by variation in the repetitive sequences in the genome include the terminal repeat, which bound both ends of the genome. This direct repeat (DR), or terminal repeat (TR), varies in size: U1102 8087 bp, HST 8231 bp, and Z29 8793 bp. These DR regions are themselves bounded by copies of human telomeric repeats, which have been postulated to play a role in latency possibly by stabilizing the genome as a mini-chromosome (see Fig. 1). The telomeric repeat bounding the left end of the repeats is heterogeneous, het, thus, at

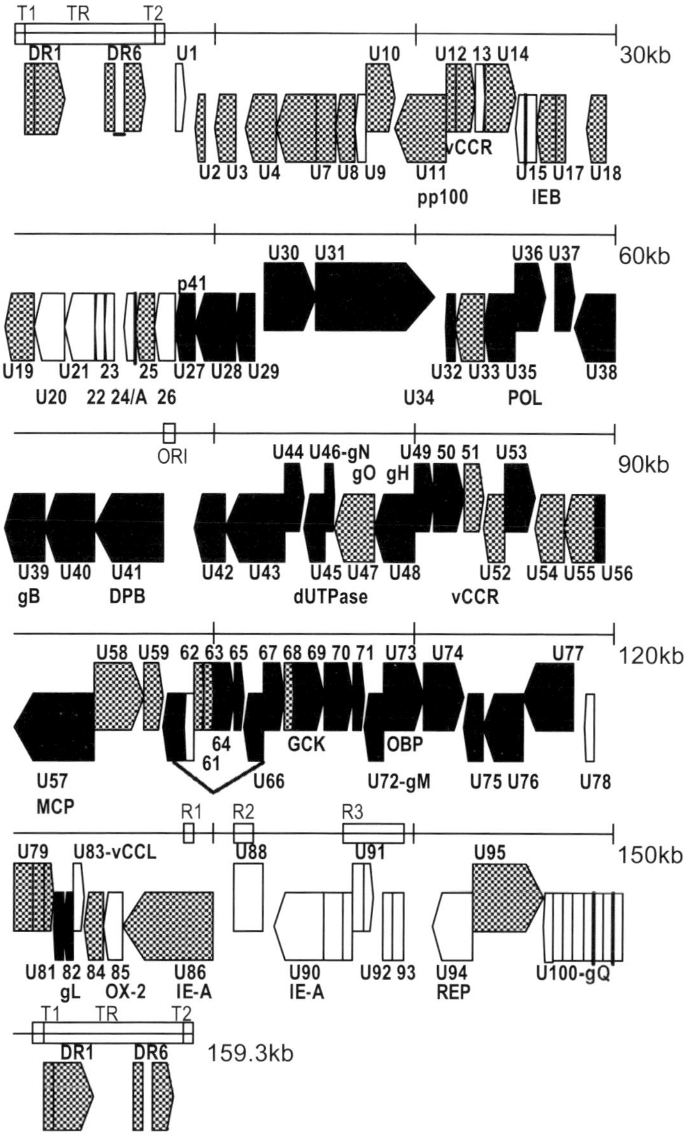

Fig. 1 Structure of HHV-6 genome with herpesvirus and betaherpesvirus conserved genes. Herpesvirus conserved genes are indicated in black while betaherpesvirus conserved genes are indicated patterned. Similarities were determined by comparisons of encoded amino acid sequences as described in the text. TR, terminal or direct repeat (DR); t, telomeric repeat; ORI, origin of lytic replication; and R1, R2, R2, repetitive sequences. Conserved or HHV-6-specific genes are indicated for reference: p41 DNA processivity factor also monoclonal antibodies used in diagnostics; pp100, major immunodominant tegument phospho-protein; U94/REP, parvovirus rep homologue and gene/replication regulatory latency gene; IEA and IEB, immediate early regulatory genes; POL, DNA polymerase; gB, gH, gL, gO, gm, gn glycoprotein; dutpase; vccr, viral chemokine receptors; MCP, major capsid protein; GCK, ganciclovir kinase, site resistance mutations; OX-2, homologue of this member of immunoglobulin gene family.

the left end of the genome and the junction of the right end of unique sequence with the right DR. Redundant copies of this repeat are also present in opposite orientation in either side of the origin of lytic replication and suggest a role of RNA copies in replication possibly through priming the lagging strand (Gompels and Macaulay, 1995; Gompels et al., 1995; Mrazek and Karlin, 1998; Dominguez et al., 1999). These repeats are adjacent to the pac repeats, directing DNA packaging, at the ends of the genomes and the junction between the DR and the unique sequences. These have been functionally defined and described in detail for both HHV-6A and HHV-6B (Thomson et al., 1994; Gompels et al., 1995; Dominguez et al., 1999; Turner et al., 2002; Borenstein et al., 2004). The origin was defined as a minimal 400 bp, but further study suggests that the functioning ori is larger, 1.3 kb containing both repeats for binding the origin-binding protein (OBP) followed by imperfect direct repeats, IDR1, 2, 3, which vary between the strains (Dykes et al., 1997; Turner et al., 2002). The ORI, is mutagenic and multiple copies have been identified in tissue culture passaged Z29 strains (Stamey et al., 1995). Pac sequences together with the origin of replication can form 'amplicons' which could be used as artificial vectors for gene delivery as described with some applications for stem cell delivery (Deng and Dewhurst, 1998; Turner et al., 2002; Borenstein et al., 2004). A series of repetitive sequence regions are found at the heterogeneous right end of the genome, R1, R2, and R3. R1 has reiterations encoding an SR domain in the IE2 homologue in HHV-6 U86 (Gompels et al., 1995). R2 contains simple TG repeats, resulting in a large reading frame open in six frames, U88, and unlikely to be coding. In HHV-6B it has been further sub-divided to R2A and R2B regions (Dominguez et al., 1999). R3 contains tandem repeats of an approximately 110 bp sequence which includes a KpnI restriction endonuclease site, and has been shown to have a role as an enhancer for U95 expression (Takemoto et al., 2001). This region in HHV-6A was refractory to cloning, possibly due to secondary structure formation from the repeated sequences there (Martin et al., 1991; Gompels et al., 1995). Not surprisingly, this right end of the genome which contains most of the repetitive sequence of the genome is heterogeneous between all strains and includes the R1, R2, R3 repeats followed by the het region and the DR repeats.

The overall composition of the genome is 43% G + C with 58% in the DR and 41% in the unique sequences. Lower G + C composition is also found at the origin for lytic replication, ORI. There is compositional polarity around the ORI as found in other organisms and particularly betaherpesviruses, this could be driven by the telomeric repeats as described above, or could be due to errors accumulated during copy repair synthesis of the lagging strand from priming with Okazaki fragments (Gompels et al., 1995; Gompels and Macaulay, 1995; Mrazek and Karlin, 1998; Dominguez et al., 1999). Other compositional biases are present across the IEA region which has marked localized CpG suppression, suggesting mutagenic effects of methylation possibly during regulation of gene expression during latent infections as described previously (Gompels et al., 1995). This localized suppression has only been observed in betaherpesviruses, while in alphaherpesviruses there is none and in gammaherpesviruses there is global suppression suggested related to distinct forms of latency (Honess et al., 1989). Thus, localized CpG suppression is a feature in betaherpesviruses, which may reflect similarities in latency control in monocytic/macrophages or bone marrow progenitor cells for this herpesvirus subgroup. The overall structural features of the HHV-6 genome show some similarities to all betaherpesviruses, but in roseoloviruses, in particular HHV-6, there are unique features as well, which are the residues of this virus specific evolution and frame its biology.

Genome rearrangements and relationship to other herpesviruses

HHV-6 together with HHV-7 form the Roseolovirus grouping of the betaherpesvirus subgroup of herpesviridae. This has also been termed as beta-2 herpesviruses, compared to the Cytomegalovirus (CMV) grouping of beta-1 herpesviruses. Recent studies, particularly on gB and polymerase sequences of extensive sets of animal herpesviruses suggest that this classification may be broadened. Examples include the roseolovirus, porcine CMV-significant in xeno-transplantation, the fatal elephant roseolovirus, and recently identified chimpanzee HHV-6 (Ehlers et al., 2001; Chmielewicz et al., 2003; Lacoste et al., 2005). Similar to other betaherpesvirus, infection is species specific, thus animal models with human viruses are restricted, although as with some of the CMV group, there may be some utility using these related (albeit distant) animal viruses for study of *in vivo* models for human infection.

The HHV-6 genome organization of conserved genes is a betaherpesvirus-specific arrangement, as first described for HHV-6A strain U1102. Both the genome organization and the encoded protein sequences share a closer relation than to the other herpesvirus lineages (Gompels and Macaulay, 1995; Dominguez et al., 1999) (Fig. 1). Seven discrete blocks of conserved genes can be identified as compared to other human herpesviruses, and these are rearranged in the alpha and gammaherpesvirus lineages of bird and mammalian herpesviruses (Gompels et al., 1995) (Fig. 2). The genome rearrangements shown by these lineage comparisons suggest that recombination has played a major role in their evolution. For example, the central conserved gene block III, appears to have recombined into repetitive sequences at the right end (relative) of the genome in the other two lineages (Gompels et al., 1995). This ancient recombination is supported by recent sequencing studies on hypervariable genes adjacent to this region in the centre of the genome or in the repetitive sequences towards the right end of the genome in HCMV where switching between phylogeny defined groups between adjacent genes are observed (Dolan et al., 2004; Mattick et al., 2004). In HHV-6, study at this same central locus, HCMV UL74/HHV-6U47, also provides evidence for strain recombination as described (Kasolo et al., 1997) and discussed below.

Genome composition and general molecular biology

HHV-6 encodes approximately 100 open reading frames, almost all of these appear coding, a few have been deleted or annotated in the updated gene list shown in

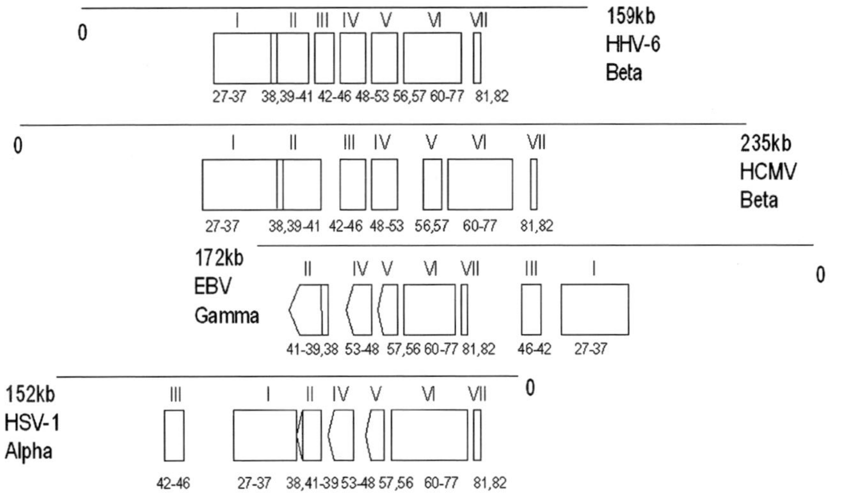

Fig. 2 Relationship between human herpesvirus genome organizations. HHV-6 shares with HCMV a betaherpesvirus specific arrangement of core-conserved genes, here collected into seven gene blocks, I–VII. The core genes are identified in Fig. 1. The numbers refer to the HHV-6 homologues of these core conserved genes. HSV and EBV represent the alpha and gammaherpesvirus lineages, respectively.

Table 1, owing to comparisons between HHV-7 and strains, spliced products identified, as well as using gene prediction software. In this chapter to highlight HHV-6-specific genes versus, Roseolovirus, betaherpesvirus-specific or herpesvirus conserved genes selected genome comparisons will be made, in particular to the other human roseolovirus, HHV-7. References are also to the prototypic beta-1 herpesvirus, HCMV, plus the initial prototypic herpesvirus, HSV, where many gene functions were first defined. HHV-7 is more compact at 144 kb or 153 kb in size for strains JI and RK, respectively, with similar long unique regions of 133 kb and variation in the DR of 5.8 and 10 kb each (Nicholas, 1996; Megaw et al., 1998). With a few notable exceptions described below, it shares all genes with HHV-6 as indicated with varying degrees of conservation of encoded amino acid sequences (Table 1, Fig. 1). In HHV-6, the open reading frames are designated from U1 to U100 with those in the direct repeats from DR1 to DR7, although recent analyses show evidence for splicing as well as expression for two of these, DR2 and DR6, which include previous DR designations as indicated (Fig. 1, Table 1) (Gompels et al., 1995). In HHV-7 similar nomenclature is used with homologous genes U2–U100 (Nicholas, 1996; Megaw et al., 1998). A few genes are lacking in HHV-7 compared to HHV-6 as discussed further below, but there are also a few HHV-7-specific genes noted H1–H7 (Gompels et al., 1995; Nicholas, 1996; Megaw et al., 1998). There is limited splicing observed in HHV-6, approximately 10% of the genes, these are primarily in the immediate early genes and selected early/late genes as noted experimentally or predicted from sequence motifs for DR1, DR6, U12, U15, U7, U66, U79, U83, U90 and U100 (Figs. 1, 2, Table 1). These are likely to be

Table 1

Features of HHV-6 strain U1102 genes

Gene[a]		Strand	Start[b]	Stop[c]	Codons	Id[d]	Properties[e]
DR1	ex1	+	501	759	689	33	HCMV US22 gene family
	ex2		843	2653			Includes DR2
DR6	ex1	+	4725	5028	395	59	HCMV US22 gene family
	ex2	+	5837	6720			Includes DR7 transactivator/transformation
U1		+	8245	8613	123	*	SR domain
U2		–	9816	8716	366	48	HCMV US22 gene family
U3		–	11276	10155	373	49	HCMV US22 gene family
U4		–	13092	11485	535	59	Related to U7 exon 2
U7	ex1	–	15921	14948	> 872	54	HCMV US22 gene family (exon 1);
	ex2	–	14858	13214			related to U4 (exon 2); includes U5
U8		–	17091	16021	356	51	HCMV US22 gene family
U9		–	17552	17241	104	*	
U10		+	17604	18914	436	50	[HCMVUL31]
U11		–	21578	18966	870	31	Structural phosphoprotein; pp100 major antigen [HCMVUL32]
U12	ex1	+	21680	21712	351	47	G protein-coupled receptor; chemokine receptor [HCMVUL33]
	ex2	+	21790	22812			
U13		+	22898	23218	106	33	
U14		+	23331	25145	604	50	HCMVUL25/UL35 gene family
U15	ex1	–	25992	25676	191	68	
	ex2	–	25602	25530			
	ex3	–	25364	25179			
U17	ex1	–	27349	27121	334	53	HCMV US22 gene family; IE-B[f] transactivator
	ex2	–	27034	26259			[HCMVUL36] includes U16 IE-B
U18		–	29389	28508	293	44	IE-B membrane glycoprotein [HCMVUL37]
U19		–	30818	29649	389	34	IE-B protein [HCMVUL38]

Table 1 (*continued*)

Gene[a]	Strand	Start[b]	Stop[c]	Codons	Id[d]	Properties[e]
U20	–	32337	31069	422	22	Probable membrane glycoprotein
U21	–	33641	32340	433	31	Probable membrane glycoprotein; HHV-7 U21 downregulates MHC class I
U22	–	34347	33739	202	*	Probable membrane glycoprotein
U23	–	35085	34375	236	p	Probable membrane glycoprotein
U24	–	35655	35392	87	28	Contains a hydrophobic domain
U24A	–	35847	35674	57	25	Contains a hydrophobic domain
U25	–	36814	35864	316	47	HCMVUS22 gene family [HCMVUL43]
U26	–	37809	36922	295	30	
U27	–	38903	37797	368	68	Processivity subunit DNA polymerase; p41 [HCMVUL44][HSVUL42]
U28	–	41434	39020	804	47	Ribonucleotide reductase large subunit [HCMVUL45] [HSVUL39]
U29	–	42356	41457	299	53	Capsid protein; component of intercapsomeric triplex [HCMVUL46] [HSVUL38]
U30	+	42325	45132	935	46	Tegument protein [HCMVUL47][HSVUL37]
U31	+	45150	51383	2077	46	Large tegument protein [HCMVUL48][HSVUL36]
U32	–	51721	51455	88	66	Capsid protein; hexon tips [HCMVUL48A][HSVUL35]
U33	–	53135	51723	470	59	Virion protein [HCMVUL49]
U34	–	53916	53086	276	56	Membrane-linked phosphoprotein; Role primary envelopment[HCMVUL50][UL34]
U35	–	54253	53933	106	58	Role in DNA packaging [HCMVUL51][HSVUL33]
U36	+	54252	55706	484	58	Role in DNA packaging [HCMV52][HSVUL32]
U37	+	55710	56504	264	62	Role primary envelopment[HCMV53][HSVUL31]
U38	–	59588	56550	1012	67	Catalytic subunit of replicative DNA polymerase [HCMVUL54][HSVUL30]
U39	–	62080	59588	830	56	Envelope glycoprotein gB [HCMVUL55][HSVUL27]

U40	–	64214	62034	726	56	Role in DNA packaging [HCMVUL57][HSVUL28]
U41	–	67620	64222	1132	68	Single-stranded conserved DNA-binding protein [HCMVUL57][HSVUL29]
U42	–	70598	69054	514	56	Post-translational regulator of gene expression [HCMVUL69][HSVUL54]
U43	–	73405	70823	860	61	Component of DNA helicase-primase complex; primase [HCVMUL70][HSVUL52]
U44	+	73470	74087	205	58	[HCMVUL71][HSVUL51]
U45	–	75218	74088	376	50	Deoxyuridine triphosphatase[HCMVUL72][HSVUL50]
U46	+	75291	75545	84	52	Membrane protein gN; complexes with gM [HCMVUL73][HSVUL49A]
U47	–	77768	75912	618	23	Membrane glycoprotein gO; complexes with gH/gL
U48	–	80118	78034	694	39	Envelope glycoprotein gH; complexes gL and gO or gQ, gH/gL/gQ binds CD46 [HCMVUL75][HSVUL22]
U49	+	80277	81035	252	52	Role in fusion [HCMVUL76][HSVUL24]
U50	+	80812	82479	555	55	Role in DNA packaging [HCMVUL77][HSVUL25]
U51	+	82574	83479	301	35	G protein-coupled receptor; chemokine receptor [HCMVUL78]
U52	–	84274	83500	258	56	[HCMVUL79]
U53	+	84281	85867	528	52	N-terminal protease domain acts in capsid maturation and is a capsid protein; C-terminal domain is the minor capsid scaffold protein [HCMVUL80][HSVUL26]
U53.5	+	85133	85867	244	51	Major capsid scaffold protein [HSVUL26.5]
U54	–	87427	86051	458	42	Virion transactivator; HCMV pp65 major tegument protein; gene duplication in HCMV [HCMVUL82/83]
55	–	88803	87505	432	33	Role in DNA synthesis HCMV, UTPase [HCMVUL84]
U56	–	89873	88983	296	65	Capsid protein; component of intercapsomeric triplex [HCMVUL85][HSVUL18]

Table 1 (*continued*)

Gene[a]		Strand	Start[b]	Stop[c]	Codons	Id[d]	Properties[e]
U57		–	93912	89875	1345	68	Major capsid protein; forms hexons and pentons [HCMVUL86][HSVUL19]
U58		+	93924	96242	772	61	[HCMVUL87]
U59		+	96239	97291	350	38	[HCMVUL88]
U61		–	98578	98234	115	*	
U62		+	98427	98684	85	45	[HCMVUL91]
U63		+	98632	99282	216	68	[HCMVUL92]
U64		+	99260	100588	442	41	Role in DNA packaging; tegument protein [HCMVUL93][HSVUL17]
U65		+	100545	101552	335	59	Tegument protein [HCMVUL94][HSVUL16]
U66	ex1	–	102486	101614	666	72	Role in DNA packaging; putative terminase [HCMVUL89EX1][HSVUL15]
	ex2	–	98415	97288	344	52	Includes U60 [HCMVUL89EX2]
U67		+	102485	103519	344	52	[HCMVUL95][HSVUL14]
U68		+	103519	103863	114	48	[HSVUL96]
U69		+	103866	105554	562	53	Ganciclovit kinase, serine threonine protein kinase; tegument protein [HCMVUL97][HSVUL13]
U70		+	105562	107028	488	52	Deoxyribonuclease; role in maturation/packaging of DNA [HCMVUL98][HSVUL12]
U71		+	106965	107198	77	53	Myristylated tegument protein; position HCMV pp28K [HSVUL11]
U72		–	108312	107279	344	59	Envelope glycoprotein gM; role in virion envelopment and trafficking membrane proteins; forms complex with gN [HCMVUL100][HSVUL10]
U73		+	108325	110667	780	58	Origin-binding protein; helicase [HSVUL9]
U74		+	110636	112624	662	41	Component of DNA helicase-primase [HCMVUL102][HSVUL8]

U75		–	113408	112659	249	45	[HCMVUL103][HSVUL7]
U76		–	115257	113317	646	59	Minor capsid protein; role in DNA packaging [HCMVUL104][HSVUL6]
U77		+	115100	117574	824	75	Component of DNA helicase-primase complex; helicase [HCMVUL105]HSVUL5]
U78		–	119038	118709	109	*	
U79	ex1	+	120164	120794	474	42	Probable role in DNA replication
	ex2		120891	121087			Includes U80[HCMVUL112/113]
	ex3		121170	121766			
U81		–	122577	121810	255	58	Uracil-DNA glycosylase [HCMVUL114][HSVUL2]
U82		–	123405	122653	250	38	Envelope glycoprotein gL; complexes with gH in gH/gL/gQ or gH/gL/gO [HCMVUL114][HSVUL1]
U83		+	123528	123821	97	*	Chemokine
U84		–	124953	123925	342	42	Spliced in [HCMVUL117]
U85		–	125853	124981	290	37	Probable membrane glycoprotein; related to OX-2
U86		–	130044	125989	1351	29	IE-A^{g} protein; includes U87; related HCMVIE2 [HCMVUL122]
U88			131034	132272	413		IE-A open in all frames, cys repeats
U90	ex1	–	136112	136054	941	28	IE-A transactivator; includes U89; positional
	ex2	–	135965	135772			homologue HCMV IE1
	ex3	–	135664	133092			
U91	ex1	+	136267	136477	153	27	Probable membrane glycoprotein
	ex2	+	136580	136830			
U92		–	138492	138052			Kpn repeats, no methionine initiator
U93		–	139124	138534			Kpn repeats, no initiator, part duplicate U92
U94		–	142867	141395	490	*	AAV-2 rep 68/78 homologue; replication inhibition; latency gene; single stranded DNA binding protein
U95		+	142942	146307	1121	25	HCMVUS22 gene family; positional homologue MCMVIE2

Table 1 (*continued*)

Gene[a]		Strand	Start[b]	Stop[c]	Codons	Id[d]	Properties[e]
U100	ex1	–	150282	149873	656	28	Envelope glycoprotein gp82/105 gQ forms a
	ex2	–	149771	149490			complex with gH/gL binds CD46
	ex3	–	149081	148746			Includes U96–U99
	ex4	–	148628	148551			
	ex5	–	148454	148347			
	ex6	–	148255	148142			
	ex7	–	148055	147895			
	ex8	–	147383	147374			
	ex9	–	147223	147095			
	ex10	–	146984	146642			
DR1	ex1	+	151735	151993	689	33	HCMVUS22 gene family
	ex2		152077	153887			Includes DR2
DR6	ex1	+	155959	156262	395	59	HCMVUS22 gene family
	ex2		157071	157954			Includes DR7 transactivator/transformation

Note: Genes specific to HHV-6 are given an asterisk *. Positional homologue indicated by 'p'.

[a]Exons (ex) are listed. Genes with counterparts in all mammalian herpesviruses are shaded. An extra G residue has been inserted at 128132 in the sequence of Gompels et al. (1995) as indicated by Nicholas (1996).

[b]First exons: from first nucleotide of first complete codon (U7) or initiation codon (other ORFs).

[c]To last nucleotide of stop codon or exon.

[d]Percentage identical amino acid residues to the HHV-7 strain RK counterpart as determined by Gap at default values; U23 proteins did not align at these settings.

[e]Properties derived from the current analysis and summaries in Gompels et al. (1995), Nicholas (1996) and numerous other herpesvirus genome sequence papers. For genes with counterparts in all mammalian herpesviruses, the HSV-1 nomenclature is given in square parentheses. For genes conserved in beta herpesviruses the HCMV homologue is also given.

[f]Immediate early B locus.

[g]Immediate early A locus. Shading indicates conserved genes in human herpesviruses; feature table adapted from Genbank submission update (Gompels and Davison, in preparation).

further identified as at least one gene has been shown to have non-consensus alternative splicing, U83 (French et al., 1999). Thus sequence comparisons between strains may be confounded from such alternative splicing.

Analyses of HHV-6 and HHV-7 coding sequences in addition to genomic organization, also place them within the betaherpesvirus subgroup of the herpesviridae. Together, they share the closest relation with HCMV, the prototype betaherpesvirus, with approximately two thirds of the genes encoding similar proteins with a third betaherpesvirus specific as shown in Fig. 1. However, this relationship is distant and can only be determined by encoded amino acid sequences as well as 'positional homologues'. Furthermore, HCMV is almost double the size, 230 kb, encoding extended glycoprotein gene families absent from HHV-6 and HHV-7. These appear to be the main genes under selection, suggesting distinct immune surveillance control (Dolan et al., 2004). HCMV is representative of beta-1 herpesviruses, cytomegaloviruses and HHV-6/-7 as beta-2 herpesviruses, roseoloviruses. There are several notable gaps in the genomic sequence with no genes identified. One is filled by the origin of lytic replication, ORI, and two others at U18 and U79, interestingly overlap with similar regions in HCMV where possible regulatory microRNAs have been identified (HCMV UL36 and UL112) (Pfeffer et al., 2005), and may be conserved in HHV-6.

There are some key genes that highlight broad similarities and some distinctions between these betaherpesviruses are reflected in their genomic composition. The virus infects and spreads by cell fusion and candidate glycoproteins which mediate this process have been identified in each virus, the conserved gH and gL complex, encoded by HHV-6 and HHV-7 U48 and U82 (Liu et al., 1993a,b; Anderson et al., 1996; Anderson and Gompels, 1999; Mori et al., 2003; Santoro et al., 2003). This is tempered by either two glycoproteins, gQ1/2 (U100) or gO (U47), which dictate different receptor interactions for the complex, either with widespread CD46 for the former, or an unidentified receptor for the latter (Mori et al., 2003; Akkapaiboon et al., 2004). Conserved and specific replication and structural genes have been identified. An origin of lytic replication has been localized and characterized plus viral genes involved in replication including enzyme targets of established anti-viral drugs, such as a viral DNA polymerase, HHV-6 and HHV-7 U38, and phosphotransferase, HHV-6 and HHV-7 U69. Unlike HCMV, HHV-6 and HHV-7 encode an 'OBP', homologue, HHV-6 and HHV-7 U73, which is also found in alphaherpesviruses (Krug et al., 2001). This suggests a difference in replication strategy from that of HCMV, although HHV-6 and HHV-7 have more complicated origins, a feature similar to HCMV rather than HSV. These differences, may affect replication strategies as well as the specificity of potential anti-virals, which target replication. About a third of HHV-6 and HHV-7 genes are specific to these two viruses and presumably reflect adaptations to their particular cellular tropisms.

In both HHV-6 and HHV-7 genome, a number of genes encoding cellular homologues have been identified. These include chemokine receptors (U12 and U51), a chemokine (specific for HHV-6, U83) and members of the immunoglobulin super-family, including an OX-2 homologue (HHV-6 and HHV-7 U85). These are either examples of convergent evolution or are cellular genes trapped via an RNA intermediate, as they lack splicing, in the distant past as they show little nucleotide sequence similarity with homologues determined only via amino acid sequence comparisons. Nonetheless, functional similarities have been detected for the chemokine and chemokine receptor genes and these may contribute an immunomodulatory role and also affect virus dissemination, thus providing HHV-6-specific enhancer functions (Isegawa et al., 1998; French et al., 1999; Zou et al., 1999; Milne et al., 2000; Luttichau et al., 2003; Dewin et al., 2006; Tadagaki et al., 2005).

As in other herpesviruses and first identified in HSV (Honess and Roizman, 1974), gene regulation appears to follow a 'cascade' regulation. This is composed of 'immediate-early' (IE) or 'alpha' genes which include regulators of virus gene expression, followed by 'early' (E) or 'beta' genes including enzymes for DNA replication, then 'late' (L) or 'gamma' genes which include structural genes for the virus particle. The IE genes are important in the switch between lytic replication and latency and include U86 and U90 sequence and positional homologues respectively of HCMV IE2 and IE1 as well as U17 also with an HCMV positional homologues (Schiewe et al., 1994; Flebbe-Rehwaldt et al., 2000; Stanton et al., 2002). In HHV-6, the spliced U90 IE1 gene is hypervariable between HHV-6A and HHV-6B strains, shows different organization between strain variants (Yamamoto et al., 1994), and has been used for genotyping.

Herpesvirus conserved and HHV-6-specific genes

The human herpesvirus conserved genes are in the core of the genome. These represent essential activities involved in primary envelopment, virus fusion, DNA replication, DNA packaging, and capsid formation (Fig. 1 and Table 1). These gene products have been used to determine evolutionary history of the herpesvirus family. Studies of these sequence relationships between the alpha, beta and gammaherpesvirus lineages representing mammalian, bird and some reptile herpesviruses, suggest this divergence is partly due to co-evolution with the host species, giving a date for this event at 400 million years ago. The increased variation over the host species generated by higher mutation/recombination rates or mechanisms is specific for each virus species. Thus, although herpesviruses show high diversity, this has been generated over a long time, compared to that of HIV which has generated similar diversity, but only over several human generations (McGeoch et al., 2000; Bowden et al., 2004; Mattick et al., 2004; McGeoch and Gatherer, 2005).

The betaherpesvirus-specific genes encode mainly at either end of the genome, additional activities in DNA replication (U55), gene regulation (IEA-U86), tegument proteins (U11, U54), immunomodulation and signalling (chemokine receptors), and many with unknown function (US22 family). The gO gene, U47, has a positional homologue in HCMV, UL74, and although they share some motifs, they are largely divergent. Roseolovirus-specific genes are also encoded at either end of the genome. These genes include U13, U15, U20, U21, U23, U24, U26, U85, U88, U90, U91, U92, U93 and U100. Of these no evidence for protein expression has been found for U88, U92 and U93, which overlap repetitive sequences. Where expression studies have been performed or there is information inferred from homologous gene products, the roseolovirus-specific genes encode glycoproteins with possible roles in immunomodulation (U21, U85), and/or virus entry/tropism (U20–U24, U91, U100). These together with the roseolovirus-specific variation present in the conserved genes, as indicated by the shared similarity with HHV-7 (Table 1), provide the fine tuning necessary for the biology of this subgroup.

The HHV-6-specific genes identified (Fig. 1, Table 1) include U1, U9, U22, U61, U78, U83 and U94. U1, U9, U61 and U78 are all small genes with no evidence for expression. While U22, U83, and U94 are expressed and again cover the fine-tuning categories, U22 a late glycoprotein, U83 a viral chemokine with a role in immunomodulation and/or virus dissemination, and U94 a latency gene with roles in gene regulation and replication inhibition (Rotola et al., 1998; French et al., 1999; Mori et al., 2000; Dhepakson et al., 2002; Turner et al., 2002; Luttichau et al., 2003; Casselli et al., 2005; Dewin et al., 2006). There are a number of small open reading frames which have been identified previously as HHV-6A or HHV-6B specific or variable in earlier studies, but these are not included in the feature table as currently there is no evidence for expression and further these are in repetitive sequences which can give sequencing artefacts as well as are prone to mutagenic recombination. However, of the HHV-6-specific genes, while U22 and U94 are well conserved between HHV-6A and HHV-6B strains, U83 is hypervariable, encodes a chemokine which can direct recruitment of cellular subsets for infection and dissemination, thus likely to add specificity in possible cellular niches between strains (French et al., 1999; Luttichau et al., 2003; Dewin et al., 2006).

Genomes and HHV-6 strains

Like other herpesviruses there are numerous HHV-6 strains that have been identified. Most analyses have been performed on a limited set of loci, and there are genes that show greater variation or more conservation, again similar to other herpesvirues analysed. The most comprehensive analyses of herpesvirus genomic differences have been on HCMV also a betaherpesvirus, where 12 distinct genomic strains have been sequenced as well as large segments of genomic analyses (Murphy et al., 2003; Dolan et al., 2004). Betaherpesviruses characteristically have extensive variation, a combination of conserved genes together with hotspots of hypervariation, presumably this reflects a betaherpesvirus-specific mode of DNA replication with consequent specific mutation/recombination rates distinct from the other herpesvirus lineages and the human host (Murphy et al., 2003; Dolan et al., 2004).

HHV-6 also encodes roseolovirus specific (OBP, U73, although related to alphaherpesvirus proteins) and HHV-6-specific (U94) replication genes, which also will influence rate of change in this virus. In analyses of 3 strains genomic sequence, the results show that similar to the recent HCMV results (not available when HHV-6 genomes originally sequenced and analyzed) (Murphy et al., 2003). HHV-6 also shows a combination of conserved genes as well as hypervariation (Fig. 3). Distinct

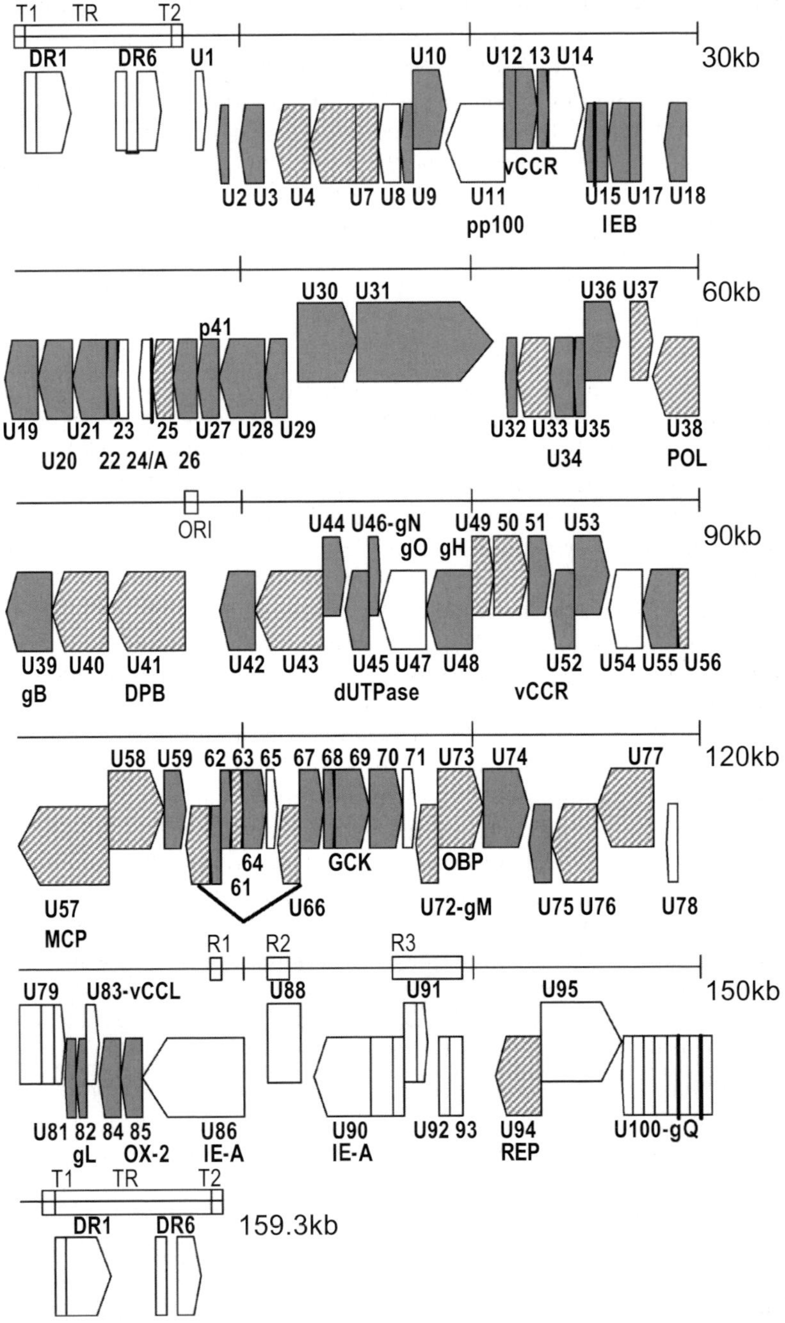

Fig. 3 Structure of HHV-6 genome with strain conserved and variant genes. Genes conserved between strains are shaded grey: patterned, 97–100% identities; solid, above 90% identities.

from HCMV, it currently appears that there are two major strain groups for HHV-6, termed variant A and variant B, whereas for HCMV, although the overall variation is higher than in HHV-6, there appear to be mosaic genomes, with some evidence for linked subsets of genes. This could also be the case for HHV-6, but the demographics and age of acquisition appear to be distinct. So for HHV-6 there appears to be congenital and early infection, perhaps first fever for infants, whereas although there can be congenital infection with HCMV, infection usually comes later in early childhood. Therefore, infection with HHV-6 could appear as Mendelian traits linked with close family groups.

To date the only population identified where there are equal prevalence of both strain groups identified is in Southern Africa (Zambia) through analyses of the gO gene, U47 (Kasolo et al., 1997). In other countries, primarily USA, Europe and Japan, where limited PCR analyses have been performed, the major childhood strain identified has been variant B. While for HCMV where variation has been analysed, representatives containing all groups of variant genes have been identified worldwide, including the gO homologue, HCMV UL74 (Dolan et al., 2004; Mattick et al., 2004). Mosaics might be possible in countries where A and B variants are equally present, and limited evidence can be observed for this in analyses of the U47 gO gene which is at a site for recombination and divergence of herpesvirus lineages (Gompels et al., 1995; Kasolo et al., 1997). Where HHV-6 sequences are examined from both blood samples of both HIV positive and negative febrile infants from Zambia (Kasolo et al., 1997; Gompels and Kasolo, unpublished results) recombination between the variants at these loci are indicated (Table 2). Here, there appears to be mosaics between A and B strains as identified using A or B specific codons. Further genomic-wide investigation will be required to address this. To fully examine this diversity and its implications for disease relationships, advances in technology may be required in order to analyse strains present only in small blood or tissue samples in acute pediatric infections from remote regions of the world. However recent data demonstrating use of saliva samples in following primary infections could aid these analyses (Zerr et al., 2005).

Inspection of the variation of the HHV-6 genomes sequenced identifies highly conserved as well as variant genes (Fig. 3). While the average identity between sequences from HHV-6A and HHV-6B strains has been reported at 90%, this includes all hypervariable genes and mismatched repetitive sequences in the repeat regions (Dominguez et al., 1999; Isegawa et al., 1999). As shown previously, and summarized in Fig. 2, most of the genome is conserved with less than 10% variation, averaging at 5%. In addition, over a quarter of the genes, 22 open reading frames (ORFs) have less than this average divergence, showing 97–100% nucleotide sequence identities, which would make these difficult or impossible to use to identify differences between strains. These include U4, U7, U25, U33, U37, U38, U40, U41, U43, U49, U50, U56, U57, U58, U63, U66, U72, U73, U76, U77 and U94. These are scattered throughout the genome and include genes with roles in gene regulation (IEB-U17), DNA replication (OBP-U73, DNA polymerase-U35), DNA packaging, glycoprotein genes, tegument genes for envelopment, receptor

Table 2

Strain differences in U47 and recombination

	U47 N-terminal coding changes											
		*	*			*	*	*	*			
Strain	1	2	3	4	5	6	7	8	9	10	11	**Variant**
Lab strain												
U1102+	S	S	S	K	N	V	P	K	Q	T	F	**A**
AJ+	—	—	—	E	I	—	—	—	—	—	—	**A**
GS+	—	—	—	E	—	—	—	—	—	I	—	**A**
Z29+	—	P	P	E	I	I	R	E	H	—	—	**B**
HST	—	P	P	E	I	I	R	E	H	—	—	**B**
European												
KF	—	—	P	E	I	I	R	E	H	—	—	**B**
MBE+	—	P	P	E	I	I	E	E	H	—	—	**B**
MAR+	—	P	P	E	I	I	E	E	H	—	—	**B**
BOU+	—	P	P	E	I	I	E	E	H	—	—	**B**
Zambian												
Zam7	—	—	—	E	—	—	—	—	H	I	—	**A**
Zam12	—	P	P	E	—	—	—	—	H	I	—	**B/A**
Zam13	—	P	P	E	I	I	R	E	H	—	—	**B**
Zam15	—	P	P	E	—	I	R	E	H	—	—	**B**
Zam18	—	P	P	E	I	I	R	—	—	—	—	**B**
Zam317	—	P	P	E	I	I	R	—	—	—	—	**B/A**
Zam325	—	—	—	E	I	L	—	—	H	—	—	**A/B**
Zam352	—	—	—	E	I	—	—	E	—	—	—	**A/B**
Zam25Ly	—	—	—	—	—	—	—	—	—	—	—	**A**
Zam3+	—	P	P	E	I	I	R	E	H	—	—	**B**
Zam5+	—	—	—	E	I	—	—	—	—	—	—	**A**
Zam14+	L	P	P	E	I	I	R	E	H	—	—	**B**
Zam23+	—	—	—	—	—	I	—	—	—	—	—	**A**
Zam25+	—	P	P	E	I	I	R	E	H	—	—	**B**
Zam27+	—	—	—	—	—	—	—	—	—	—	L	**A**
Zam31+	—	P	P	E	I	I	R	E	H	—	—	**B**
Zam50+	—	P	P	E	I	I	R	—	H	—	—	**B**
Zam59+	—	—	—	—	—	—	—	—	—	—	L	**A**

Note: * indicates variant specific coding change, + indicates HIV positive. Coding changes 1–11 correspond to U47 amino acids 215/218/219/220/239/240/247/248/256/259/ in strain U1102 and 182/185/186/187/206/207/214/215/223/226/ in strain Z29 from start methionine. Lab strains data from Genbank genomes (Gompels et al., 1995; Dominguez et al., 1999; Isegawa et al., 1999), MBE, MAR, BOU strains (Aubin et al., 1991, 1993), other strains sequences and Zambian HIV-samples from Gompels et al. (1993) and Kasolo et al. (1997). Zambian HIV+ sequences are from Kasolo and Gompels (unpublished).

signalling and structural genes (Fig. 3). They overlap with the genes conserved in all mammalian herpesviruses, but also have genes distinct to betaherpesviruses (US22 family; U33 virion protein) and HHV-6 (U94 rep and latency gene). Similar findings have now been demonstrated for HCMV strains in whole genome analyses

(Murphy et al., 2003; Dolan et al., 2004) where a core of genes is conserved. However, there are also hotspots for variation between the strains that have also been found in HCMV. Some of these areas overlap, for example, regions adjacent to the repeats, particularly at the right end in this genomic orientation, this repeat-rich region may drive accumulation of mutations here, as repetitive sequences through recombination mismatches, for example can generate change. These variable regions include genes for expression regulation (IEA-U86/U90), glycoproteins (gO-U47, gQ-U100) and immunomodulation (U83) which may contribute to fine tuning of cellular niches *in vivo*, immune evasion or possibly adaptations to allelic variation present in host populations. Similar results have been identified in HCMV where hotspots for variation also include specific glycoproteins and immunomodulatory genes (Murphy et al., 2003; Dolan et al., 2004).

Acknowledgements

Gompels acknowledges the support of the BBSRC and Wellcome Trust; Kasolo thanks the Wellcome Trust for an international career development award.

References

Ablashi D, Agut H, Berneman Z, Campadelli-Fiume G, Carrigan D, Ceccerini-Nelli L, Chandran B, Chou S, Collandre H, Cone R, Dambaugh T, Dewhurst S, Diluca D, Foatomasi L, Fleckenstein B, Frenkel N, Gallo R, Gompels U, Hall C, Jones M, Lawrence G, Martin M, Montagnier L, Neipel F, Nicholas J, Pellett P, Razzaque A, Torrelli G, Thomson B, Salahuddin S, Wyatt L, Yamanishi K. Arch Virol 1993; 129: 1–4.

Akkapaiboon P, Mori Y, Sadaoka T, Yonemoto S, Yamanishi K. J Virol 2004; 78: 7969–7983.

Anderson RA, Gompels UA. J Gen Virol 1999; 80(Pt 6): 1485–1494.

Anderson RA, Liu DX, Gompels UA. Virology 1996; 217: 517–526.

Borenstein R, Singer O, Moseri A, Frenkel N. J Virol 2004; 78: 4730–4743.

Bowden R, Sakaoka H, Donnelly P, Ward R. Infect Genet Evol 2004; 4: 115–123.

Caselli E, Bracci A, Galvan M, Boni M, Rotola A, Bergamini C, Cermelli C, Dal Monte P, Gompels UA, Cassai E, Di Luca D. Virol 2005 Dec 17; [Epub ahead of print].

Cha TA, Tom E, Kemble GW, Duke GM, Mocarski ES, Spaete RR. J Virol 1996; 70: 78–83.

Chmielewicz B, Goltz M, Lahrmann KH, Ehlers B. Xenotransplantation 2003; 10: 349–356.

Deng H, Dewhurst S. J Virol 1998; 72: 320–329.

Dewin DR, Catusse J, Gompels UA. J Immunol 2006; 176: 544–556.

Dhepakson P, Mori Y, Jiang YB, Huang HL, Akkapaiboon P, Okuno T, Yamanishi K. J Gen Virol 2002; 83: 847–854.

Di Luca D, Mirandola P, Ravaioli T, Dolcetti R, Frigatti A, Bovenzi P, Sighinolfi L, Monini P, Cassai E. J Med Virol 1995; 45: 462–468.

Dolan A, Cunningham C, Hector RD, Hassan-Walker AF, Lee L, Addison C, Dargan DJ, McGeoch DJ, Gatherer D, Emery VC, Griffiths PD, Sinzger C, Mcsharry BP, Wilkinson GW, Davison AJ. J Gen Virol 2004; 85: 1301–1312.

Dominguez G, Dambaugh TR, Stamey FR, Dewhurst S, Inoue N, Pellett PE. J Virol 1999; 73: 8040–8052.
Downing RG, Sewankambo N, Serwadda D, Honess R, Crawford D, Jarrett R, Griffin BE. Lancet 1987; 2: 390.
Dykes C, Chan H, Krenitsky DM, Dewhurst S. J Gen Virol 1997; 78(Pt 5): 1125–1129.
Ehlers B, Burkhardt S, Goltz M, Bergmann V, Ochs A, Weiler H, Hentschke J. J Gen Virol 2001; 82: 475–482.
Flebbe-Rehwaldt LM, Wood C, Chandran B. J Virol 2000; 74: 11040–11054.
French C, Menegazzi P, Nicholson L, Macaulay H, Diluca D, Gompels UA. Virology 1999; 262: 139–151.
Gautheret-Dejean A, Aubin JT, Poirel L, Huraux JM, Nicolas JC, Rozenbaum W, Agut H. J Clin Microbiol 1997; 35: 1600–1603.
Gompels UA. Roseoloviruses: Human Herpesviruses 6 and 7. In: Principles and Practice of Clinical Virology (Zuckerman AJ, Banatvala JE, Pattison JR, Griffiths PD, Schoub BD, editors). 5th ed. Chichester: Wiley; 2004.
Gompels UA, Carrigan DR, Carss AL, Arno J. J Gen Virol 1993; 74(Pt 4): 613–622.
Gompels UA, Luxton J, Knox KK, Carrigan DR. Lancet 1994; 343: 735–736.
Gompels UA, Macaulay HA. J Gen Virol 1995; 76: 451–458.
Gompels UA, Nicholas J, Lawrence G, Jones M, Thomson BJ, Martin ME, Efstathiou S, Craxton M, Macaulay HA. Virology 1995; 209: 29–51.
Hall CB, Caserta MT, Schnabel KC, Boettrich C, Mcdermott MP, Lofthus GK, Carnahan JA, Dewhurst S. J Pediatr 2004; 145: 472–477.
Hall CB, Long CE, Schnabel KC, Caserta MT, Mcintyre KM, Costanzo MA, Knott A, Dewhurst S, Insel RA, Epstein LG. N Engl J Med 1994; 331: 432–438.
Honess RW, Gompels UA, Barrell BG, Craxton M, Cameron KR, Staden R, Chang YN, Hayward GS. J Gen Virol 1989; 70: 837–855.
Honess RW, Roizman B. J Virol 1974; 14: 8–19.
Inoue Y, Yasukawa M, Fujita S. J Virol 1997; 71: 3751–3759.
Isegawa Y, Mukai T, Nakano K, Kagawa M, Chen J, Mori Y, Sunagawa T, Kawanishi K, Sashihara J, Hata A, Zou P, Kosuge H, Yamanishi K. J Virol 1999; 73: 8053–8063.
Isegawa Y, Ping Z, Nakano K, Sugimoto N, Yamanishi K. J Virol 1998; 72: 6104–6112.
Kasolo FC, Mpabalwani E, Gompels UA. J Gen Virol 1997; 78: 847–855.
Katsafanas GC, Schirmer EC, Wyatt LS, Frenkel N. Proc Natl Acad Sci USA 1996; 93: 9788–9792.
Kempf W, Adams V, Wey N, Moos R, Schmid M, Avitabile E, Campadelli-Fiume G. Proc Natl Acad Sci USA 1997; 94: 7600–7605.
Kidd IM, Clark DA, Ait-Khaled M, Griffiths PD, Emery VC. J Infect Dis 1996; 174: 396–401.
Kondo K, Kondo T, Okuno T, Takahashi M, Yamanishi K. J Gen Virol 1991; 72: 1401–1408.
Kondo K, Kondo T, Shimada K, Amo K, Miyagawa H, Yamanishi K. J Med Virol 2002a; 67: 364–369.
Kondo K, Shimada K, Sashihara J, Tanaka-Taya K, Yamanishi K. J Virol 2002b; 76: 4145–4151.
Kondo K, Sashihara J, Shimada K, Takemoto M, Amo K, Miyagawa H, Yamanishi K. J Virol 2003; 77: 2258 2264.
Krug LT, Inoue N, Pellett PE. Virology 2001; 288: 145–153.

Lacoste V, Verschoor EJ, Nerrienet E, Gessain A. J Gen Virol 2005; 86: 2135–2140.
Liu DX, Gompels UA, Foa-Tomasi L, Campadelli-Fiume G. Virology 1993a; 197: 12–22.
Liu DX, Gompels UA, Nicholas J, Lelliott C. J Gen Virol 1993b; 74: 1847–1857.
Lopez C, Pellett P, Stewart J, Goldsmith C, Sanderlin K, Black J, Warfield D, Feorino P. J Infect Dis 1988; 157: 1271–1273.
Lusso P, Malnati M, De Maria A, Balotta C, Derocco SE, Markham PD, Gallo RC. J Immunol 1991; 147: 685–691.
Luttichau HR, Clark-Lewis I, Jensen PO, Moser C, Gerstoft J, Schwartz TW. J Biol Chem 2003; 278: 10928–10933.
Martin ME, Thomson BJ, Honess RW, Craxton MA, Gompels UA, Liu MY, Littler E, Arrand JR, Teo I, Jones MD. J Gen Virol 1991; 72: 157–168.
Mattick C, Dewin D, Polley S, Sevilla-Reyes E, Pignatelli S, Rawlinson W, Wilkinson G, Dal Monte P, Gompels UA. Virology 2004; 318: 582–597.
McGeoch DJ, Dolan A, Ralph AC. J Virol 2000; 74: 10401–10406.
McGeoch DJ, Gatherer D. J Virol 2005; 79: 725–731.
Megaw AG, Rapaport D, Avidor B, Frenkel N, Davison AJ. Virology 1998; 244: 119–132.
Milne RS, Mattick C, Nicholson L, Devaraj P, Alcami A, Gompels UA. J Immunol 2000; 164: 2396–2404.
Mori Y, Dhepakson P, Shimamoto T, Ueda K, Gomi Y, Tani H, Matsuura Y, Yamanishi K. J Virol 2000; 74: 6096–6104.
Mori Y, Yang X, Akkapaiboon P, Okuno T, Yamanishi K. J Virol 2003; 77: 4992–4999.
Mrazek J, Karlin S. Proc Natl Acad Sci USA 1998; 95: 3720–3725.
Murphy E, Yu D, Grimwood J, Schmutz J, Dickson M, Jarvis MA, Hahn G, Nelson JA, Myers RM, Shenk TE. Proc Natl Acad Sci USA 2003; 100: 14976–14981.
Nicholas J. J Virol 1996; 70: 5975–5989.
Pfeffer S, Sewer A, Lagos-Quintana M, Sheridan R, Sander C, Grasser FA, Van Dyk LF, Ho CK, Shuman S, Chien M, Russo JJ, Ju J, Randall G, Lindenbach BD, Rice CM, Simon V, Ho DD, Zavolan M, Tuschl T. Nat Methods 2005; 2: 269–276.
Rapp JC, Krug LT, Inoue N, Dambaugh TR, Pellett PE. Virology 2000; 268: 504–516.
Roffman E, Frenkel N. J Infect Dis 1991; 164: 617–618.
Rotola A, Ravaioli T, Gonelli A, Dewhurst S, Cassai E, Di Luca D. Proc Natl Acad Sci USA 1998; 95: 13911–13916.
Salahuddin SZ, Ablashi DV, Markham PD, Josephs SF, Sturzenegger S, Kaplan M, Halligan G, Biberfeld P, Wong-Staal F, Kramarsky B, Gallo RC. Science 1986; 234: 596–601.
Santoro F, Greenstone HL, Insinga A, Liszewski MK, Atkinson JP, Lusso P, Berger EA. J Biol Chem 2003; 278: 25964–25969.
Schiewe U, Neipel F, Schreiner D, Fleckenstein B. J Virol 1994; 68: 2978–2985.
Secchiero P, Flamand L, Gibellini D, Falcieri E, Robuffo I, Capitani S, Gallo RC, Zauli G. Blood 1997; 90: 4502–4512.
Stamey FR, Dominguez G, Black JB, Dambaugh TR, Pellett PE. J Virol 1995; 69: 589–596.
Stanton R, Fox JD, Caswell R, Sherratt E, Wilkinson GW. J Gen Virol 2002; 83: 2811–2820.
Tadagaki K, Nakano K, Yamanishi K. J Virol 2005; 79: 7068–7076.
Takahashi K, Sonoda S, Higashi K, Kondo T, Takahashi H, Takahashi M, Yamanishi K. J Virol 1989; 63: 3161–3163.
Takemoto M, Shimamoto T, Isegawa Y, Yamanishi K. J Virol 2001; 75: 10149–10160.

Tanaka-Taya K, Sashihara J, Kurahashi H, Amo K, Miyagawa H, Kondo K, Okada S, Yamanishi K. J Med Virol 2004; 73: 465–473.
Tedder RS, Briggs M, Cameron CH, Honess R, Robertson D, Whittle H. Lancet 1987; 2: 390–392.
Thomson BJ, Dewhurst S, Gray D. J Virol 1994; 68: 3007–3014.
Turner S, Diluca D, Gompels U. J Virol Methods 2002; 105: 331–341.
Van Loon NM, Gummuluru S, Sherwood DJ, Marentes R, Hall CB, Dewhurst S. Clin Infect Dis 1995; 21: 1017–1019.
Ward KN, Thiruchelvam AD, Couto-Parada X. J Med Virol 2005; 76: 563–570.
Yamamoto T, Mukai T, Kondo K, Yamanishi K. J Clin Microbiol 1994; 32: 473–476.
Yamanishi K, Okuno T, Shiraki K, Takahashi M, Kondo T, Asano Y, Kurata T. Lancet 1988; 1: 1065–1067.
Yasukawa M, Inoue Y, Ohminami H, Sada E, Miyake K, Tohyama T, Shimada T, Fujita S. J. Virol 1997; 71: 1708–1712.
Yasukawa M, Inoue Y, Ohminami H, Terada K, Fujita S. J Gen Virol 1998; 79: 143–147.
Zerr DM, Meier AS, Selke SS, Frenkel LM, Huang ML, Wald A, Rhoads MP, Nguy L, Bornemann R, Morrow RA, Corey L. N Engl J Med 2005; 352: 768–776.
Zhang Y, De Bolle L, Aquaro S, Van Lommel A, De Clercq E, Schols D. J Virol 2001; 75: 10511–10514.
Zou P, Isegawa Y, Nakano K, Haque M, Horiguchi Y, Yamanishi K. J Virol 1999; 73: 5926–5933.

Human Herpesvirus-6: General Virology, Epidemiology and Clinical Pathology
Perspectives in Medical Virology
Gerhard Krueger and Dharam Ablashi (Editors)

DOI 10.1016/S0168-7069(06)12004-2

Proteins of HHV-6

Yasuko Mori
Laboratory of Virology and Vaccinology, National Institute of Biomedical Innovation, 7-6-8, Saito-Asagi, Ibaraki, Osaka 567-0085, Japan

Introduction

HHV-6 belongs to the genus *Roseolavirus* of β-herpesviruses (Pellett and Dominguez, 2001). HHV-6 is classified into variants A and B (HHV-6A and HHV-6B) on the basis of distinct genetic, immunological and biological characteristics. HHV-6 genome contains open reading frames (ORFs) U1-U100 and flanking 8–9 kb terminal direct repeats (DRs) at either end. HHV-6B contains 119 ORFs and HHV-6A contains 110 ORFs (Dockrell, 2003). Here, the proteins of HHV-6 ever reported are shown.

IE proteins

HHV-6 gene transcription follows a similar pattern that characterizes herpesvirus, with immediate-early (IE), early and late proteins expressed (Dockrell, 2003). IE proteins are the first proteins expressed following viral entry, independent of *de novo* protein synthesis and play a crucial role in the initiation of infection, and the establishing productive infections, regulating reactivation from latency and evading immune recognition.

HHV-6 immediate-early A locus (IE-A) locates in the position analogous to the human cytomegalovirus (HCMV) major IE (MIE) locus that is well-known to play critical roles in viral infection. Similarly to HCMV MIE, HHV-6 IE-A consists of two genetic units, IE1 and IE2, corresponding to ORFs U90–U89 and U90–U86/87, respectively (Papanikolaou et al., 2002). However, the HHV-6 IE-A locus exhibits limited sequence homology with the HCMV MIE locus. IE2 proteins derived from the U86/87 region with apparent molecular mass of 100, 85 and 55 kDa are detected in HHV-6-infected cells 3 days after infection, while IE1 proteins with apparent molecular mass greater than 170 kDa are detectable as early as 8 h (Papanikolaou et al., 2002). Mapping of the IE2 proteins suggests differential splicing and alternative translation initiation in the IE2 genetic unit. The IE2 proteins show a mixed cytoplasmic and nuclear localization pattern. Thus, it is detected (8–48 h) as intranuclear granules, while at later time points (72–120 h) the IE2 protein coalesces into a few large immunoreactive patches. Transfection of cells with an IE2 expression vector alone does not reproduce the patch-like distribution, suggesting that other viral proteins are necessary for this process to occur (Papanikolaou et al., 2002). In addition, the 437 amino acid carboxyl-terminus domain of IE2 binds to a DNA fragment containing the putative IE-A promoter. Cotransfection experiments in T cells indicate that IE2 can induce the transcription of a complex promoter, such as the human immunodeficiency virus (HIV)-long terminal repeat (LTR), as well as simpler promoters, whose expression is driven by a unique set of responsive elements (CRE, NFAT and NF-kB) (Gravel et al., 2003). Moreover, minimal promoters having a single TATA box or no defined eukaryotic regulatory elements are significantly activated by IE2, suggesting that IE2 is likely to play an important role in initiating the expression of several HHV-6 genes.

Furthermore, the heterogeneous nuclear ribonucleoprotein K (hnRNP K) and the beta subunit of casein kinase 2 (CK2beta) interacts with HHV-6 IE2, indicating that these interactions may affect viral and cellular RNA transcription and translation in viral replication (Shimada et al., 2004).

Analyses of HHV-6 IE gene expression have revealed that the IE1 gene of the HHV-6A and HHV-6B variants exhibits a higher degree of sequence variation than other regions of the genome and no obvious similarity to its positional analogue in HCMV.

The IE1B (HHV-6B) transcript consists of five exons (3720 nucleotides), three of which are coding for the IE1 protein (Gravel et al., 2002). The 1078-amino acid-long IE1B protein is 62% identical and 75% similar to the 941-amino acid IE1 from HHV-6A. IE1B protein can be detected at 4 h p.i.I, and it is distributed as small intranuclear structures. The maximal number of IE1 bodies (approximately 10–12/nucleus) is detected at 12 h p.i., after which the IE1 bodies condense into 1–3 larger entities by 24–48 h p.i. (Gravel et al., 2002). During infection, the IE1B protein is phosphorylated on serine and threonine residues. HHV-6A IE1 can also be detected at 4 h p.i. as small dots, and accumulates at later PI (Fig. 1).

IE1 proteins form a stable interaction with, PML-bodies [also known as ND10 or nuclear promyelocytic leukemia protein (PML) oncogenic domains (PODS)] (Stanton et al., 2002). Remarkably, PML bodies remain structurally intact and associate with the IE1 protein throughout lytic HHV-6 infection, unlike other herpesviruses, suggesting that HHV-6 appears to have no requirement or mechanism to induce PML-body dispersal during lytic replication. In addition, IE1 is

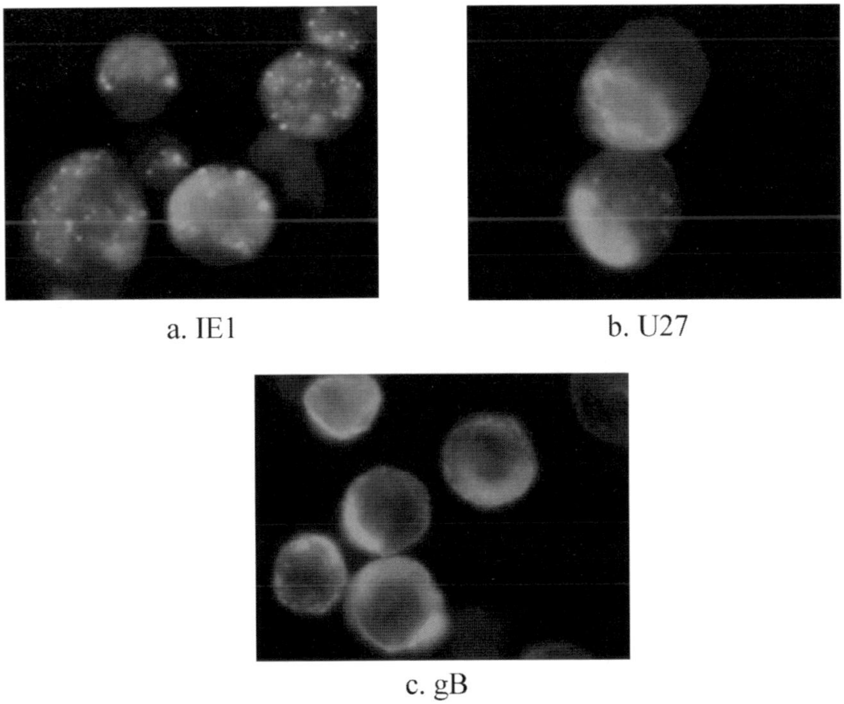

Fig. 1 Indirect immunofluorescence assay of HHV-6-infected cells. The HSB-2 cells infected with HHV-6A (strain GS) were stained with monoclonal antibody for IE1 (a) U27 (b) or gB (c) at 86 h p.i. (a) IE1 locates in nucleus with punctuated pattern. (b) U27 locates in nucleus like forming replication compartment. (c) gB locates in the cytoplasm. (for colour version: see colour section on page 349).

covalently modified by conjugation to the small ubiquitin-like protein, SUMO-1. Overexpression of SUMO-1 in cell lines results in substantially enhanced levels of IE1 expression; thus, sumoylation may bestow stability to the protein (Stanton et al., 2002). However, SUMOylation-deficient mutants of IE1 co-localize with PODs as efficiently as wild type IE1, indicating that POD targeting is independent of IE1 SUMOylation status (Gravel et al., 2004).

U16/U17

HHV-6 U16 is located in the immediate-early B (IE-B) region of the genome. The ORF in one cDNA is generated by splicing together in frame ORFs U17 and U16, and the second cDNA included ORFs U16 and U15. A third differentially spliced cDNA (U16+) is identified by 5′ rapid amplification of cDNA ends (Flebbe-Rehwaldt et al., 2000). At least two potential transcription initiation sites are shown to be used to express the transcripts encoding U17 and U16 gene products. The U17/U16 spliced gene products are expressed at IE times after infection, but a multiply spliced gene product encoded by U16 is expressed as a late gene. The U17/U16 and the U16+ gene products transactivate the HIV LTR. Thus, while there are similarities to the HCMV UL36–UL38 gene family, some of the IE-B U17/U16 transcripts are unique to HHV-6 (Flebbe-Rehwaldt et al., 2000).

U53, proteinase

HHV-6 U53 encodes its own proteinase, which is essential for capsid maturation, DNA packaging and the ultimate formation of new virus particles (Tigue et al., 1996). The mature proteinase consists of 230 residues, but is synthesized in the form of a precursor, which has an additional 298 residues attached to the C terminus of the mature enzyme. Autolytic removal of these residues, which themselves constitute a form of the viral assembly protein) releases the N-terminal proteinase in its mature form; processing takes place at two locations positioned, respectively, at the C terminus of the proteinase. Autoprocessing of the precursor form of HHV-6 proteinase at two sites (termed M and R) is required to generate the mature enzyme, which could represent targets for novel antivirals against HHV-6 (Tigue and Kay, 1998a,b). Despite sharing 40% identity with other betaherpesvirus proteinases such as human cytomegalovirus proteinase, the one-chain HHV-6 enzyme is distinguished from these two-chain proteinases by the absence of an internal autocatalytic cleavage site.

U69, kinase

HHV-6 U69 gene product (pU69) is the presumed functional homologue of HCMV UL97-encoded kinase (pUL97), which converts ganciclovir into its monophosphate metabolite in HCMV-infected cells (Ansari and Emery, 1999; Manichanh et al., 2001; De Bolle et al., 2002; Safronetz et al., 2003). pU69, like pUL97, is expressed as a nuclear protein. The insertion of U69 into baculovirus confers sensitivity to ganciclovir in insect cells (Ansari and Emery, 1999). The efficiency of ganciclovir phosphorylation induced by HHV-6 is relatively poor. The ganciclovir is phosphorylated in human cells infected with pU69-expressing recombinant vaccinia viruses (rVVs), although the levels of phosphorylated ganciclovir metabolites are approximately 10-fold lower than those observed with pUL97. The limited phosphorylation of ganciclovir by pU69 may contribute to its modest antiviral activity against HHV-6 in certain cell systems (De Bolle et al., 2002).

Cellular homologues

Herpesviruses provide examples of viral piracy of host genes, which may play roles in immune evasion. HHV-6 encodes several chemokine and chemokine receptor homologues (Isegawa et al., 1998; Menotti et al., 1999; Zou et al., 1999; Milne et al., 2000; Bradel-Tretheway et al., 2003; Luttichau et al., 2003).

U83 encodes a functional chemokine (Zou et al., 1999). Although the gene has relatively little sequence similarity to human chemokine genes, the protein expressed has the typical cysteine residues of a chemokine, transduces signals that involve calcium fluxes and induces chemotactic activation. The recombinant U83 protein is capable of inducing transient calcium mobilization in THP-1 cells and of chemotactically activating THP-1 cells (Zou et al., 1999). Furthermore, the U83 has been found to cause calcium mobilization as efficiently through the CCR2 receptor (Luttichau et al., 2003), suggesting that the U83 protein might play an important role in HHV-6 propagation *in vivo* by activating and trafficking mononuclear cells to sites of viral replication, or U83 during reactivation of the virus in for example monocyte-derived microglia could perhaps be involved in the pathogenesis of the CCR2-dependent disease, multiple sclerosis.

HHV-6 contains two genes, U12 and U51 that encode putative homologues of cellular G-protein-coupled receptors (GCR). The U12 gene is expressed late in infection from a spliced mRNA. U12 functionally encoded a calcium-mobilizing receptor for beta-chemokines such as regulated upon activation, normal T expressed and secreted (RANTES), macrophage inflammatory proteins 1α and 1β (MIP-1α and MIP-1β) and monocyte chemoattractant protein 1, but not for the α-chemokine interleukin-8, suggesting that the chemokine selectivity of the U12 product is distinct from that of the known mammalian chemokine receptors (Isegawa et al., 1998).

U51 gene defines a new family of betaherpesvirus-specific genes encoding multiple transmembrane glycoproteins with similarity to G protein-coupled receptors, in particular, human chemokine receptors. When synthesized in transient expression systems, U51 intracellular trafficking is regulated in a cell-type-dependent fashion. In human monolayer HEK-293 and 143tk− cells, U51 accumulates predominantly in the endoplasmic reticulum and fails to be transported to the cell surface. In contrast, in T-lymphocytic cell lines J-Jhan, Molt-3 and Jurkat, U51 is successfully transported to the plasma membrane. The transport of U51 to the cell surface requires a cell-specific function present in activated T lymphocytes and T-cell lines (Menotti et al., 1999).

On the other hand, U51 stably expressed in cell lines shows specific binding of the CC chemokine RANTES and competitive binding with other beta chemokines, such as eotaxin; monocyte chemoattractant proteins 1, 3 and 4; as well as the HHV-8 chemokine vMIPII. In epithelial cells already secreting RANTES, U51 expression induces specific transcriptional downregulation of RANTES (Milne et al., 2000).

U94, AAV-2 rep homologue

HHV-6A and HHV-6B, respectively encode homologue (U94) of the parvovirus nonstructural gene, ns1 or rep (Thomson et al., 1991). Interestingly, this homologue is encoded just by HHV-6, but not HHV-7. The U94 transcript is spliced to remove a 2.6-kb intron and is expressed at very low levels relative to other HHV-6B genes, reaching approximately 10 copies/cell 3 days after infection (Rapp et al., 2000). HHV-6 rep locates in the nucleus in HHV-6-infected T cells, and is first detected at 24 h p.i. and accumulates to higher levels by 72 h p.i, and even when the late protein glycoprotein H 8 (gH) is detected in nearly 90% of HHV-6-infected cells, the rep protein is detected in only a small percentage of them, indicating that the rep protein may also be expressed at very low levels in HHV-6-infected cells. Furthermore, the rep protein binds to single-stranded DNA (Dhepakson et al., 2002) and binds to a transcription factor, human TATA-binding protein, through its N-terminal region (Mori et al., 2000).

HHV-6 rep is transcribed in latency-infected lymphocytes and contributes to the maintenance of latency: stable expression of rep results in susceptibility to infection but low-level viral replication and gene transcription in the absence of cytopathic effects (Rotola et al., 1998).

HHV-6 rep gene cloned under control of HCMV immediate-early promoter complements replication of a rep-deficient AAV-2 genome (Thomson et al., 1994). In cotransfection experiments with heterologous promoters linked to the CAT reporter gene, HHV-6 rep activates the HIV LTR in fibroblast cell lines but not in T-cells. In contrast, AAV-2 rep inhibits HIV LTR activity in both fibroblast and T-cell lines. The effect of HHV-6 and AAV-2 rep genes on the HIV LTR is independent of the NF-kappa B, Sp1 and TATA box elements. HHV-6 rep may be a multifunctional regulatory protein with properties related to, but distinct from, those of AAV-2 rep (Thomson et al., 1994).

Envelope glycoproteins

The genes, U39 and U48 encode the conserved herpesvirus glycoproteins, glycoprotein B (gB) and glycoprotein H (gH), which are found on virions of most herpesviruses (Cirone et al., 1994).

U39, gB

There is 39% amino acid identity with the gB of HHV-6 and HCMV. HHV6-CMV gB peptide homology is evident through the entire sequence, but is especially strong in the amino-terminal portion of CMV gB, which contains linear and conformational epitopes recognized by CMV-neutralizing antibodies. All 10 cysteine residues of HHV6 gB match corresponding residues of CMV gB. Sequence data suggest strong structural similarity and possible immunologic cross-reactivity of gB from the two viruses (Chou and Marousek, 1992). The neutralizing monoclonal antibodies (mabs) specific for HHV-6 gB inhibit infection and prevent cell–cell fusion (Takeda et al., 1997; Mori et al., 2002).

The primary HHV-6A gB translation product of about 112 kDa is proteolytically cleaved to form disulfide-bond-linked subunits of 64 and 58 kDa, for HHV-6B, the corresponding products are about 102, 59 and 50 kDa, respectively (Foa-Tomasi et al., 1992; Campadelli-Fiume et al., 1993; Ellinger et al., 1993). HHV-6 gB proteins locate in cytoplasm (Fig. 1).

U48 and U82, gH and gL

Formation of the glycoprotein gH/gL heterooligomer has important implications for understanding the pathology of HHV-6-associated disease because this complex is essential for infectivity and fusogenic cell-to-cell spread (Josephs et al., 1991; Liu et al., 1993a,b; Qian et al., 1993; Anderson et al., 1996; Takeda et al., 1997; Anderson and Gompels, 1999). Definition of the HHV-6 gH domain involved in protein–protein interactions is addressed by targeting regions defined by conserved cysteines identified by alignment of gH amino acid sequences representative of all herpesvirus subfamilies. The N-terminus of HHV-6 gH includes a 230-amino-acid domain required for interaction with HHV-6 gL encompassing residues conserved specifically among betaherpesviruses. HCMV homologues, UL75 (gH) or UL115 (gL), can substitute for HHV-6 glycoproteins and participate in heterologous complex formation (Anderson et al., 1996). Furthermore, the region that governs this heterologous gL binding also maps to the N-terminal portion of HHV-6 gH. Surprisingly, further deletion of HHV-6 gH to 145-amino acid-domain residues abolishes complex formation with HHV-6 gL, but allows interaction with HCMV gL (Anderson et al., 1996). Anti-fusion monoclonal antibodies specific for HHV-6 gH inhibits infection and prevent cellular spread by syncytia formation (Foa-Tomasi et al., 1991). Reactivity of these monoclonal antibodies with gH deletion mutants suggests a conserved C-terminal fusion-associated domain (Anderson and Gompels, 1999).

U100, gQ

U100 gene, the glycoprotein Q (gQ) that is unique to the genus *Roseolavirus* of human herpesviruses (Gompels et al., 1995; Dominguez et al., 1999; Isegawa et al., 1999). The U100 gene is subject to differential splicing, and a number of envelope-expressed polypeptides result. In contrast to the other glycoprotein-encoding genes, U100 of HHV-6A and HHV-6B demonstrate only 72.1% sequence identity (Isegawa et al., 1999). This glycoprotein may therefore have a role in the differential effects of HHV-6A and HHV-6B infections. Along with gB and gH, the gQ contains epitopes recognized by neutralizing antibodies and therefore represents a target for variant-specific neutralizing antibodies (Pfeiffer et al., 1993, 1995).

Recently, it was found that HHV-6 gH/gL complex associates with the 80-kDa form of gQ (gQ–80 K) that is found on the viral envelope. Besides gQ–80 K, the gQ gene encodes an additional product whose mature molecular mass is 37 kDa (gQ–37 K) and that is derived from a different transcript. Therefore, they were designated gQ–80 K as gQ1 and gQ–37 K as gQ2 (Mori et al., 2003a; Akkapaiboon et al., 2004). Furthermore, gQ2 also interacts with the gH/gL/gQ1 complex in HHV-6-infected cells and in virions (Akkapaiboon et al., 2004).

HHV-6 employs the complement regulator CD46 (membrane cofactor protein) as a receptor for fusion and entry into target cells (Santoro et al., 1999). Like other known herpesviruses, HHV-6 encodes multiple glycoproteins, several of which have been implicated in the entry process. The gH has been reported to be the viral component responsible for binding to CD46. Antibodies to CD46 co-immunoprecipitated an approximately 110-kDa protein band specifically associated with HHV-6-infected cells, which was identified as gH (Santoro et al., 2003). CD46 also was co-immunoprecipitated with anti-gH antibody (Santoro et al., 2003). Recently, gH/gL/gQ1/gQ2 complex has been found to bind to human CD46 (Mori et al., 2003b; Akkapaiboon et al., 2004).

U47, gO

HHV-6 encodes U47 gene, which is a positional homologue of the HCMV glycoprotein O (gO) gene (Gompels et al., 1995; Isegawa et al., 1998; Dominguez et al., 1999). The U47 gene also encodes a third component of the HHV-6 gH/gL containing envelope complex (Mori et al., 2004). A monoclonal antibody against HHV-6 gO reacted in immunoblots with protein species migrating at 120–130 and 74–80 kDa in lysates of HHV-6-infected cells and with a 74- to 80-kDa protein species in purified virions (Mori et al., 2004). The 80-kDa form of gO is co-immunoprecipitated with an anti-gH mab, but an anti-gQ mab, which co-immunoprecipitates gH, does not co-precipitate gO. Furthermore, the gH/gL/gO complex does not bind to human CD46, indicating that the complex is not a ligand for CD46 (Mori et al., 2004). These findings suggest that the viral envelope contains at least two kinds of tripartite complexes, gH/gL/gQ1/gQ2 and gH/gL/gO, and that the gH/gL/gO complex may play a role different from that of gH/gL/gQ1/gQ2 during viral infection. The predicted HHV-6A and HHV-6B gO gene products have 76.8% amino acid identity, which is much lower than the similarity between other glycoproteins, such as gB, gH, gL and gM, suggesting that the gH/gL/gO complex may confer different biological properties on the variants that cause them to target different cells (Mori et al., 2004).

U11

U11 encodes the major structural antigen p100 and 100 K, a phosphoprotein, which differs between HHV-6A and HHV-6B, with only 80.1% sequence identity (Neipel et al., 1992; Isegawa et al., 1999). The peotein is a tegment protein and is the dominant antigen recognized by HHV-6-specific IgG.

U14

U14 encodes 75 kDa protein, and interacts with the tumor suppressor p53. U14 localizes to the dot-like structures observed in both nucleus and cytoplasm in HHV-6B infected cells by 18 h p.i. U14 is virion protein, and p53 are incorporated into virions with U14 (Takemoto et al., 2005).

U27

U27 encodes 41-kDa early antigen called p41, and exhibits nuclear localization (Fig. 1) and DNA-binding activity (Agulnick et al., 1993). U27 encodes DNA polymerase processivity factor, and binds to 110-kDa protein, which is HHV-6 DNA polymerase (Pol-6). An *in vitro* DNA synthesis assay using primed M13 single-stranded DNA template demonstrates that p41 not only increase the DNA synthesis activity of Pol-6 but also allow Pol-6 to synthesize DNA products corresponding to full-length M13 template (7249 nucleotides). By contrast, Pol-6 alone can only synthesize DNA of <100 nucleotides. The functional interaction between Pol-6 and p41 appears to be specific because they cannot be physically or functionally substituted *in vitro* by their herpes simplex virus 1 homologues. Moreover, as revealed by mutational analysis, both the N and C termini of Pol-6 contribute to its binding to p41. In the case of p41, the N terminus is required for increasing DNA synthesis but not binding to Pol-6, whereas the C terminus is totally dispensable.

Perspective

It is not so easy to handle virus, including the inability to prepare high-titered cell-free virus and lack of a useful animal-model system. Nonetheless, several interesting studies of HHV-6 proteins have been done. HHV-6 can infect and replicate in T cells, therefore, HHV-6-encoded genes and gene products would have the novel functions different from the other herpesviruses. We anticipate that further observations for roles of HHV-6 proteins in virus infection will be identified.

References

Agulnick AD, Thompson JR, Iyengar S, Pearson G, Ablashi D, Ricciardi RP. J Gen Virol 1993; 74(Pt 6): 1003–1009.

Akkapaiboon P, Mori Y, Sadaoka T, Yonemoto S, Yamanishi K. J Virol 2004; 78: 7969–7983.

Anderson RA, Gompels UA. J Gen Virol 1999; 80(Pt 6): 1485–1494.

Anderson RA, Liu DX, Gompels UA. Virology 1996; 217: 517–526.

Ansari A, Emery VC. J Virol 1999; 73: 3284–3291.

Bradel-Tretheway BG, Zhen Z, Dewhurst S. J Virol Methods 2003; 111: 145–156.

Campadelli-Fiume G, Guerrini S, Liu X, Foa-Tomasi L. J Gen Virol 1993; 74(Pt 10): 2257–2262.

Chou S, Marousek GI. Virology 1992; 191: 523–528.

Cirone M, Campadelli-Fiume G, Foa-Tomasi L, Torrisi MR, Faggioni A. AIDS Res Hum Retrov 1994; 10: 175–179.

De Bolle L, Michel D, Mertens T, Manichanh C, Agut H, De Clercq E, Naesens L. Mol Pharmacol 2002; 62: 714–721.

Dhepakson P, Mori Y, Jiang YB, Huang HL, Akkapaiboon P, Okuno T, Yamanishi K. J Gen Virol 2002; 83: 847–854.
Dockrell DH. J Med Microbiol 2003; 52: 5–18.
Dominguez G, Dambaugh TR, Stamey FR, Dewhurst S, Inoue N, Pellett PE. J Virol 1999; 73: 8040–8052.
Ellinger K, Neipel F, Foa-Tomasi L, Campadelli-Fiume G, Fleckenstein B. J Gen Virol 1993; 74(Pt 3): 495–500.
Flebbe-Rehwaldt LM, Wood C, Chandran B. J Virol 2000; 74: 11040–11054.
Foa-Tomasi L, Boscaro A, di Gaeta S, Campadelli-Fiume G. J Virol 1991; 65: 4124–4129.
Foa-Tomasi L, Guerrini S, Huang T, Campadelli-Fiume G. Virology 1992; 191: 511–516.
Gompels UA, Nicholas J, Lawrence G, Jones M, Thomson BJ, Martin ME, Efstathiou S, Craxton M, Macaulay HA. Virology 1995; 209: 29–51.
Gravel A, Dion V, Cloutier N, Gosselin J, Flamand L. J Gen Virol 2004; 85: 1319–1328.
Gravel A, Gosselin J, Flamand L. J Biol Chem 2002; 277: 19679–19687.
Gravel A, Tomoiu A, Cloutier N, Gosselin J, Flamand L. Virology 2003; 308: 340–353.
Isegawa Y, Mukai T, Nakano K, Kagawa M, Chen J, Mori Y, Sunagawa T, Kawanishi K, Sashihara J, Hata A, Zou P, Kosuge H, Yamanishi K. J Virol 1999; 73: 8053–8063.
Isegawa Y, Ping Z, Nakano K, Sugimoto N, Yamanishi K. J Virol 1998; 72: 6104–6112.
Josephs SF, Ablashi DV, Salahuddin SZ, Jagodzinski LL, Wong-Staal F, Gallo RC. J Virol 1991; 65: 5597–5604.
Liu DX, Gompels UA, Foa-Tomasi L, Campadelli-Fiume G. Virology 1993a; 197: 12–22.
Liu DX, Gompels UA, Nicholas J, Lelliott C. J Gen Virol 1993b; 74(Pt 9): 1847–1857.
Luttichau HR, Clark-Lewis I, Jensen PO, Moser C, Gerstoft J, Schwartz TW. J Biol Chem 2003; 278: 10928–10933.
Manichanh C, Olivier-Aubron C, Lagarde JP, Aubin JT, Bossi P, Gautheret-Dejean A, Huraux JM, Agut H. J Gen Virol 2001; 82: 2767–2776.
Menotti L, Mirandola P, Locati M, Campadelli-Fiume G. J Virol 1999; 73: 325–333.
Milne RS, Mattick C, Nicholson L, Devaraj P, Alcami A, Gompels UA. J Immunol 2000; 164: 2396–2404.
Mori Y, Akkapaiboon P, Yang X, Yamanishi K. J Virol 2003a; 77: 2452–2458.
Mori Y, Akkapaiboon P, Yonemoto S, Koike M, Takemoto M, Sadaoka T, Sasamoto Y, Konishi S, Uchiyama Y, Yamanishi K. J Virol 2004; 78: 4609–4616.
Mori Y, Dhepakson P, Shimamoto T, Ueda K, Gomi Y, Tani H, Matsuura Y, Yamanishi K. J Virol 2000; 74: 6096–6104.
Mori Y, Seya T, Huang HL, Akkapaiboon P, Dhepakson P, Yamanishi K. J Virol 2002; 76: 6750–6761.
Mori Y, Yang X, Akkapaiboon P, Okuno T, Yamanishi K. J Virol 2003b; 77: 4992–4999.
Neipel F, Ellinger K, Fleckenstein B. J Virol 1992; 66: 3918–3924.
Papanikolaou E, Kouvatsis V, Dimitriadis G, Inoue N, Arsenakis M. Virus Res 2002; 89: 89–101.
Pellett PE, Dominguez G. In: Knipe DM, Howley PM, editors. Fields Virology. vol. 2. 4th ed., Chapter 80. Philadelphia: Lippincott, Williams & Wilkins; 2001; pp. 69–2784.
Pfeiffer B, Berneman ZN, Neipel F, Chang CK, Tirwatnapong S, Chandran B. J Virol 1993; 67: 4611–4620.
Pfeiffer B, Thomson B, Chandran B. J Virol 1995; 69: 3490–3500.
Qian G, Wood C, Chandran B. Virology 1993; 194: 380–386.
Rapp JC, Krug LT, Inoue N, Dambaugh TR, Pellett PE. Virology 2000; 268: 504–516.

Rotola A, Ravaioli T, Gonelli A, Dewhurst S, Cassai E, Di Luca D. Proc Natl Acad Sci USA 1998; 95: 13911–13916.

Safronetz D, Petric M, Tellier R, Parvez B, Tipples GA. J Med Virol 2003; 71: 434–439.

Santoro F, Greenstone HL, Insinga A, Liszewski MK, Atkinson JP, Lusso P, Berger EA. J Biol Chem 2003; 278: 25964–25969.

Santoro F, Kennedy PE, Locatelli G, Malnati MS, Berger EA, Lusso P. Cell 1999; 99: 817–827.

Shimada K, Kondo K, Yamanishi K. Microbiol Immunol 2004; 48: 205–210.

Stanton R, Fox JD, Caswell R, Sherratt E, Wilkinson GW. J Gen Virol 2002; 83: 2811–2820.

Takeda K, Haque M, Sunagawa T, Okuno T, Isegawa Y, Yamanishi K. J Gen Virol 1997; 78(Pt 9): 2171–2178.

Takemoto M, Koike M, Mori Y, Yonemoto S, Sasamoto Y, Kondo K, Uchiyama Y, Yamanishi K. Human herpesvirus 6 open reading frame U14 protein and cellular p53 interact with each other and are contained in the virion. J Virol 2005; 79: 13037–13046.

Thomson BJ, Efstathiou S, Honess RW. Nature 1991; 351: 78–80.

Thomson BJ, Weindler FW, Gray D, Schwaab V, Heilbronn R. Virology 1994; 204: 304–311.

Tigue NJ, Kay J. FEBS Lett 1998a; 441: 467–946.

Tigue NJ, Kay J. J Biol Chem 1998b; 273: 26441–26446.

Tigue NJ, Matharu PJ, Roberts NA, Mills JS, Kay J, Jupp R. J Virol 1996; 70: 4136–4141.

Zou P, Isegawa Y, Nakano K, Haque M, Horiguchi Y, Yamanishi K. J Virol 1999; 73: 5926–5933.

Human Herpesvirus-6: General Virology, Epidemiology and Clinical Pathology
Perspectives in Medical Virology
Gerhard Krueger and Dharam Ablashi (Editors)

DOI 10.1016/S0168-7069(06)12005-4

Biological Features of HHV-6

Lorenzo Dagna[a,b], Fabio Santoro[a], Paolo Lusso[a]
[a]*Unit of Human Virology, Department of Biological and Technological Research (DIBIT), San Raffaele Scientific Institute, 20132 Milan, Italy*
[b]*Universita' Vita-Salute San Raffaele School of Medicine, San Raffaele Scientific Institute, 20132 Milan, Italy*

Introduction

The biology of human herpesvirus-6 (HHV-6) is complex and unique. A major factor in determining this complexity is the existence of two viral variants or subgroups, defined HHV-6A and 6B, which exhibit different biological features, possibly also reflected in the different disease associations described *in vivo*. Together with HHV-7, HHV-6 is the only human herpesvirus to show a distinct tropism for cells of the T-lymphoid lineage. However, in line with the nearly ubiquitous distribution of its cellular receptor, CD46, HHV-6 has in fact a much broader cellular tropism, although in many cell types the viral cycle is restricted and does not proceed to completion. Productive infection almost invariably results in cytopathic effects and cell death. However, HHV-6 may also cause important phenotypic and functional changes when it establishes nonproductive infection, as seen with mononuclear phagocytic cells and dendritic cells (DC). The biological features of HHV-6 have been extensively characterized using *in vitro* models, while *in vivo* studies are still limited, owing to the lack of suitable and easily accessible animal model systems. The recent development of an *ex vivo* infection model in structurally intact human lymphoid tissue has provided a new tool for investigating the complex virus–cell interactions in a physiologically relevant system. In this chapter, we summarize the body of knowledge accrued over the past 20 years on the biology of HHV-6.

HHV-6 cellular receptor

The early events of the interaction between HHV-6 and the host cell have remained poorly understood for a long time, except for the recognition that the CD4 molecule is not involved in the cellular receptor mechanism for HHV-6 (Lusso et al., 1989) and that specific viral glycoproteins, i.e. the gH/gL complex, gB, and gQ are critically involved in such interactions, as shown by the ability of monoclonal antibodies (mAbs) against these proteins to inhibit HHV-6 infection and syncytia formation (Foa-Tomasi et al., 1991; Gompels et al., 1991; Liu et al., 1993; Qian et al., 1993; Anderson and Gompels, 1999). The development of an HHV-6 envelope-mediated cell fusion assay based on vaccinia virus technology has provided a critical tool for investigating the nature of the HHV-6 cellular receptor. It was using this assay that Santoro and colleagues, in 1999, identified CD46 as a critical cell-surface receptor mediating fusion and entry for both HHV-6 variants A and B (Santoro et al., 1999). This conclusion was based on several complementary lines of evidence: first, CD46 is selectively downmodulated from the target cell surface during the course of infection; second, CD46-specific mAbs block both the infectivity and the fusogenic activity of both HHV-6 variants; third, HHV-6-mediated fusion is also inhibited by a soluble form of CD46; fourth, gain-of-function studies demonstrated that recombinant expression of human CD46 rendered otherwise insensitive nonhuman cells susceptible to both HHV-6-induced fusion and entry (Santoro et al., 1999). Subsequent studies have permitted mapping of the HHV-6-binding region within the short consensus repeat (SCR)-2 and -3 domains of CD46 (Greenstone et al., 2002), while the viral glycoprotein directly implicated in the interaction with CD46 was identified as gH (Santoro et al., 2003). The critical requirement of CD46 for fusion induced by both HHV-6 variants has been clearly documented in different cell types and experimental systems. However, in a recent study by Mori et al. (2002), one HHV-6B strain, HST, was found to be unable to induce CD46-mediated syncytia formation after short-term virion binding to the external cell membrane (the so-called 'fusion from without'). This phenomenon may result from a lower CD46-binding affinity of this particular strain or, more generally, of HHV-6B, insufficient to mediate this purely extracellular fusion process. Alternatively, it cannot be excluded that at least some strains of HHV-6B utilize another, still undefined, cellular receptor for entry.

CD46, also designated membrane cofactor protein, is a type-1 glycoprotein expressed on the surface of all nucleated human cells examined to date (Liszewski et al., 1991; Seya et al., 1995). It is a member of a family of glycoproteins that act as regulators of complement activation (RCA). Their physiological function is essential because it prevents the spontaneous activation of complement on autologous cells. Interestingly, CD46 has also been identified as a primate-specific cellular receptor for other viruses, including vaccine strains of measles virus (MV) (Dorig et al., 1993; Naniche et al., 1993). Multiple isoforms of CD46 display MV receptor function (Manchester et al., 1994). However, the relevance of CD46 for the entry of clinical isolates of MV is controversial (Buckland and Wild, 1997; Sakata et al.,

1998). Such isolates seem to use an alternative receptor molecule, SLAM (CDw150), which is expressed on a subset of T and B cells (Tatsuo et al., 2000). CD46 also serves as a cellular receptor or attachment protein for other human pathogens, including some adenovirus strains, various pathogenic Neisseriae, and group A Streptococci. It is noteworthy that other members of the RCA family function as viral receptors, including CD21 for Epstein–Barr virus (EBV) (Ablashi et al., 1976; Fingeroth et al., 1984) and CD55 for several echoviruses (Bergelson et al., 1994) and coxsackie viruses (Bergelson et al., 1995). These findings suggest the existence of common biological or structural features that favor the exploitation of this class of molecules by infectious agents.

The discovery that HHV-6 exploits an almost ubiquitous molecule as a cell surface receptor may help to explain its ability to enter a variety of cell types *in vitro* (see below) as well as the diverse clinical manifestations to which HHV-6 has been linked. However, it was also demonstrated that CD46 alone may be insufficient to enable HHV-6 fusion/entry, since some human T-cell lines are apparently nonpermissive for HHV-6-induced fusion despite high levels of CD46 expression. One interpretation of this finding is that fusion requires an additional factor besides CD46, possibly a coreceptor, which is not consistently expressed in all human cell types. This second factor or receptor might contribute to the restriction in tissue tropism exhibited by HHV-6 *in vivo*. A related complexity is that both variants A and B of HHV-6 use CD46 as a receptor but display different tropism for infection of certain human cell lines. For example, the immature T-cell line HSB-2 is selectively susceptible to HHV-6A, but not to HHV-6B (Ablashi et al., 1991). Restriction at the entry stage (for example, due to different coreceptor usage) and/or subsequent post-entry replication blocks (as seen with differentiated neural stem cells; see below) may both independently contribute to the different biological properties displayed by the two variants as well as, possibly, to their different pathologic effects.

Species specificity

HHV-6 has a restricted range of susceptible species, essentially limited to humans and selected nonhuman primates. Antibodies to HHV-6 or to a closely related simian herpesvirus have been demonstrated in monkeys (Higashi et al., 1989), and a simian HHV-6 homolog has been recently identified in mandrill and drill monkeys (Lacoste et al., 2000) as well as chimpanzees (Lacoste et al., 2005). Although an established rodent model of HHV-6 infection is not presently available, preliminary experiments in human CD46-transgenic mice have shown some degree of susceptibility to HHV-6A (P. Lusso et al., unpublished). An efficient small-animal model of HHV-6 infection was developed using severe combined immunodeficiency (SCID)-hu *Thy/Liv* mice, in which human fetal liver and thymic tissues are implanted under the renal capsule (Gobbi et al., 1999). These mice can be successfully infected with HHV-6A and -6B, but the infection remains confined to the human graft. *In vitro*, HHV-6 was shown to replicate efficiently in cells from selected nonhuman primate species (Lusso et al., 1990, 1994). Experimental infection attempts *in vivo* have documented replication of HHV-6A (strain GS) in pig-tailed and cynomolgus macaques (P. Lusso et al., unpublished), while HHV-6B (strain HST) was found to infect cynomolgus macaques and African green monkeys (Yalcin et al., 1992).

Cellular tropism of HHV-6

In spite of its initial designation as 'human B-lymphotropic virus' (HBLV) (Salahuddin et al., 1986), both HHV-6A and -6B variants replicate most efficiently in $CD4^+$ T-lymphocytes both *in vitro* (Lusso et al., 1988) and *in vivo* (Takahashi et al., 1989). However, a wide range of cell types is susceptible to either productive or nonproductive HHV-6 infection (De Bolle et al., 2005).

Target cells in vitro

The finding that both HHV-6 variants use the ubiquitous molecule CD46 as a membrane receptor is compatible with a broad cellular tropism. Productive infection, however, is limited to a small range of cells most likely by intracellular restriction factors acting beyond the viral entry step. Studies on the cellular tropism of HHV-6, and particularly of the A variant, have suggested that this virus is broadly 'immunotropic' as it infects several cells implicated in the generation of effective immune responses. Thereby, HHV-6 may affect, directly or indirectly, both the cellular and humoral arms of the immune system (see below). There is universal consensus that the primary target cells for HHV-6 infection both *in vivo* and *in vitro* are $CD4^+$ T-Cells (Lusso et al., 1988; Takahashi et al., 1989). When activated mononuclear cells obtained from different sources (e.g. cord or peripheral blood, thymus, tonsils, lymph node) are exposed *in vitro* to HHV-6, the vast majority of the infected cells display the phenotype of activated $CD4^+$ T-cells ($CD2^+$, $CD4^+$, $CD5^+$, $CD7^+$, $CD26^+$, $CD38^+$, $CD71^+$). In SCID-hu *Thy/Liv* mice, thymocytes at different maturation stage are productively infected (Gobbi et al., 1999). In studies performed using human lymphoid tissue blocks *ex vivo* (see below), both naïve ($CD45RA^+62L^+$) and memory T-cells were shown to be productively infected with HHV-6A (Grivel et al., 2003). However, there are some important differences in the *in vitro* cellular tropism of the two major viral subgroups (A and B). Besides the differential ability to infect human T- and B-cell lines (Table 1), HHV-6A efficiently infects different types of cytotoxic effectors, such as $CD8^+$ T-cells (Lusso et al., 1991a,b), natural killer (NK) cells (Lusso et al., 1993), and $\gamma\delta$ T-cells (Lusso et al., 1995). Because such cells are involved in the mechanisms of antiviral defense *in vivo*, this strategy may allow HHV-6 to counteract the protective immune surveillance of the host and thereby establish persistent infection. By contrast, HHV-6B seems to have a more restricted cellular tropism and, in particular, to infect rather inefficiently cytotoxic effector cells (Lusso et al., 1991a; Grivel et al., 2003).

Table 1

Tropism of HHV-6A and -6B for primary and immortalized lymphoid cells

	Subgroup A (GS-like)	Subgroup B (Z29-like)	References
Primary blood cells			
$CD4^+$ T-cells	+	+	Lusso et al. (1988); Takahashi et al. (1989)
$CD8^+$ T-cells	+	±	Lusso et al. (1991a)
NK cells	+	–	Lusso et al. (1993)
γδ T-cells	+	–	Lusso et al. (1995)
B cells	–	–	Lusso et al. (1988)
Cell lines			
HSB-2	+	–	Ablashi et al. (1991)
Molt-3	±	+	Ablashi et al. (1991)
SupT1	+	+	Ablashi et al. (1991)
Jurkat/J-JHAN	+	+	Lusso et al. (1991b)
Hut78	–	–	Ablashi et al. (1987)
CEM	+	+	Ablashi et al. (1987); Santoro et al., (unpublished)
EBV^+ LCL	+	–	Ablashi et al. (1988)

The cells of the mononuclear phagocytic system represent another important target of HHV-6A and 6B both *in vitro* and *in vivo* (Kondo et al., 1991; Burd and Carrigan, 1993), and have been suggested as one of the possible HHV-6 reservoirs *in vivo* (Kondo et al., 1991). However, there is some controversy on the ability of these cells to support productive HHV-6 infection. Smith et al. (2003) have reported that after exposure to infectious HHV-6 stocks, primary macrophages (differentiated in culture from peripheral blood monocytes) never show any detectable expression of U16/17, U60/61, and U89/90 mRNA, nor viral DNA accumulation or cytopathic effects. By contrast, other studies with *in vitro*-differentiated macrophages have reported at least a transient HHV-6 replication (Kondo et al., 1991) as well as HHV-6-induced cytopathic effects (Burd et al., 1993). In another study, freshly isolated monocytes were shown to be resistant to HHV-6, but their differentiation with interleukin (IL)-15 induced a greater susceptibility to productive infection (Arena et al., 2000). The reasons for these discrepancies are still unclear, although they may be related to the variable experimental conditions used for growing and differentiating primary monocytes *in vitro*. Regardless of the completion of the viral lytic cycle, it is unquestionable that HHV-6 can induce important phenotypic and functional alterations in mononuclear phagocytic cells (see below).

Similar to monocyte/macrophages, blood-derived DC can be infected *in vitro* with both HHV-6A and -6B, although there is again controversy on the productive nature of such infection. In one study (Smith et al., 2003), no signs of productive infection were detected in differentiated DC. By contrast, Kakimoto and colleagues (Kakimoto et al., 2002) reported a widespread expression of late viral antigens in cultured DC, even though this was not accompanied by any detectable cytopathic effect, suggesting that the virus was unable to effectively complete its lytic cycle.

Unlike T cells, which are promptly infected after contact with HHV-6, B cells seem to be infectable only after their immortalization with EBV (Ablashi et al., 1988), suggesting that EBV infection induces the expression of cell surface or soluble factors that facilitate HHV-6 infection. However, persistent nonproductive infection with HHV-6, in the absence of EBV, has been reported in a Burkitt's lymphoma cell line with an immature B-cell phenotype (Bandobashi et al., 1997). Other types of interactions between HHV-6 and EBV have been documented *in vitro* (Flamand et al., 1993).

HHV-6 was shown to productively infect *in vitro* human endothelial cells from different sources in the absence of cytopathic effects, leading to the suggestion that endothelial cells may function as another *in vivo* reservoir for HHV-6 (Caruso et al., 2002). HHV-6 also induces endothelial cells to secrete high amount of chemokines, such as IL-8 and monocyte chemoattractant protein (MCP-1). It is tempting to speculate that immune cells attracted by these proinflammatory chemokines may then be infected by the virus after contact with endothelial cells and/or during transendothelial migration, and become themselves carriers of infectious virus, thus playing a potential role in virus dissemination (Caruso et al., 2002).

Consistent with the described neurotropism of HHV-6 *in vivo*, primary neural cells can be productively infected *in vitro*, although their efficiency in sustaining viral replication is generally low. It has been shown that human fetal astrocytes can be productively infected with both HHV-6 variants, and that the progeny virus is able to reinfect both primary T cells and astrocytes (He et al., 1996). A more recent study, however, suggested that only HHV-6A variant can establish a productive lytic infection in human progenitor-derived astrocytes from fetal brain tissue, whereas HHV-6B induces a nonproductive infection (Donati et al., 2005). Similar observations were made using differentiated human neural stem cells, in which HHV-6A (GS), but not HHV-6B (Z29), was shown to productively infect both astroctyes and, to a lesser extent, neurons, causing syncytia formation and down-modulation of lineage-specific markers; also in this study, HHV-6B DNA was shown to persist in neural cells in the absence of any sign of virus expression or cytopathic effect (Foglieni et al., submitted).

Finally, several immortalized human cell lines can be infected by HHV-6, including lines of T-cell (Ablashi et al., 1988; Cermelli et al., 1997), epithelial (Chen et al., 1994), fibroblastic (Luka et al., 1990), monocytoid (Arena et al., 1997), immature hemopoietic (Furlini et al., 1996), megakaryoblastoid and neuroblastoma (Ablashi et al., 1988), hepatocytic (Inagi et al., 1996), and astrocytoma (Yoshikawa et al., 2002a) origin. Although the *in vivo* relevance of these observations remains unknown, these models confirm the broad cellular tropism of HHV-6.

Cellular and tissue host range in vivo

Only a few studies have carefully investigated the tissue host range of HHV-6 *in vivo*. During acute primary infection, HHV-6B has been identified primarily in $CD4^+$ T-cells (Takahashi et al., 1989), which not only represent the primary target for virus replication, but may also function as a reservoir for HHV-6 latency. However, high levels of HHV-6 DNA have also been detected in circulating monocytes during primary infection, and virus isolation was successfully achieved from these cells; moreover, they were shown to express U79/80 mRNA, which is an index of active viral replication (Kondo et al., 2002a). Conversely, in convalescent patients, monocytes were shown to harbor HHV-6 DNA in the absence of signs of productive infection (Kondo et al., 1991). Bone marrow progenitor cells ($CD34^+$) are also susceptible to HHV-6 infection: they can harbor latent HHV-6 that can be longitudinally transmitted to differentiated blood cells of different lineages (Luppi et al., 1999; Andre-Garnier et al., 2004), including monocytes/macrophages and DC (Kondo et al., 1991; Burd and Carrigan, 1993; Kakimoto et al., 2002). Exposure to HHV-6 was shown to exert dramatic suppressive effects on the maturation and growth of normal human bone marrow precursor cells committed toward different lineages (Knox and Carrigan, 1992), including the macrophage lineage (Burd et al., 1993). These observations are consistent with the reported association of HHV-6 infection with graft failure in patients who received bone marrow transplantation (Drobyski et al., 1993).

Endothelial cells seem to represent another important target for infection *in vivo* as suggested by endothelial cell activation and/or damage secondary to HHV-6 reactivation, which may predispose to the development of thrombotic microangiopathy (Matsuda et al., 1999; Takatsuka et al., 2003). Finally, in healthy subjects, HHV-6 DNA has been demonstrated *ex vivo* in several tissues including tonsils, salivary glands, thyroid, brain, liver, myocardium, kidney, and thyroid (Fox et al., 1990; Corbellino et al., 1993; Luppi et al., 1994a; Chan et al., 2001; Grivel et al., 2001; Ozaki et al., 2001; Roush et al., 2001; Ishikawa et al., 2002; Donati et al., 2003; Harma et al., 2003). Nevertheless, it is unclear to what extent the detection of viral DNA reflects the ability of HHV-6 to infect tissue-specific resident cells or, alternatively, infiltrating lymphoid and monocytic cells.

Biological effects of HHV-6 on the host cell

In vitro studies have documented a variety of effects of HHV-6 on its host cell, which can significantly impair its functon and survival (Table 2).

Cytopathic effects

When HHV-6 establishes productive infection, it typically behaves as a cytopathic virus that induces widespread cell death. However, the exact mechanisms of HHV-6 cytopathicity are not yet definitely elucidated. After exposure to HHV-6, primary

Table 2

Biological effects of HHV-6A and -6B in infected T-lymphoid cells

	Subgroup A (GS-like)	Subgroup B (Z29-like)	References
Cytopathic effects	+	+	Lusso et al. (1988)
Latent infection	+	+	
Cell immortalization	–	–	
Downregulation of CD46	+	±	Santoro et al. (1999)
Downregulation of CD3	+	+	Lusso et al. (1988)
Induction/upregulation of CD4	+	±	Lusso et al. (1991a)
Induction of RANTES	+	+	Grivel et al. (2003)

human mononuclear cells show dramatic cytomorphological changes that start to appear within 3–4 days: the cells become enlarged and lose their structured, blastic shape becoming rounded and refractile. These enlarged T lymphocytes are locked in the G_2/M phase of the cell cycle, as indicated by the accumulation of p53, cyclin B_1 Cyclin B, cyclin A, and tyrosine^{15}-phosphorylated cdk1 (De Bolle et al., 2004), and usually survive in culture for 3–4 days before dying. Although the majority of these cells are not syncytial, membrane fusion resulting in the formation of giant multinucleated cells does also occur, particularly when continuous $CD4^+$ T-cell lines, like SupT1, are employed as targets. This phenomenon, which is usually more prominent with HHV-6A than with HHV-6B, has been exploited for the establishment of fusion inhibition tests *in vitro* (Santoro et al., 1999).

Evidence has been provided that the infection of primary $CD4^+$ is associated with widespread induction of apoptosis both *in vitro* and *in vivo* (Inoue et al., 1997; Yasukawa et al., 1998). By contrast, studies with continuous cell lines have documented an aberrant cytoplasmic accumulation of the tumor-suppressor and anti-apoptotic protein p53 upon HHV-6 infection (Takemoto et al., 2004; Zhen et al., 2005). Because the elevation of p53 begins early after the onset of infection and does not occur using UV-inactivated virus, it is likely to depend on the expression of the immediate-early genes (Takemoto et al., 2004). In this model, the vast majority of HHV-6-infected cells do not bind annexin V, indicating that they are not undergoing apoptosis (Zhen et al., 2005). These conflicting observations are at present difficult to reconcile.

Latent infection

Like other herpesviruses, HHV-6 establishes latent infection *in vivo* and thereby can persist in the host indefinitely after primary infection. The best-characterized *in vitro* model of latent infection is represented by long-term (30 days) cultured macrophages, which after exposure to HHV-6 survive a transient period of low-level productive infection (Kondo et al., 1991). Latency-associated HHV-6 transcripts from the IE1/IE2 locus have been identified in latently infected macrophages both *in vitro* and *in vivo* (Kondo et al., 2002b). A second *in vitro* model of HHV-6 latency was described in papillomavirus-immortalized cervical epithelial cells, in which high numbers of viral genome copies were shown to persist in an episomal form for prolonged periods of time in the absence of any sign of productive infection (Chen et al., 1994). Latent HHV-6 infection was also reported in an EBV-negative Burkitt's lymphoma cell line (Bandobashi et al., 1997).

Although circulating monocytes and epithelial cells of the bronchial and salivary glands have been suggested as possible *in vivo* reservoirs (Krueger et al., 1990; Kondo et al., 1991), there is still uncertainty regarding the exact sites of viral persistence and latency. Long-lived resting memory $CD4^+$ T cells may also be a primary reservoir among peripheral blood cells. In rare cases, chromosomally integrated HHV-6 has been documented in peripheral blood mononuclear cells from patients with different clinical conditions as well as from normal subjects (Luppi et al., 1993, 1994b, 1998; Torelli et al., 1995; Daibata et al., 1999). The pathological consequences of this phenomenon are still at present unclear.

Cellular transformation

Unlike EBV and other *γ-herpesvirinae*, HHV-6 does not directly cause immortalization of its target cells, at least *in vitro*. Although some early studies had documented a transforming effect of isolated HHV-6 genes *in vitro* (Razzaque et al., 1993; Kashanchi et al., 1997), there is at present little clinical ground to support these observations. Indeed, despite several claims, the putative involvement of HHV-6 in a series of neoplastic disorders has not been substantiated.

Immune modulation

The broad immunotropism of HHV-6, particularly of variant A (Table 1), may dramatically affect, directly or indirectly, the function of the cellular and humoral arms of the immune system. As discussed above, both variants have a primary tropism for $CD4^+$ T cells, which are pivotal in the orchestration of the immune responses. Variant A also efficiently infects different types of cytotoxic effector cells such as $CD8^+$ T lymphocytes, NK cells, and γδ T lymphocytes. Moreover, both mononuclear phagocytic cells and DC can be infected, albeit usually in a nonproductive fashion, and the infection results in dramatic phenotypic and functional alterations.

In accordance with the above observations, multiple lines of clinical and experimental evidence suggest that HHV-6 may be an immunosuppressive agent in its own right. One such hint comes from the SCID-hu *Thy/Liv* mouse model in which infection with either HHV-6 subgroup A or B results in a rapid destruction of the thymic grafts with dramatic thymocyte depletion affecting all major intrathymic cell populations (Gobbi et al., 1999). Consistent with these experimental observations, disseminated coinfection with HHV-6A and -6B has been etiologically linked with thymic atrophy and progressive immunodeficiency in a child who showed no transcriptional downregulation of DC-SIGN in infected immature DC (Niiya et al., 2004). The alteration of the expression of DC-SIGN, which is involved in DC rolling and transmigration into periphery and T-lymphocyte activation, may affect their ability to initiate and sustain specific immune responses. Altogether, these data suggest that the interference with the functional maturation of DC is a potential mechanism of HHV-6-mediated immunosuppression.

Another intriguing mechanism that might be exploited by HHV-6 to evade the immune system has been recently proposed (Kemper et al., 2003), although these results have not yet been reproduced in other studies. Coengagement of CD3 and the HHV-6 receptor CD46 in the presence of IL-2 was shown to induce a T-regulatory 1(Tr1)-specific cytokine phenotype in human $CD4^+$ T cells (Kemper et al., 2003). These CD3/CD46 stimulated IL-10-producing $CD4^+$ T cells proliferate strongly, suppress activation of bystander T cells, and acquire a memory phenotype. If confirmed, the CD46-mediated induction of Tr1 cells may provide a further explanation for the choice of CD46 as a cellular receptor by many microbial agents.

Human lymphoid tissue ex vivo*: a new model for the study of HHV-6 pathogenesis*

The study of the pathogenesis of HHV-6 infection has been significantly hampered by the lack of physiologically relevant models. The systems most commonly used to grow HHV-6 *in vitro*, such as cord blood- and peripheral blood-derived mononuclear cells cultured in suspension, require the cells to be maximally stimulated with polyclonal activators in order to sustain HHV-6 replication, a condition that is unlikely to occur *in vivo*. Recently, a new *ex vivo* model for the study of the cellular tropism and pathogenic effects of HHV-6 has been established using structurally intact lymphoid tissue blocks in the absence of exogenous stimulation (Grivel et al., 2003). This system, which was previously used to study HIV infection (Glushakova et al., 1995; Grivel and Margolis, 1999) as well as the interactions between HIV-1 and HHV-6 (Grivel et al., 2001), is physiologically relevant, since it better preserves the structural and morphological features of the lymphoid tissue, in particular the complex interactions between lymphocytes and stromal cells. HHV-6 replicates vigorously in *ex vivo* lymphoid tissue, with peak levels at day 10–14 post infection, resulting in significant cell depletion, particularly of $CD4^+$ T lymphocytes. In accordance with *in vitro* observations in PBMC grown in suspension (Lusso et al., 1991b), HHV-6A was able to efficiently infect both $CD4^+$ and $CD8^+$ T lymphocytes, whereas HHV-6B infection of CD8 T cells was less efficient. As a proof of the physiological relevance of this model, and at variance with the typical pattern seen in suspension cultures, the infection does not propagate to all the target cells present in the tissue blocks, with only about 5–30% of T lymphocytes productively infected by the virus at any given time. Moreover, at variance with the observations in SCID-hu *Thy/Liv* mice, bystander uninfected cells are not depleted in this model (Gobbi et al., 1999).

Infection with HHV-6 had profound phenotypic and functional effects on human lymphoid tissue. HHV-6A was shown to induce a dramatic and widespread downmodulation of the expression of CD46, which affects both infected and bystander cells; in contrast, the effect of HHV-6B on CD46 expression was less marked. As discussed below, this marked reduction of CD46 expression may result in an increased susceptibility to lymphoid tissue damage secondary to spontaneous activation of the complement cascade. Besides CD46, a second surface molecule that is dramatically downmodulated in HHV-6-infected lymphoid tissue is CD3, confirming previous observations in suspension T-cell cultures (Lusso et al., 1991b). However, at variance with CD46, CD3 is downmodulated only in HHV-6-infected cells but remains unaltered in bystander cells, suggesting a differential mechanism of viral modulation of these cell surface markers. Also, as previously documented using PBMC grown in suspension (Lusso et al., 1991a), the CD4 glycoprotein is upregulated in HHV-6-infected lymphoid tissue without bystander effects.

Measurement of cytokine release in HHV-6-infected lymphoid tissue showed only marginal effects on most cytokines, except for a dramatic increase in RANTES production. This phenomenon appears to be specific and cannot be simply explained as a consequence of generalized cell activation. In addition, analysis of RANTES expression at the single-cell level showed that uninfected cells also upregulate their RANTES production in the context of HHV-6-infected lymphoid tissue. This increased secretion of RANTES in HHV-6-infected tissues may profoundly influence the physiology of the immune responses, as well as, more specifically, the replication of the different biological variants of HIV-1.

Dysregulation of complement activation

As discussed above, HHV-6 binds to the SCR-2 and -3 domains of CD46 (Greenstone et al., 2002), which are the same as those involved in the complement-regulatory function of CD46. This suggests that HHV-6 may functionally impair the protective effect of CD46 against the activation of autologous complement and the consequent cellular damage, as demonstrated in some *in vitro* models (Schnorr et al., 1995). Aberrant activation of human complement and widespread complement-mediated cytotoxicity have been documented in differentiated neural stem cells exposed to HHV-6 (Foglieni et al., submitted for publication). Of note, the effect was observed using UV light-inactivated virus, confirming that the mere binding to CD46, in the absence of productive infection, is sufficient to alter the complement regulatory function of the receptor.

References

Ablashi DV, Easton JM, Guegan JH. Biomedicine 1976; 24: 286.

Ablashi DV, Salahuddin SZ, Josephs SF, Imam F, Lusso P, Gallo RC, Hung C, Lemp J, Markham PD. Nature 1987; 329: 207.

Ablashi DV, Josephs SF, Buchbinder A, Hellman K, Nakamura S, Llana T, Lusso P, Kaplan M, Dahlberg J, Memon S, Gallo RC. J Virol Methods 1988; 21: 29.
Ablashi DV, Balachandran N, Josephs SF, Hung CL, Krueger GR, Kramarsky B, Salahuddin SZ, Gallo RC. Virology 1991; 184: 545.
Anderson RA, Gompels UA. J Gen Virol 1999; 80(Pt 6): 1485.
Andre-Garnier E, Milpied N, Boutolleau D, Saiagh S, Billaudel S, Imbert-Marcille BM. J Gen Virol 2004; 85: 3333.
Arena A, Liberto MC, Capozza AB, Foca A. New Microbiol 1997; 20: 13.
Arena A, Liberto MC, Iannello D, Capozza AB, Foca A. New Microbiol 1999; 22: 293.
Arena A, Merendino RA, Bonina L, Iannello D, Stassi G, Mastroeni P. New Microbiol 2000; 23: 105.
Bandobashi K, Daibata M, Kamioka M, Tanaka Y, Kubonishi I, Taguchi H, Ohtsuki Y, Miyoshi I. Blood 1997; 90: 1200.
Bergelson JM, Chan M, Solomon KR, St John NF, Lin H, Finberg RW. Proc Natl Acad Sci USA 1994; 91: 6245.
Bergelson JM, Mohanty JG, Crowell RL, St John NF, Lublin DM, Finberg RW. J Virol 1995; 69: 1903.
Buckland R, Wild TF. Virus Res 1997; 48: 1.
Burd EM, Carrigan DR. Virus Res 1993; 29: 79.
Burd EM, Knox KK, Carrigan DR. Blood 1993; 81: 1645.
Caruso A, Rotola A, Comar M, Favilli F, Galvan M, Tosetti M, Campello C, Caselli E, Alessandri G, Grassi M, Garrafa E, Cassai E, Di Luca D. J Med Virol 2002; 67: 528.
Cermelli C, Pietrosemoli P, Meacci M, Pecorari M, Sabbatini AM, Colombari B, Portolani M. New Microbiol 1997; 20: 187.
Chan PK, Ng HK, Hui M, Cheng AF. J Med Virol 2001; 64: 42.
Chen M, Popescu N, Woodworth C, Berneman Z, Corbellino M, Lusso P, Ablashi DV, DiPaolo JA. J Virol 1994; 68: 1173.
Corbellino M, Lusso P, Gallo RC, Parravicini C, Galli M, Moroni M. Lancet 1993; 342: 1242.
Daibata M, Taguchi T, Nemoto Y, Taguchi H, Miyoshi I. Blood 1999; 94: 1545.
De Bolle L, Hatse S, Verbeken E, De Clercq E, Naesens L. FEBS Lett 2004; 560: 25.
De Bolle L, Van Loon J, De Clercq E, Naesens L. J Med Virol 2005; 75: 76.
Donati D, Akhyani N, Fogdell-Hahn A, Cermelli C, Cassiani-Ingoni R, Vortmeyer A, Heiss JD, Cogen P, Gaillard WD, Sato S, Theodore WH, Jacobson S. Neurology 2003; 61: 1405.
Donati D, Martinelli E, Cassiani-Ingoni R, Ahlqvist J, Hou J, Major EO, Jacobson S. J Virol 2005; 79: 9439.
Dorig RE, Marcil A, Chopra A, Richardson CD. Cell 1993; 75: 295.
Drobyski WR, Dunne WM, Burd EM, Knox KK, Ash RC, Horowitz MM, Flomenberg N, Carrigan DR. J Infect Dis 1993; 167: 735.
Fingeroth JD, Weis JJ, Tedder TF, Strominger JL, Biro PA, Fearon DT. Proc Natl Acad Sci USA 1984; 81: 4510.
Flamand L, Gosselin J, D'Addario M, Hiscott J, Ablashi DV, Gallo RC, Menezes J. J Virol 1991; 65: 5105.
Flamand L, Stefanescu I, Ablashi DV, Menezes J. J Virol 1993; 67: 6768.
Flamand L, Gosselin J, Stefanescu I, Ablashi D, Menezes J. Blood 1995; 85: 1263.
Flamand L, Stefanescu I, Menezes J. J Clin Invest 1995; 97: 1373.

Flamand L, Romerio F, Reitz MS, Gallo RC. J Virol 1998; 72: 8797.
Foa-Tomasi L, Boscaro A, di Gaeta S, Campadelli-Fiume G. J Virol 1991; 65: 4124.
Fox JD, Briggs M, Ward PA, Tedder RS. Lancet 1990; 336: 590.
French C, Menegazzi P, Nicholson L, Macaulay H, DiLuca D, Gompels UA. Virology 1999; 262: 139.
Furlini G, Vignoli M, Ramazzotti E, Re MC, Visani G, La P. Blood 1996; 87: 4737.
Glushakova S, Baibakov B, Margolis LB, Zimmerberg J. Nat Med 1995; 1: 1320.
Gobbi A, Stoddart CA, Malnati MS, Locatelli G, Santoro F, Abbey NW, Bare C, Linquist-Stepps V, Moreno MB, Herndier BG, Lusso P, McCune JM. J Exp Med 1999; 189: 1953.
Gompels UA, Carss AL, Saxby C, Hancock DC, Forrester A, Minson AC. J Virol 1991; 65: 2393.
Greenstone HL, Santoro F, Lusso P, Berger EA. J Biol Chem 2002; 277: 39112.
Grivel JC, Margolis LB. Nat Med 1999; 5: 344.
Grivel JC, Ito Y, Faga G, Santoro F, Shaheen F, Malnati MS, Fitzgerald W, Lusso P, Margolis L. Nat Med 2001; 7: 1232.
Grivel JC, Santoro F, Chen S, Faga G, Malnati MS, Ito Y, Margolis L, Lusso P. J Virol 2003; 77: 8280.
Harma M, Hockerstedt K, Lautenschlager I. Transplantation 2003; 76: 536.
Hasegawa H, Utsunomiya Y, Yasukawa M, Yanagisawa K, Fujita S. J Virol 1994; 68: 5326.
He J, McCarthy M, Zhou Y, Chandran B, Wood C. J Virol 1996; 70: 1296.
Higashi K, Asada H, Kurata T, Ishikawa K, Hayami M, Spriatna Y, Yamanishi K. J Gen Virol 1989; 70(Pt 12): 3171.
Inagi R, Guntapong R, Nakao M, Ishino Y, Kawanishi K, Isegawa Y, Yamanishi K. J Med Virol 1996; 49: 34.
Inoue Y, Yasukawa M, Fujita S. J Virol 1997; 71: 3751.
Isegawa Y, Ping Z, Nakano K, Sugimoto N, Yamanishi K. J Virol 1998; 72: 6104.
Ishikawa K, Hasegawa K, Naritomi T, Kanai N, Ogawa M, Kato Y, Kobayashi M, Torii N, Hayashi N. J Gastroenterol 2002; 37: 523.
Kakimoto M, Hasegawa A, Fujita S, Yasukawa M. J Virol 2002; 76: 10338.
Kashanchi F, Araujo J, Doniger J, Muralidhar S, Hoch R, Khleif S, Mendelson E, Thompson J, Azumi N, Brady JN, Luppi M, Torelli G, Rosenthal LJ. Oncogene 1997; 14: 359.
Kemper C, Chan AC, Green JM, Brett KA, Murphy KM, Atkinson JP. Nature 2003; 421: 388.
Kikuta H, Nakane A, Lu H, Taguchi Y, Minagawa T, Matsumoto S. J Infect Dis 1990; 162: 35.
Knox KK, Carrigan DR. J Infect Dis 1992; 165: 925.
Knox KK, Pietryga D, Harrington DJ, Franciosi R, Carrigan DR. Clin Infect Dis 1995; 20: 406.
Kondo K, Kondo T, Okuno T, Takahashi M, Yamanishi K. J Gen Virol 1991; 72(Pt 6): 1401.
Kondo K, Kondo T, Shimada K, Amo K, Miyagawa H, Yamanishi K. J Med Virol 2002a; 67: 364.
Kondo K, Shimada K, Sashihara J, Tanaka-Taya K, Yamanishi K. J Virol 2002b; 76: 4145.
Krueger GR, Wassermann K, De Clerck LS, Stevens WJ, Bourgeois N, Ablashi DV, Josephs SF, Balachandran N. Lancet 1990; 336: 1255.
Lacoste V, Mauclere P, Dubreuil G, Lewis J, Georges-Courbot MC, Rigoulet J, Petit T, Gessain A. J Virol 2000; 74: 11993.

Lacoste V, Verschoor EJ, Nerrienet E, Gessain A. J Gen Virol 2005; 86: 2135.
Liszewski MK, Post TW, Atkinson JP. Ann Rev Immunol 1991; 9: 431.
Liu DX, Gompels UA, Foa-Tomasi L, Campadelli-Fiume G. Virology 1993; 197: 12.
Luka J, Okano M, Thiele G. J Clin Lab Anal 1990; 4: 483.
Luppi M, Marasca R, Barozzi P, Ferrari S, Ceccherini-Nelli L, Batoni G, Merelli E, Torelli G. J Med Virol 1993; 40: 44.
Luppi M, Barozzi P, Maiorana A, Marasca R, Torelli G. J Infect Dis 1994a; 169: 943.
Luppi M, Barozzi P, Marasca R, Torelli G. Leukemia 1994b; 8(Suppl 1): S41.
Luppi M, Barozzi P, Morris CM, Merelli E, Torelli G. Lancet 1998; 352: 1707.
Luppi M, Barozzi P, Morris C, Maiorana A, Garber R, Bonacorsi G, Donelli A, Marasca R, Tabilio A, Torelli G. J Virol 1999; 73: 754.
Lusso P, Markham PD, Tschachler E, di Marzo Veronese F, Salahuddin SZ, Ablashi DV, Pahwa S, Krohn K, Gallo RC. J Exp Med 1988; 167: 1659.
Lusso P, Gallo RC, DeRocco SE, Markham PD. Lancet 1989; 1: 730.
Lusso P, Markham PD, DeRocco SE, Gallo RC. J Virol 1990; 64: 2751.
Lusso P, De Maria A, Malnati M, Lori F, DeRocco SE, Baseler M, Gallo RC. Nature 1991a; 349: 533.
Lusso P, Malnati M, De Maria A, Balotta C, DeRocco SE, Markham PD, Gallo RC. J Immunol 1991b; 147: 685.
Lusso P, Malnati MS, Garzino-Demo A, Crowley RW, Long EO, Gallo RC. Nature 1993; 362: 458.
Lusso P, Secchiero P, Crowley RW. AIDS Res Hum Retroviruses 1994; 10: 181.
Lusso P, Garzino-Demo A, Crowley RW, Malnati MS. J Exp Med 1995; 181: 1303.
Luttichau HR, Clark-Lewis I, Jensen PO, Moser C, Gerstoft J, Schwartz TW. J Biol Chem 2003; 278: 10928.
Manchester M, Liszewski MK, Atkinson JP, Oldstone MB. Proc Natl Acad Sci USA 1994; 91: 2161.
Matsuda Y, Hara J, Miyoshi H, Osugi Y, Fujisaki H, Takai K, Ohta H, Tanaka-Taya K, Yamanishi K, Okada S. Bone Marrow Transplant 1999; 24: 919.
Mayne M, Cheadle C, Soldan SS, Cermelli C, Yamano Y, Akhyani N, Nagel JE, Taub DD, Becker KG, Jacobson S. J Virol 2001; 75: 11641.
Milne RS, Mattick C, Nicholson L, Devaraj P, Alcami A, Gompels UA. J Immunol 2000; 164: 2396.
Mori Y, Seya T, Huang HL, Akkapaiboon P, Dhepakson P, Yamanishi K. J Virol 2002; 76: 6750.
Naniche D, Varior-Krishnan G, Cervoni F, Wild TF, Rossi B, Rabourdin-Combe C, Gerlier D. J Virol 1993; 67: 6025.
Niiya H, Azuma T, Jin L, Uchida N, Inoue A, Hasegawa H, Fujita S, Tohyama M, Hashimoto K, Yasukawa M. J Gen Virol 2004; 85: 2639.
Ozaki Y, Tajiri H, Tanaka-Taya K, Mushiake S, Kimoto A, Yamanishi K, Okada S. J Clin Microbiol 2001; 39: 2173.
Qian G, Wood C, Chandran B. Virology 1993; 194: 380.
Razzaque A, Williams O, Wang J, Rhim JS. Virology 1993; 195: 113.
Roush KS, Domiati-Saad RK, Margraf LR, Krisher K, Scheuermann RH, Rogers BB, Dawson DB. Am J Clin Pathol 2001; 116: 648.
Sakata H, Kurita M, Murakami Y, Nagasawa S, Watanabe M, Ueda S, Matsumoto M, Sato T, Kobune F, Seya T. Biol Pharm Bull 1998; 21: 1121.

Salahuddin SZ, Ablashi DV, Markham PD, Josephs SF, Sturzenegger S, Kaplan M, Halligan G, Biberfeld P, Wong-Staal F, Kramarsky B, Gallo RC. Science 1986; 234: 596.
Santoro F, Kennedy PE, Locatelli G, Malnati MS, Berger EA, Lusso P. Cell 1999; 99: 817.
Santoro F, Greenstone HL, Insinga A, Liszewski MK, Atkinson JP, Lusso P, Berger EA. J Biol Chem 2003; 278: 25964.
Schnorr JJ, Dunster LM, Nanan R, Schneider-Schaulies J, Schneider-Schaulies S, ter Meulen V. Eur J Immunol 1995; 25: 976.
Seya T, Hara T, Iwata K, Kuriyama S, Hasegawa T, Nagase Y, Miyagawa S, Matsumoto M, Hatanaka M, Atkinson JP. Int Immunol 1995; 7: 727.
Smith A, Santoro F, Di Lullo G, Dagna L, Verani A, Lusso P. Blood 2003; 102: 2877.
Smith AP, Paolucci C, Di Lullo G, Burastero SE, Santoro F, Lusso P. J Virol 2005; 79: 2807.
Takahashi K, Sonoda S, Higashi K, Kondo T, Takahashi H, Takahashi M, Yamanishi K. J Virol 1989; 63: 3161.
Takatsuka H, Wakae T, Mori A, Okada M, Fujimori Y, Takemoto Y, Okamoto T, Kanamaru A, Kakishita E. Bone Marrow Transplant 2003; 31: 475.
Takemoto M, Mori Y, Ueda K, Kondo K, Yamanishi K. J Gen Virol 2004; 85: 869.
Tatsuo H, Ono N, Tanaka K, Yanagi Y. Nature 2000; 406: 893.
Torelli G, Barozzi P, Marasca R, Cocconcelli P, Merelli E, Ceccherini-Nelli L, Ferrari S, Luppi M. J Med Virol 1995; 46: 178.
Vossen MT, Westerhout EM, Soderberg-Naucler C, Wiertz EJ. Immunogenetics 2002; 54: 527.
Yalcin S, Mukai T, Kondo K, Ami Y, Okawa T, Kojima A, Kurata T, Yamanishi K. J Gen Virol 1992; 73(Pt 7): 1673.
Yasukawa M, Inoue Y, Ohminami H, Terada K, Fujita S. J Gen Virol 1998; 79(Pt 1): 143.
Yoshikawa T, Asano Y, Akimoto S, Ozaki T, Iwasaki T, Kurata T, Goshima F, Nishiyama Y. J Med Virol 2002a; 66: 497.
Yoshikawa T, Ihira M, Asano Y, Tomitaka A, Suzuki K, Matsunaga K, Kato Y, Hiramitsu S, Nagai T, Tanaka N, Kimura H, Nishiyama Y. J Med Virol 2002b; 66: 82.
Zhen Z, Bradel-Tretheway B, Sumagin S, Bidlack JM, Dewhurst S. J Virol 2005; 79: 11914.
Zou P, Isegawa Y, Nakano K, Haque M, Horiguchi Y, Yamanishi K. J Virol 1999; 73: 5926.

PART II: DIAGNOSIS AND EPIDEMIOLOGY

Human Herpesvirus-6: General Virology, Epidemiology and Clinical Pathology
Perspectives in Medical Virology
Gerhard Krueger and Dharam Ablashi (Editors)

DOI 10.1016/S0168-7069(06)12006-6

Strain Variants of HHV-6

Dharam V. Ablashi[a,b]
[a]*HHV-6 Foundation, 285 San Ysidro Road, Santa Barbara, CA 93108, USA*
[b]*Department of Microbiology & Immunology, School of Medicine, Georgetown University, Washington, DC, USA*

Introduction

After the first isolation of human herpesvirus-6 (HHV-6) in 1986 (Salahuddin et al., 1986), two other isolates were reported (Downing et al., 1987; Lopez et al., 1988). At that time no attention was paid to whether there were variations in these isolates, although these isolates were from HIV-1 and HIV-2 AIDS patients. The original HBLV isolate reported was from 6 to 8 AIDS patients with lymphoma and other lymphoproliferative disorders (Salahuddin et al., 1986), suggesting that perhaps HHV-6, can be more frequently reactivated in AIDS and immunosuppressed patients. Besides these isolates, in the early phase of HHV-6 research Agut et al. (1988) reported isolation of related viruses from an Ivory Coast patient with T-cell leukemia who was simultaneously infected with HTLV-1 and HIV-2. Becker et al. (1989) also isolated HHV-6 from AIDS patients from South Africa. Levy et al. (1990) isolated HHV-6 variant B from the saliva of an HIV-infected individual. All these HHV-6 isolates were from HIV-infected patients, and were not fully characterized.

Krueger et al. (1987) reported the first evidence of HHV-6 in Germany, and their isolate, CO1 was later identified as HHV-6A. In fact, from 1987 to 1994, Krueger and his associates identified and characterized five more HHV-6 isolates (CO1–CO6) from chronic fatigue syndrome (CFS) patients and from patients with vascular diseases (Krueger et al., 1990, 1991; Eichler et al., 1994). These were later

Table 1

Categorization of different HHV-6 strains using restriction analysis polymorphisms

Group A	Group B
GS	Z-29
Davilla	HST, KBT, KSM, SUZ
U1102	OK
AJ	KF, BA
SIE, TAN	SF
CO1, CO2, CO3, CO5, CO6	VW, AW, BOU, MAR, BLE, MBE

found to be HHV-6A (Table 1). These isolates were not from HIV positive or AIDS patients.

Kikuta et al. (1989) made interesting observations that there was polymorphism of HHV-6 DNA in five Japanese patients with *Exanthem subitum*. Perhaps this drew attention to the fact that there may be biologic, molecular and immunologic variations in HHV-6 isolates. In 1991, two independent reports were published that indicated that HHV-6 strains fall into two groups (Ablashi et al., 1991; Schirmer et al., 1991). Furthermore, other reports supported that there are genomic and antigenic variations in HHV-6 isolates (Wyatt et al., 1990; Balachandran et al., 1992; Dewhurst et al., 1992a,b; Gompels et al., 1993). Based on these data, it was suggested that HHV-6 isolates should be divided into Groups A and B (Ablashi et al., 1993). The ICTV discussed the variations in HHV-6 isolates at a special session at the 16th International Herpesvirus Workshop in July 1991 and at the first International Herpesvirus Symposium in Japan in 1992. A nomenclature of HHV-6 strain groups was based on tropism on $CD4^+$ T-lymphocytes, but each group was characterized by strain, GS or Z-29 (USA) or U1102 (Europe). These three strains were therefore considered prototypes of variant A (GS & U1102) and variant B (Z-29). The second property considered was the use of monoclonal antibodies specific to variant A or B. The human sera reacted with both variants, and as of now there is no serologic test that can differentiate whether antibody (IgG or IgM) is directed to A or B variant. The third property that clearly distinguishes the two groups is the restriction enzyme endonuclease site. This was based on the first studies performed on strains GS and Z-29 (Josephs et al., 1988) and strains U1102 and Z-29 (Frenkel et al., 1990). Nucleotide sequence analysis also supported the existence of variants A and B of HHV-6. It indicated that these two classes are very closely related and that each contains some degree of heterogenicity. Some of the isolates belonging to HHV-6 variants A and B are identified in Table 1. Molecular studies from various laboratories have clearly shown differences between variants A and B that made some investigators believe that these two variants are two very distinct viruses and should be reclassified (Ablashi et al., 1993). The question of assigning a new herpesvirus number to either of the two groups was considered. There was a lot of discussion on renaming variant A HHV-6, and variant B HHV-8. (There was no HHV-8 at that time.) Some felt that, since HHV-6A isolates did not carry diseases like *Exanthem subitum*, variant B should be called HHV-6 and variant A renamed HHV-8. Ablashi and a few others objected saying that, since GS and some other HHV-6A isolates were discovered first, in 1986 and 1987, variant A should be called HHV-6. Since more data was needed on the epidemiology of variants A and B, although there was already enough data on the molecular biology and pathogenesis, it was decided that this matter should be brought back to the ICTV to reconsider reclassifying HHV-6 some time in the near future.

Epidemiology

The prevalence rate reported on HHV-6 suggests that 97% of children in the USA have been infected with HHV-6B (Campadelli-Fiume et al., 1999; Clark, 2000; Krueger and Ablashi, 2003). The prevalence rates of HHV-6A and B in Africa, however, showed that 44% of children were infected with HHV-6A (Kasolo et al., 1997). As with those infected with HHV-6B, these children exhibited febrile illness, rash and other symptoms generally observed with HHV-6 infection (Yamanishi et al., 1988). Most of the studies in children in Japan reveal that HHV-6B infection is more prevalent. Another study on HHV-6 infection in febrile infants 90 days of age and younger showed that HHV-6 DNA was detected in the range of 70–169,000 DNA copies/ml, in 10.6% of these infants. HHV-6 variants A and B infection were seen in these children with or without fever (Byington et al., 2002). It is evident that there are geographical differences in HHV-6A and B distribution and that some areas of the world, such as Japan, basically lack HHV-6A infection, whereas, in Africa, variants A and B are nearly evenly distributed.

Data have also been obtained in adults showing that in 60% of men (mean age 68 years) studied for HHV-6 reactivation, the rate was higher in critically ill patients than in healthy volunteers (54/101 vs. 0/50, $P = 0.001$). Reactivation of HHV-6 (5066 DNA copies/ml) in blood leukocytes was of HHV-6A (Razonable et al., 2002). The epidemiology of HHV-6 is not well understood. It is possible that infection of HHV-6A usually comes later in life. There are no serologic assays, as yet, that differentiate between A and B antibody distribution to assess the true prevalence rate of variant A. Although salivary glands are considered a reservoir for HHV-6, the majority of isolates from saliva are HHV-6B. There is more HHV-6A in the lung, but the mode of transmission of HHV-6A needs to be investigated further. Until we have serological assays directed for HHV-6A and B, the prevalence of HHV-6 will remain confusing. More discussion on the epidemiology of HHV-6 is covered in another chapter in the book.

Molecular biology

The genomic architecture shared by HHV-6A and B and HHV-7 is unique (Braun et al., 1997; Campadelli-Fiume et al., 1999; Clark, 2000) among human herpesviruses, and resembles that of channel catfish virus. The unit length of the HHV-6A and B molecule ranges approximately between 162 and 170 kb and is composed of

143 kb unique, long segment bracketed by direct repeats of DR_L (left) and DR_R (right) that can vary in length from 8 to 13 kb, upon passage *in vitro*. The complete sequence of HHV-6A (U1102) has been determined (Gompels et al., 1995). The u segment is 143–147 bp in length, and is flanked at each terminus by an 8.087 by DR, for an overall length of 159.321. HHV-6A mostly contains 119 open reading frames (ORFs), 9 of which are absent in HHV-6B strain. Splicing (U1102) is predicted, resulting in 97 unique genes, 88 of which have counterparts in HHV-6B. In comparing HHV-6B (Z-29) to HHV-6A (U1102), 9 ORFs in variant A do not have a counterpart in variant B as a result of either the lack of an initiation codon truncation or frameshift mutation. Several HHV-6B ORFs give rise to proteins longer than their HHV-6A counterparts that are dispersed throughout the genome. Conserved genes among six herpesviruses can be organized into seven gene blocks, the arrangement of which is subfamily specific and there is an additional block specific to *β*-herpesviruses. The genes of HHV-6A and B variants belonging to conserved gene blocks have greater than 94% of aa identity. Genes specific in the Roseola virus genus of β-herpesviruses (HHV-6, HHV-7) include 20–24 (predicted glycoproteins), U26, U85 (OX-2 glycoprotein homologous) and U100. A comparison of HHV-6A and HHV-6B genomes reveals that these two genomes are colinear with an overall nucleotide sequence identity of 90%. The regions of significant variation include DR and a 24 kb segment located at the right of U86 (except for U94). The unique region spanning ORFs U86–U100 has 72% nucleotide sequence identity and the U regions differ more than 10%. The divergence in particular genes between variants A and B throughout the genomes are important, but particularly at the right end of U. This may be critical in defining the biological differences between A and B (Gompels et al., 1993, 1995; Inoue et al., 1994; Braun et al, 1997; Clark, 2000; Krueger et al., 2003; DeBolle et al., 2005). In-depth molecular differences in both variants are covered in other chapters.

Biological aspects of variants A and B

HHV-6 is predominantly a T-lymphotropic virus, but it has the ability to infect other cell types (Ablashi et al., 1988; Braun et al., 1997). Both HHV-6A and B replicate efficiently in activated peripheral blood or cord blood mononuclear cells. HHV-6, especially HHV-6A, is more cell associated. Isolating HHV-6B from peripheral blood from patients with *Exanthem subitum* or from transplant patients is much easier than isolating HHV-6A. Although HHV-6A DNA copy numbers in peripheral blood mononuclear cells are reasonably high, isolation of the virus is difficult. This is why there are very few HHV-6A isolates reported (Table 1). Earlier studies by Ablashi et al. (1988) showed that HHV-6A replicates well in $CD38^+$ immature T-cells (HSB_2). HHV-6B, on the other hand, prefers to replicate in mature T-cells (MOLT-3). Both variants have been shown to infect productively, both *in vivo* and *in vitro*, mature T-cells ($CD4^+$–$CD8^-$ and $CD3^+$–$CD4^+$) (Clark, 2000), supporting the finding that HHV-6 can be isolated only from the non-adherent mononuclear cells from *Exanthem subitum*. DNA has, however, been detected in both adherent and non-adherent cells by PCR (in mononuclear fractions during the acute phase of illness). Macrophages are persistently infected with HHV-6 following primary infection and serve as a reservoir for virus reactivation. Interestingly, both variants enter the cell through interaction with CD46, which is the receptor for HHV-6 (Braun et al., 1997; Clark, 2000; DeBolle et al., 2005).

HHV-6 infects a wide variety of cell types including diploid lung fibroblasts, neural cells, megakaryocytes and NK cells. In most of these cell lines there is a low level of viral replication. It has been shown that HHV-6A induces $CD4^+$ receptor on $CD8^+$ cells, making them infectable with HIV. It has also been shown that B-lymphocytes are not infectable with HHV-6A, but Epstein–Barr virus (EBV)-positive B-lymphocytes can be infected with HHV-6A. It was reported that HHV-6A, not HHV-6B, activates EBV and similarly HHV-6A also activates HHV-8 (Flamand et al., 1993; Cuomo et al., 1995; Zeng et al., 2005). It was also shown that HHV-6A induces cell membrane receptor expression that predisposes them to superinfection by other viruses, such as EBV and HIV (Krueger et al., 1990; Schonnebeck et al., 1991). HHV-6B (Z-29) grew best in HTLV-1 transformed cell line MOLT-4, which is not infectable with HHV-6A. The propagation and replication of HHV-6A and B isolates have been studied. HHV-6A (GS U1102) replicate quite well in HSB_2, J Jhan and $supT_1$ cells, whereas HHV-6B grows well in MOLT-3, MOLT-4 and MT-4. Both HHV-6B (SF) and strain HST grew well in MT-4 and HPB-ALL cell lines. Like J Jhan, another continuous cell line, SupT1, can be used to propagate HHV-6A GS, u1102 and HHV-6B (Z-29, OK). Thus far, using continuous T-cell lines has not been very useful in providing HHV-6 isolation. Since HHV-6 is cell associated, especially variant A, the yield of cell-free virus in the supernatant ranges between 10^{-3} and 10^{-4}/ml. It has been observed that glial cells can be infected with HHV-6A more efficiently than with B. Infected cells undergo viral DNA synthesis and the cytopathic effects are quite noticeable. This was not true when glial cells were infected with Z-29 (HHV-6B), but primary astrocytes are infectable with HHV-6A and B (Clark, 2000; DeBolle et al., 2005). HHV-6 late antigens can also be localized in oligodendrocyte sections from plaque from multiple sclerosis (MS) patients. DiLuca et al. (1996) also reported distribution of HHV-6A in various human tissues. More critical studies of the cell biology of HHV-6A and B are needed to further the understanding of the pathogenesis of HHV-6 variants.

Reactivity of monoclonal antibodies to HHV-6 variants A and B

Iyengar et al. (1991) was able to show that HHV-6A isolates reacted to a late protein monoclonal antibody gp110, whereas HHV-6B isolates showed no activity. Similarly, gp82/105 showed specificity to variant A (Balachandran et al., 1989). It has also been observed that the immediate-early protein monoclonal antibody against IE antigen from HHV-6A can detect the IE antigen in infected cells between 4 and 8 h after infection by HHV-6 GS or U1102 strains (Arsenault et al., 2003). The same antibody fails to react with HHV-6B (Z-29). Another late protein monoclonal antibody to P101 produced at the Center for Disease Control (CDC) by Dr. Pellett only reacted with HHV-6B (Z-29) infected cells. OHV-3 (P98) monoclonal antibody produced by Dr. Yamanishi's group in Japan was specific for HHV-6B isolates. Thus, monoclonal antibodies are very useful in diagnostics, to identify the HHV-6A or B infections in tissues and cells, and in developing immunologic assays to detect HHV-6A or B antigens in clinical specimens. These antibodies are very useful in developing antigenemia assays for detecting HHV-6 reactivation in transplant recipients (Lautenschlager et al., 2000). We do not know whether the recombinant or peptide-based ELISA to detect antibody to early protein (P41) would be equally specific and sensitive as the one developed from native protein. The use of variants A- and B- specific monoclonal antibodies to extract native proteins from infected cells will surely be an asset. This would definitely enhance our understanding of the role of HHV-6A and B variants in the pathology of various illnesses.

Disease associations of HHV-6A and -6B variants

Roseola infantum

It was previously thought that primary infection of HHV-6 in infants and young children lead to *roseola* and febrile illnesses. This infection was attributed to HHV-6B (Yamanishi et al., 1988). Some of the symptoms identified include diarrhea, vomiting, seizure, nasal congestion, rash and high fever. Similar symptoms were noted in Africa in children and the DNA analysis from the peripheral blood from these children revealed HHV-6A infections (Hidaka et al., 1997; Kasolo et al., 1997; Randhawa et al., 1997). It is, therefore, obvious that both variants A and B participate in infecting growing children with identical pathologies. Hall et al. (1994) also observed that one third of congenital infection was due to HHV-6A. Later on these children still retained HHV-6A DNA. There have not been longitudinal follow-up studies to assess whether these children with congenital HHV-6A infection could be at risk of developing a central nervous system (CNS) illness such as multiple sclerosis.

CNS disease

One complication of CNS disease with HHV-6 could be seizures. CNS involvement may lead to meningitis and encephalitis or encephalopathy (Suga et al., 1993). The CNS is most likely one of the sites of persistence of virus following primary infection. Both variants have been detected in the cerebral spinal fluids (CSF). There is more recent evidence that HHV-6A may have a greater neurotropism (Aberle et al., 1996; Hall et al., 1998). In this study 84 of 660 CSF samples from children under 3 years of age with or without febrile illness were HHV-6 positive by PCR. Seven children had primary infection, and in the remaining 77, HHV-6 was detected in periferal blood mononuclear cells (PBMC) and CSF in 30 (39%). In CSF alone, 47 (61%) HHV-6A was detected more frequently than in PBMCs. In children who had HHV-6 in PBMCs and CSF, variant B was detected in the former and HHV-6A in the latter. There is considerable evidence showing that severe meningitis was caused by HHV-6B in an immune-competent woman who was treated with ganciclovir (Birnbaum et al., 2005). Cases of encephalitis have been reported frequently with HHV-6 infection as a result of complications. In some cases, variant A has been detected and in other cases variant B has been reported (Clark, 2000; DeBolle et al., 2005). In the early studies as well as in some more recent studies, attempts have been made to identify the variant (Isaacson et al., 2005). Whitley et al. (2005) believes that rare cases of HHV-6 infections of the CNS exist. At the present time only PCR (Taqman) assay is utilized that may not be sensitive enough, because if the HHV-6 persists in low DNA copy numbers the assay may not detect it. New, more sensitive serological assays are needed to detect active infection in patients with low-grade infection.

Epilepsy and HHV-6B

Donati et al. (2003) reported up to 23,079 copies of HHV-6 DNA in the hippocampal sections and was subtyped as HHV-6B. Expression of HHV-6 was confirmed by western blot analysis and by immunohistochemistry using HHV-6 monoclonal antibody (gp116/64/54). The study concluded that HHV-6B is present in specimens from a subset of patients with mesial temporal-lobe epilepsy (MTLE). The infection is localized in astrocytes in the absence of inflammation. Thus, HHV-6 seems to play a role in the development of MTLE. Another study by Uesugi et al. (2000) reported that HHV-6 and HSV was detected by PCR in surgical tissue from temporal-lobe epileptic patients. HHV-6-induced mild encephalitis/meningitis most frequently causes temporal-lobe epilepsy. In four out of the six patients, the mesial temporal lobe was positive for HHV-6. This study fails to type the HHV-6 variant. Further investigations are necessary in order to study the role of HHV-6 in epilepsy, since it is a serious disease and affects a great number of individuals, including children.

Transplantation and HHV-6 variants

Similar to the other HHV, HHV-6 has the ability to enhance pathogenicity in the immunocompromised host including transplant recipients. Over 24,000 stem cells and cord blood cell transplants are been performed in the USA. HHV-6 reactivation rate in transplant patients is extremely high and its active infection leads to enhancement of CMV disease, pneumonitis, and even rejection of the graft (Wainwright et al., 2001). All evidence that exists about kidney, heart, liver and bone marrow transplants (there may be a few exceptions) points out that it is HHV-6B which is involved (Singh et al., 1995; Randhawa et al., 1997; Zerr et al., 2005; Lautenschlager, et al., 2000). Nitsche et al. (2001) often detected HHV-6A DNA in the plasma but not in the blood leukocytes of patients after bone marrow transplantation. More attention should be paid to controlling HHV-6 reactivation in the patients so that there are no added complications.

HHV-6 (variants A and B) and multiple sclerosis

MSis characterized as a T-cell mediated, autoimmune pathogenesis process in genetically predisposed individuals. Although the origin of MS is unclear, viral agents are suspected in its pathogenesis. Earlier studies by Challoner et al. (1995) revealed HHV-6B variant in MS plaques and HHV-6 localized in oligodendrocytes, in later studies, Alvarez-Lafuente et al. (2004) showed that a subset of MS patients with relapsed and remitting MS (RRMS) had active HHV-6 infection. They are, therefore, at greater risk for exacerbation with RRMS. Another study by Rotola et al. (2004) showed that 20% of patients have an active HHV-6A infection in the early stages of MS. Akhyani et al. (2000) found DNA in the serum of MS patients that was HHV-6A. Soldan et al. (2000) showed that 67% of MS patients had lymphoproliferative response to HHV-6A, compared to healthy controls who had a 78% lymphoproliferative response to HHV-6B. Many other studies conclude that there are CSF and other tissues from MS with a high prevalence of HHV-6A (Akhyani et al., 2000). Kim et al. (2000) reported that HHV-6A genomic sequences were present in MS patients' PBMCs, whereas, no HHV-6 genome was detected in the healthy controls. Since A and B variant-specific monoclonal antibodies exist, in-depth study of variant types in MS tissues can be performed to assess the prevalence of HHV-6A. Also, with more sensitive PCR probes, the frequency of HHV-6A can be documented more consistently. It is most likely that the PBMCs of the MS patients may carry HHV-6 variant B (Ablashi et al., 1998, 2000).

HHV-6A and chronic fatigue syndrome (CFS)

According to the CDC, the incidence of CFS in USA is approximately 400/100,000. It is now believed that subsets of CFS patients have active HHV-6 infection, based on early antigen-antibody assay and PCR (Patnaik et al., 1995; Ablashi et al., 2000). The PCR data show that approximately 35% of CFS patients who exhibit CNS disease or cognitive disorders have active HHV-6A infection. Although the frequency of HHV-6 DNA detected by PCR is low, due to the insensitivity of the assay, the evidence shows that in the USA, all HHV-6 DNA positive CSF samples were infected by HHV-6A, and the majority of plasma samples positive by nested or Taqman PCR were also variant A infected. DiLuca et al. (1995) showed in his study that all CFS patients harbor HHV-6A DNA. A similar study was reported by Yalcin et al. (1994), but here she found both variants. Isolate CO1 (Table 1) was isolated from the PBMCs of a CFS patient in Germany and was identified as variant A (Ablashi et al., 1991). It is evident that of the subsets of CFS patients carrying HHV-6 active infection, a few are infected with HHV-6 variant A because no typing of variants was attempted (Buchwald et al., 1992; Wagner et al., 1996), suggesting that HHV-6 plays a role in the pathogenesis of these patients. This is supported by the use of antiviral therapy, which relieves the patients' symptoms. The specific serological assays to detect variant-specific antibodies and more sensitive molecular probes for PCR would be helpful in strengthening the association of HHV-6A in the pathogenesis of CFS.

Concluding remarks

The data reviewed in this chapter describe various aspects of HHV-6 variants. Based on the findings, the following conclusions can be drawn:

1. The HHV-6A and B variants are two distinct viruses and should be reclassified according to the nomenclature of herpesviruses;
2. The existing epidemiology of HHV-6 is confusing because there are no serological tests to assess the prevalence rates of variants A and B in various parts of the world.
3. More specific reagents and assays need to be developed to differentiate between active and latent infections, so that the infection can be controlled in the early stages.
4. Earlier reports and current studies have not been able to show clear-cut variant association with various diseases and disorders, or few attempts have been made to do so. More emphasis must be placed to type the variants in these diseases.
5. The consequences of HHV-6 infection and its concurrent complications are still being underestimated.

References

Aberle SW, Mandl CW, Kunz C, Popow-Kraupp TJ. Clin Microbiol 1996; 34: 3223–3225.

Ablashi DV, Agut H, Berneman Z, Campadelli-Fiume G, Carrigan D, Ceccarini-Nelli L, Balachandran N, Chou S, Collandre H, Cone R, Dambaugh T, Dewhurst S, DiLuca D, Foa-Tomasi L, Fleckenstein B, Frenkel N, Gallo R, Gompels U, Hall C, Jones H, Lawrence G, Martin M, Montagnier L, Neipel F, Nicholas J, Pellett P, Razzaque A, Torvelli G, Thomson B, Salahuddin S, Wyatt L, Yamanishi K. Arch Virol 1993; 129: 363–366.

Ablashi DV, Balachandran N, Josephs SF, Hung CL, Krueger GRF, Kramarsky B, Salahuddin SZ, Gallo RC. Virology 1991; 184: 545–552.

Ablashi DV, Eastman HB, Owen CB, Roman MM, Friedman J, Zabriskie JB, Peterson DL, Pearson GR, Whitman JE. J Clin Virol 2000; 16: 179–191.

Ablashi DV, Lapps W, Kaplan M, Whitman JE, Richert JR, Pearson GR. Mult Scler 1998; 4: 490–496.

Ablashi DV, Lusso P, Hung CL, Salahuddin SZ, Josephs SF, Llana T, Kramarsky B, Biberfeld P, Markham PD, Gallo RC. Int J Cancer 1988; 42: 787–791.

Agut H., Guetard D., Collandre H., Dauguet C., Montagnier L., Miclea J.M., Baurmann H., GessainA, Lancet, I, 1988, 712

Akhyani N, Berti R, Brennan MB, Soldan SS, Eaton JM, McFarland HF, Jacobson S. J Infect Dis 2000; 182: 1321–1325.

Alvarez-Lafuente R, De las Heras V, Bartolome M, Picazo JJ, Arroyo R. Arch Neurol 2004; 61: 1523–1527.

Arsenault S, Gravel A, Gosselin J, Flamand L. J Clin Virol 2003; 3: 284–290.

Balachandran N, Amelese RF, Zhou WW, Chang CK. J Virol Methods 1989; 63: 2835–2840.
Balachandran N, Tirawatnapong S, Pfeiffer B, Ablashi DV. J Infect Dis 1992; 37: 29–34.
Becker WB, Engelbrecht S, Becker ML, Piek C, Robson BA, Wood L, Jacobs P. Lancet 1989; I: 41.
Birnbaum T, Padovan CS, Sporer B, Rupprecht TA, Ausserer H, Jaeger G, Pfister HW. Clin Infect Dis 2005; 40: 887–889.
Braun DK, Dominguez G, Pellett PE. Clin Microbiol Rev 1997; 10: 521–556.
Buchwald D, Cheney PR, Peterson DL, Henry B, Wormsley SB, Geiger A, Ablashi DV, Salahuddin SZ, Saxinger C, Biddle R, Kilkinis R, Jolesz FA, Folks T, Balachandran N, Gallo RC, Komaroff AL. Ann Intern Med 1992; 116: 103–113.
Byington CL, Zerr DM, Taggart EW, Nguy L, Hillyard DR, Carroll KC, Corey L. Pediatr Infect Dis J 2002; 21: 996–999.
Campadelli-Fiume G, Mirandola P, Menotti L. Emerg Infect Dis 1999; 5: 353–566.
Challoner PB, Smith KT, Parker JD, MacLeod DL, Coulter SN, Rose TM, Schultz ER, Bennett JL, Garber RL, Chang M. Proc Natl Acad Sci USA 1995; 92: 7440–7444.
Clark DA. Rev Med Virol 2000; 10: 55–73.
Cuomo L, Angeloni A, Zompetta C, Cirone M, Calogero A, Frati L, Ragona G, Faggioni A. AIDS Res Hum Retroviruses 1995; 11: 1241–1245.
DeBolle L, Naesens L, DeClerq E. Clin Microbiol Rev 2005; 18: 217–245.
Dewhurst S, Chandran B, McIntyre K, Schnabel K, Hall CB. Virology 1992a; 190: 490–4933.
Dewhurst S, McIntyre K, Schnabel K, Hall CB. J Clin Microbiol 1992b; 31: 416–418.
DiLuca D, Mirandola P, Ravaioli T, Bigoni B, Cassai E. Infect Agents Dis 1996; 5: 203–214.
DiLuca D, Zorzenon M, Mirandola P, Colle R, Botta GA, Cassai EJ. Clin Microbiol 1995; 33: 1660–1661.
Donati D, Akhyani N, Fogdell-Hahn A, Cermelli C, Cassiani-Ingoni R, Vortmeyer A, Heiss JD, Cogen P, Gaillard WD, Sato S, Theodore WH, Jacobson S. Neurology 2003; 61: 1405–1411.
Downing RG, Sewankambo N, Serwadda D, Honess R, Crawford D, Jarrett R, Griffin BE. Lancet 1987; II: 390.
Eichler F, Krueger GRF. In Vivo 1994; 8: 565–576.
Flamand L, Stefanescu I, Ablashi DV, Menezes JJ. Virol 1993; 677: 6768–6777.
Frenkel N, Shirmer EC, Katsafanas G, June CHJ. Virol 1990; 64: 4598–4602.
Gompels UA, Carrigan DR, Carss AL, Arno J. J Gen Virol 1993; 74: 613–622.
Gompels UA, Nicholas J, Lawrence G, Jones M, Thomson J, Martin ED, Efstathiou S, Craxton M, Macaulay HA. Virology 1995; 209: 29–51.
Hall CB, Caserta MT, Schnabel KC, Long C, Epstein LG, Insel RA, Dewhurst S. Clin Infect Dis 1998; 26: 132–137.
Hall CB, Long CE, Schnabel KC, Caserta MT, McIntyre KM, Costanzo MA, Knolt A, Dewhurst S, Insel RA, Epstein G. N Engl J Med 1994; 331: 432–438.
Hidaka Y, Kusuhara K, Takabayushi A. Clin Inft Dis 1997; 24: 1022.
Inoue N, Dambaugh TR, Pellett PE. Infect Agents Dis 1994; 2: 343–360.
Isaacson E, Glaser CA, Forghani B, Amad Z, Wallace M, Armstrong RW, Exner MM, Schmid S. Clin Infect Dis 2005; 40: 890–893.
Iyengar S, Levine PH, Ablasi DV, Neequaye J, Pearson GR. Int J Cancer 1991; 49: 551–557.
Josephs SF, Ablashi DV, Salahuddin SZ, Kramarsky B, Franza BR, Pellett P, Buchbinder A, Memon S, Wong-Staal F, Gallo RC. J Virol Methods 1988; 21: 179–190.
Kasolo FC, Mpabalwani E, Gompels UAJ. Gen Virol 1997; 78: 847–855.

Kikuta H, Matsumoto S, Josephs SF, Gallo RC. J Infect Dis 1989; 550–551.
Kim SS, Lee KS, Park JH, Kim MY, Shin WS. Eur Neurol 2000; 43: 170–173.
Krueger GRF, Ablashi DV. Intervirology 2003; 46: 257–269.
Krueger GRF, Koch B, Ablashi DV. Lancet 1987; II: 36.
Krueger GRF, Sander C, Hoffman A, Barth A, Koch B, Braun B. In Vivo 1991; 5: 17–26.
Krueger GRF, Schonneck M, Braun M. AIDS Res Hum Retrovirus 1990; 6: 148.
Lautenschlager I, Kimmo L, Krister H. Transplantation 2000; 69: 2561–2566.
Levy JA, Ferro F, Lennette ET, Oshiro L, Poulin L. Virology 1990; 178: 113–121.
Lopez C, Pellet P, Goldsmith C, Sanderlin K, Black J, Warfield D, Feorino P. J Infect Dis 1988; 157: 1271–1273.
Nitsche A, Muller CW, Radonic A, Landt O, Ellerbrok H, Pauli G, Siegert W. J Infect Dis 2001; 183: 130–133.
Patnaik M, Komaroff AL, Conley E, Ojo-Amaize EA, Peter JB. J Infect Dis 1995; 172: 1364–1367.
Randhawa PS, Jenkins FJ, Nalesnik MA, Martens J, Williams PA, Ries A, Pham S, Demetris AJ. J Surg Pathol 1997; 21: 847–853.
Razonable RR, Fanning C, Brown RA, Espy MI, Rivero A, Wilson J, Kremer W, Smith TF, Paya CV. J Infect Dis 2002; 185: 110–113.
Rotola A, Merlotti I, Caniatti L, Caselli E, Granieri E, Tola MR, DiLuca D, Cassai E. Mult Scler 2004; 10: 348–354.
Salahuddin SZ, Ablashi DV, Markham PD, Josephs SF, Sturzenegger S, Kaplan M, Halligan G, Biberfeld P, Wong-Staal F, Kramarsky B, Gallo RC. Science 1986; 234: 596–601.
Schirmer EC, Wyatt S, Yamanishi K, Rodriguez WJ, Frenkel N. Proc Natl Acad Sci USA 1991; 88: 5922–5926.
Schonnebeck M, Krueger GRF, Braun M, Koch B, Ablashi DV, Balachandran N. In Vivo 1991; 5: 255–264.
Singh N, Carrigan DR, Gayowski T, Singh J, Marino JR. Transplantation 1995; 60: 1355–1357.
Soldan SS, Thomas BA, Leist TP, Newton-Juhng KN, Henry BA, McFarland HF, Jacobson S. Ann Neurol 2000; 47: 306–313.
Suga S, Yoshikawa T, Asano Y, Kozawa T, Nakashima T, Kobayashi I, Yamamoto H, Kyata T. Ann Neurol 1993; 33: 597–603.
Uesugi H, Schimizu J, Maehara T, Arai N, Nakayama H. Psychiatry Clin Neurosci 2000; 54: 589–593.
Wagner M, Krueger GRF, Ablashi DV, Whitman JE. J CFS 1996; 2: 3–16.
Wainwright MS, Martin PL, Morse RP, Lacaze M, Provencale JM, Coleman RE, Morgan MA, Hulette C, Kurtzberg J, Bushnell C, Epstein 1, Lewis DV. Ann Neurol 2001; 50: 612–619.
Whitley RJ, Lakeman FD. Clin Infect Dis 2005; 40: 894–895.
Wyatt LS, Balachandran N, Frenkel N. J Infect Dis 1990; 162: 852–857.
Yalcin S, Kuratsune H, Yamaguchi K, Kitani T, Yamanishi K. Microbiol Immunol 1994; 38: 587–590.
Yamanishi K, Okuno T, Shiraki K, Takahashi M, Kondo T, Asano Y, Kurata T. Lancet 1988; 1: 1065–1067.
Zeng LC, Huang Y, Huang Z, Qian L, Tang C, Qin D. Am J Pathol 2005; 166: 173–183.
Zerr DM, Corey L, Kim HW, Huang ML, Nguy L, Boeckh M. Clin Infect Dis 2005; 40: 932–940.

Human Herpesvirus-6: General Virology, Epidemiology and Clinical Pathology
Perspectives in Medical Virology
Gerhard Krueger and Dharam Ablashi (Editors)

DOI 10.1016/S0168-7069(06)12007-8

Serologic Testing for Acute and Chronic Infection

Henri Agut[a], Agnes Gautheret-Dejean[a,b]
[a]*Department of Virology, EA 2387, Pierre et Marie Curie Paris 6 University, CERVI, Pitie-Salpetriere Hospital, 83 boulevard de l'Hopital, 75013 Paris, France*
[b]*Department of Virology, EA 2387, Pierre et Marie Curie Paris 6 University, CERVI, Pitie-Salpetriere Hospital, 83 boulevard de l'Hopital, 75013 Paris, France*

Introduction

The detection of virus-specific antibodies in a body fluid, mainly serum, or the demonstration of a significant increase in antibody titer is commonly used in viral diagnosis. HHV-6 primary infection and the following chronic lifelong infection result in the synthesis of specific antibodies. These antibodies can be detected by means of diverse techniques (Fig. 1) but the interpretation of serologic results in terms of primary infection, reactivation, or reinfection may be problematic. This is due to the fact that HHV-6 infection is ubiquitous in the population, has a worldwide distribution, is acquired primarily within the first two years of life, and establishes itself as a permanent latent infection. Therefore, any child above two years of age and any adult is expected to have detectable HHV-6-specific IgG antibodies in their serum, provided they have mounted a significant immune response to the infection. This occurs in most subjects but some patients may fail to recognize HHV-6 antigens in a relevant way, in particular, in case of immunodepression. The kinetics of antibody response and the search for IgM antibodies may be useful for investigating acute infections. However, a rise in antibody titer and/or the presence of IgM cannot be considered specific for current or very recent infection, since these features can be observed during the reactivations of latent virus. Moreover, the current assays of

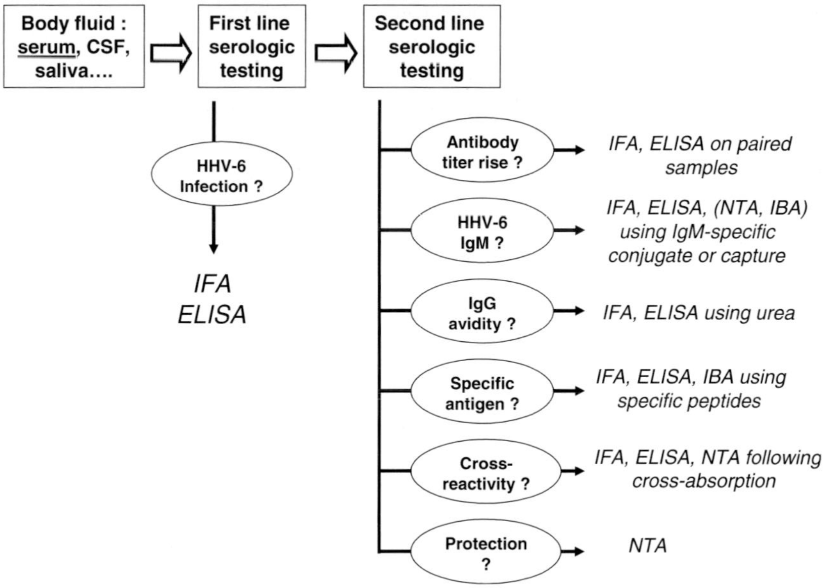

Fig. 1 Overview on the current approaches for HHV-6 serologic testing. Serum (or another body fluid) may be assayed for HHV-6-specific antibodies according to a two-step procedure: first line analyses are dedicated to the detection of HHV-6 infection, second-line analyses to the detailed characterization of the current stage of infection. IFA, immunofluorescence antibody assay (includes anti-complement immunofluorescence assay, ACIF); ELISA, enzyme-linked immunosorbent assay; IBA, immunoblot assay; NTA, neutralization assay.

HHV-6 serology have to stand other drawbacks such as the inability to distinguish HHV-6A from HHV-6B infections, persistence of maternal antibodies in young infants, and cross-reactivity with other betaherpesviruses. Despite these limitations, serologic testing remains a valuable tool for the determination of immune status to HHV-6 and epidemiologic studies regarding this virus. In the future, the identification of highly specific HHV-6 epitopes will improve both the specificity and standardization of serologic assays. Combined with the direct detection and quantitation of virus, serology may contribute to a better understanding of HHV-6 pathology and further developments of antiviral strategies.

Serologic assays

Techniques

Immunofluorescence assays

Indirect immunofluorescence antibody assays (IFA) were the first ones to be used for the detection of HHV-6 antibodies (Salahuddin et al., 1986; Linde et al., 1988;

Lopez et al., 1988) and remain still widely employed. In these tests, HHV-6-infected cells are fixed on a glass slide, a serum dilution is added and a fluorochrome-conjugated anti-immunoglobulin antibody is then applied to detect the binding of serum antibodies to specific antigens. When illuminated with ultraviolet light, the number of fluorescent foci as well as the characteristic pattern of cell staining observed with the microscope constitute the main parameters to be taken into account for the result. The staining of uninfected cells with a counterstain partly quenching non-specific fluorescence is also important, in particular, to check that the ratio of infected to uninfected cells is in agreement with the known characteristics of cell preparation. Infected cells consist of primary cells such as cord blood mononuclear cells or continuous cell lines such as HSB2 cells previously inoculated with a reference HHV-6 strain. These infected cells are used for the preparation of slides after a significant cytopathic effect (CPE) has been observed. A variation on conventional IFA is an anticomplement immunofluorescence assay (ACIF) in which human complement is applied to slides after the serum has been removed and before the conjugate is added; antigen–antibody–complement complex is detected by a fluorochrome-conjugated anti-C3 antibody (Lopez et al., 1988; Okuno et al., 1989; Robert et al., 1990). ACIF is considered to provide a lower background signal and a higher specific one than classical IFA. Although ACIF and IFA probably do not detect exactly the same HHV-6-specific antibodies, their results seem to be well correlated. As a general requirement, an extensive washing step after each exposure to a specific reagent (serum, complement, conjugate) is necessary to lower non-specific signal and reduce the rate of false-positive results. Nevertheless, this does not prevent the binding of cross-reactive antibodies, in particular those directed against other betaherpesviruses (Adler et al., 1993; Foa-Tomasi et al., 1994); as indicated below, more complex procedures are required to circumvent this phenomenon.

Enzyme immunoassays

Enzyme-linked immunosorbent assays (ELISA) have been developed in recent years for the diagnosis of HHV-6 infection (Saxinger et al., 1988; Chou and Scott, 1990; Sloots et al., 1996). These assays generally use either crude lysate of infected cells or purified virus obtained from cell culture supernatant as antigens. The antigen preparation is coated on polystyrene plates or any other appropriate surface, the serum sample is then added after what an enzyme-conjugated anti-human immunoglobulin is applied; finally, a chromogenic reaction catalyzed by the enzyme permits to detect and, to some extent, quantify the complex formed by the reference antigens and corresponding serum antibodies. ELISA is usually known to be highly sensitive, simple, rather inexpensive and susceptible of being automated, some remarkable qualities which justify its extensive use for the diagnosis of many viral infections. In the case of HHV-6, the availability of commercialized ELISA kits has been less widespread than in other domains of clinical virology. As an overall consequence, ELISA is less used than IFA, despite its theoretical higher first reaction step designed for the capture of IgM is known to improve assay specificity, and has been included successfully in HHV-6 ELISA (Nielsen and Vestergaard, 2002). An IBA has been developed to detect HHV-6-specific IgM antibodies against the immunodominant virion protein 101 K and led to relevant results when applied to serum pairs from children who had recently seroconverted (LaCroix et al., 2000). However, cross-reactive anti-HHV-6 IgM responses have been reported for children experiencing a primary HHV-7 infection in the absence of any previous HHV-6 infection (Yoshida et al., 2002b).

In parallel to the search for IgM assumed to reflect either primary infection or reactivation, another approach is to study virus-specific IgG subclass responses with the goal of finding a pattern specific for either latency or reactivation. Recently, two different immune isotype responses have been reported for HHV-6: one was the restriction to IgG1 in the latent phase of infection while the other was the presence of both IgG1 and IgG4 in reactivation (Carricart et al., 2004).

In addition, the distinction between primary and recurrent HHV-6 infection can be addressed with an antibody avidity test (Ward et al., 1993a). The principle is that low-avidity antibody present during the primary infection can be eluted easily by a mild denaturing agent such as urea, whereas high-avidity antibody synthesized during past or recurrent infection cannot be eluted. This can be applied to both IFA and ELISA: the sera whose antibody titer is significantly reduced in the presence of urea as compared to the control in the absence of denaturing agent is considered to contain low-avidity antibody. This strategy proved to be useful for demonstrating recurrent or chronic HHV-6 infection in transplant recipients, leukemia patients, HIV-positive subjects, as well as primary infection in children (Ward et al., 1993a,b; Marodi et al., 1998; Salonen et al., 2002).

Serologic findings

Seroprevalence studies

Without any doubt, serologic testing is the best and simplest tool for investigating the prevalence of infection. Seroepidemiological studies show that HHV-6 infection is very common in humans and geographically widespread (Table 1). The frequency of HHV-6 seropositivity slightly varies according to the country, the population studied, and the serologic assay employed in the study. However, the order of magnitude remains unchanged in all cases and the data are fully convergent. About 90% of the population has detectable HHV-6-specific antibodies at any age and in any country. The seropositivity in young infants corresponds to maternal antibody and, from six months old to the synthesis of their own antibodies following early primary infection. The titer of antibody decreases within the first six months reflecting the progressive loss of maternal antibody and, much later, at the end of life, witnessing a possible physiological decay of immune responses (Brown et al., 1988; Yanagi et al., 1990).

Table 1

Prevalence of HHV-6 antibodies determined in distinct populations using different assays

Country	Assay[a]	Studied population	Prevalence (%)	Reference
United Kingdom	IFA	Children 11 months	63	Briggs et al. (1988)
		Children > 15 years	65	
		Adults	63	
USA	ELISA	Adults	81–97	Saxinger et al. (1988)
Japan	ACIF	Children and adults	79	Okuno et al. (1989)
USA	IFA	Children 6–11 months	36	Levy et al. (1990)
		Children 3–5 years	95	
		Adults 21–40 years	90	
		Adults 62–88 years	35	
France	ACIF	Adults	60	Robert et al. (1990)
Japan	IFA	Children and adults	95	Yanagi et al. (1990)
Canada	ELISA	Children and adults	94	Parker et al. (1993)
Slovakia	IFA	Adults	64	Rajcani et al. (1994)
France	IFA	Adults	65	Ranger-Rogez et al. (1995)
Ghana	IFA	Children and adults	98	Cleghorn et al. (1995)
Trinidad and Jamaica	IFA	Children and adults	98	Cleghorn et al. (1995)
Denmark	ELISA	Children 12–15 months	78	Nielsen et al. (1996)
		Children > 3 years	100	
USA	IFA	Adult women	100	Baillargeon et al. (2000)
Eritrea	IFA	Children < 1 year	93	Tolfvenstam et al. (2000)
		Children 1–5 years	98	
		Children > 5 years	98	
Japan	IFA	Adults	96	Ihira et al. (2002)

[a]IFA, immunofluorescence antibody assay; ELISA, enzyme-linked immunosorbent assay; ACIF, anti-complement immunofluorescence assay.

Primary infection

HHV-6 infection is acquired primarily within the first two years of life, with a peak between 9 and 21 months (Zerr et al., 2005). All infants possess maternal antibody at birth and progressively lose it within their first month. The serologic evidence for primary infection is a seroconversion if the child was found HHV-6-negative initially, a significant rise (at least a fourfold increase) of antibody titer, and/or the detection of specific IgM. At least one of these results was reported for each published case of child experiencing a primary infection (Yamanishi et al., 1988; Suga et al., 1992; Black et al., 1996a; Nielsen and Vestergaard, 2002; Yoshida et al., 2002b; Zerr et al., 2005). A particular aspect of primary infection is that following an intrauterine transmission: in this case, specific IgM may be detected in the cord blood, a result which strongly supports the diagnosis of congenital infection (Dunne and Demmler, 1992). However, serologic testing may fail in the diagnosis of infection, due to several possible reasons. The presence of maternal antibody may occult seroconversion. The significant rise of antibody titer as well as the presence of IgM may correspond to a reactivation. Conversely, although a real primary infection has occurred as evidenced by virus isolation and/or IgG seroconversion, IgM may be not detected (Suga et al., 1992). Moreover, HHV-6 and HHV-7 primary infections may induce serologic cross-reactivity with the risk of false-positive results (Yoshida et al., 2002b).

Latent infection and reactivations

The determination of serologic status is the easiest way to know whether a subject has previously been infected with HHV-6 and now harbours latent virus. In contrast, the serologic diagnosis of an acute infection associated with viral reactivation is much more inaccurate. As mentioned above, neither the increase of antibody titer nor the presence of IgM is strictly pathognomonic of primary infection or reactivation (Suga et al., 1992). This is even more the case of immunocompromised patients whose adaptability for producing specific antibodies may be severely impaired.

Despite these well-known limitations, serologic testing has extensively been used for the investigation of HHV-6 infection in children and adults presenting suggestive clinical symptoms. In renal transplant recipients, serum antibody titer has been found to increase in parallel to virus isolation (Okuno et al., 1990; Herbein et al., 1996). The virus isolation from blood leukocytes generally occurred within the two to four weeks following transplant, whereas the increase of antibody titer was observed later on, one to two months after the graft (Yoshikawa et al., 1992). In particular, HHV-6 infection has been found in patients experiencing kidney rejection but the causative role of the virus in that process remains debated. Similarly, a significant rise of antibody titer has been detected in bone marrow and liver transplant recipients (Yoshikawa et al., 1991; Sloots et al., 1996; Ihira et al., 2001).

Central nervous system diseases

HHV-6 is involved in the pathogenesis of several neurological diseases (Kimberlin and Whitley, 1998): febrile seizures, meningitis, encephalitis, and, possibly MS. Serologic studies showed that HHV-6 seropositivity was significantly associated with seizures in infants, 12–15 months old, presenting with acute illnesses to the emergency department of a pediatric hospital (Hall et al., 1994). Whether HHV-6 is a risk for the subsequent development of epilepsy remains a matter of discussion (Eeg-Olofsson, 2003). As far as patients with meningitis or encephalitis are concerned, an intrathecal production of HHV-6 IgM or IgG has been found in 10 (20%) of 50 subjects presenting these syndromes as compared to none (0%) of the control patients (Patnaik and Peter, 1995b). This intrathecal antibody synthesis was not observed for other viruses, suggesting that the results about HHV-6 were quite specific and pointed to a causative role of the virus in these central nervous system diseases.

The possible role of HHV-6 in MS has emerged from both serologic and virologic findings. Regarding serologic results, patients with MS have been shown to have a higher serum titer of HHV-6 antibody than control patients (Sola et al., 1993; Wilborn et al., 1994; Caselli et al., 2002). Increased IgM responses have also been detected in MS patients compared with relevant controls (Soldan et al., 1997). It must be acknowledged that other studies have not found such differences (Enbom et al., 1999; Xu et al., 2002). Although the implication of HHV-6 in MS remains controversial, there is no doubt that serologic studies have pointed out a subset of MS patients exhibiting particular HHV-6 antibody responses, which may stimulate further investigations in that domain.

Pending questions and perspectives

Serologic testing is an easy procedure for viral diagnosis. The collection and storage of serum samples are easy, the immediate and delayed testing are both equally feasible making emergency analyses as well as prospective and retrospective studies possible. HHV-6 serologic testing exhibits these convenient properties. It has provided major insights both into the epidemiology and physiology of HHV-6 infection in the past two decades.

However this diagnostic approach has numerous drawbacks. The high seroprevalence of infection in the population is a basic obstacle for interpreting any seropositive result in terms of disease or acute infection. A single serum specimen may be not sufficient in such case while it is for the determination of immune status and epidemiologic studies. Many difficulties concentrate on the question of acute infection in relationship with clinical symptoms. As previously mentioned, the presence of IgM and the significant increase of antibody titer may be absent whereas HHV-6 infection is truly active; conversely, these findings may be present while the reactivation of HHV-6 is modest and clinically irrelevant. Therefore the positive predictive value of such serologic findings, which has not been investigated accurately up to date, might be unacceptably low for medical diagnosis. This also explains why the current possibilities of detecting and quantifying the virus itself in biological samples have been considered as a major breakthrough for the monitoring of HHV-6 infections in clinical settings. Another requirement is the standardization of the different IFA and ELISA tests, some commercialized and some other homemade, which should allow all scientists and physicians interested in HHV-6 to share and compare their results without any ambiguity.

So far, no serologic test is capable to discriminate between variant A- and variant B-specific antibodies. It is a real limitation but it is possible that this question is outdated now. The direct detection and characterization of viral DNA by means of molecular techniques have shown that many subjects, if not most, harbor both variants within their body (Fillet et al., 1995; Cone et al., 1996). In that context, it seems extremely difficult to consider a differential diagnosis of HHV-6 variant infection based on serology, given that both the variant-specific antibody sets are present together in most cases and probably cross-react with each other extensively. The identification of viral genomes using PCR approaches there appears as a better alternative.

A recurrent question is that of cross-reactivity with other viruses, particularly the two other human betaherpesviruses, HCMV and HHV-7. This includes at least two distinct phenomena: the production of antibodies reacting with common epitopes shared by two or three distinct viruses and the heterotypic production of antibodies to a given virus when infection by another one occurs. The rise of antibody to HHV-6 in patients with primary HCMV infection has repeatedly been reported (Irving et al., 1988; Chou and Scott, 1990; Ward et al., 1991). In this situation, a cross-reactivity between the glycoproteins gB of HCMV and gp116 of HHV-6 has been demonstrated, the reactivity to gp116 being removed by absorption of sera with gB (Adler et al., 1993). Symmetrically, HCMV antibody rise has been reported during active HHV-6 infection and the reactivity to HHV-6 could be removed following absorption with HCMV-infected cells (Sutherland et al., 1991). Some cases of seroconversion to HHV-7 have been associated with a simultaneous rise in HHV-6 antibody titer and, experimentally, some polyvalent sera raised to HHV-7 showed a cross-reactivity to HHV-6 (Foa-Tomasi et al., 1994). Further studies have indicated that the IgM response to HHV-7 primary infection was also directed to HHV-6 (Yoshida et al., 2002b). In addition, HHV-6 and HHV-7 cross-reactive antibodies have been detected in transplant recipients as demonstrated by using a cross-absorption IFA (Yoshikawa et al., 2001). These results were also found in the general population with a variable degree of cross-reactivity and a different efficiency in the removal of this activity by absorption with other viruses, depending on individual serum specimens (Black et al., 1996b). The current picture is far from clear although the general procedure of cross-absorption with heterologous antigens may be recommended as an attempt to clarify particularly intriguing clinical cases.

Future developments are expected to address some of these points. It seems that the use of panels of well-characterized peptides will fulfill the need for standardization and improve the specificity of HHV-6 serologic assays. This will require numerous convergent efforts from the community of virologists and probably lead to the combination of serologic testing with direct viral detection within integrated diagnostic strategies.

Acknowledgments

The authors deeply thank David Boutolleau for helpful discussions and Marie-Christine Papuchon for preparation of the manuscript.

References

Adler SP, McVoy M, Chou S, Hempfling S, Yamanishi K, Britt W. J Infect Dis 1993; 168: 1119–1126.

Asada H, Yalcin S, Balachandra K, Higashi K, Yamanishi K. J Clin Microbiol 1989; 27: 2204–2207.

Asano Y, Yoshikawa T, Suga S, Yazaki T, Ozaki T, Saito Y, Hatano Y, Takahashi M. J Med Virol 1990; 32: 119–123.

Baillargeon J, Piper J, Leach CT. J Clin Virol 2000; 16: 149–157.

Balachandran N, Amelse RE, Zhou WW, Chang CK. J Virol 1989; 63: 2835–2840.

Black JB, Durigon E, Kite-Powell K, de Souza L, Curli SP, Afonso AM, Theobaldo M, Pellett PE. Clin Infect Dis 1996a; 23: 1156–1158.

Black JB, Schwarz TF, Patton JL, Kite-Powell K, Pellett PE, Wiersbitzky S, Bruns R, Muller C, Jager G, Stewart JA. Clin Diagn Lab Immunol 1996b; 3: 79–83.

Briggs M, Fox J, Tedder RS. Lancet 1988; 1: 1058–1059.

Brown NA, Sumaya CV, Liu CR, Ench Y, Kovacs A, Coronesi M, Kaplan MH. Lancet 1988; 2: 396.

Carricart SE, Bustos D, Biganzoli P, Nates SV, Pavan JV. J Clin Virol 2004; 31: 266–269.

Caselli E, Boni M, Bracci A, Rotola A, Cermelli C, Castellazzi M, Di Luca D, Cassai E. J Clin Microbiol 2002; 40: 4131–4137.

Chen H, Pesce AM, Carbonari M, Ensoli F, Cherchi M, Campitelli G, Sbarigia D, Luzi G, Aiuti F, Fiorilli M. Eur J Epidemiol 1992; 8: 217–221.

Chokephaibulkit K, Brunell PA, Vimal V, Long C, Schnabel K, Hall CB. Clin Diagn Lab Immunol 1997; 4: 687–691.

Chou SW, Scott KM. J Clin Microbiol 1990; 28: 851–854.

Cleghorn FR, Maybank KA, Jack N, Pate E, Mingle J, Levine PH, Manns A. Ann Epidemiol 1995; 5: 497–500.

Cone RW, Huang ML, Hackman RC, Corey L. J Clin Microbiol 1996; 34: 877–881.

Dahl H, Linde A, Sundqvist VA, Wahren B. J Virol Methods 1990; 29: 313–323.

Dunne Jr. WM, Demmler GJ. Lancet 1992; 340: 121–122.

Eeg-Olofsson O. Virological and immunological aspects of seizure disorders. Brain Dev 2003; 25: 9–13.

Enbom M, Wang FZ, Fredrikson S, Martin C, Dahl H, Linde A. Clin Diagn Lab Immunol 1999; 6: 545–549.

Fillet AM, Raphael M, Visse B, Audouin J, Poirel L. Agut H. J Med Virol 1995; 45: 106–112.

Foa-Tomasi L, Avitabile E, Ke L, Campadelli-Fiume G. J Gen Virol 1994; 75(Pt 10): 2719–2727.

Hall CB, Long CE, Schnabel KC, Caserta MT, McIntyre KM, Costanzo MA, Knott A, Dewhurst S, Insel RA, Epstein LG. N Engl J Med 1994; 331: 432–438.

Herbein G, Strasswimmer J, Altieri M, Woehl-Jaegle ML, Wolf P, Obert G. Clin Infect Dis 1996; 22: 171–173.

Ihira M, Yoshikawa T, Ishii J, Nomura M, Hishida H, Ohashi M, Enomoto Y, Suga S, Iida K, Saito Y, Nishiyama Y, Asano Y. J Med Virol 2002; 67: 534–537.

Ihira M, Yoshikawa T, Suzuki K, Ohashi M. Suga S, Asonuma K, Tanaka K, Asano Y. Microbiol Immunol 2001; 45: 225–232.

Irving WL, Cunningham AL, Keogh A, Chapman JR. Antibody to both human herpesvirus 6 and cytomegalovirus. Lancet 1988; 2: 630–631.

Iyengar S, Levine PH, Ablashi D, Neequaye J, Pearson GR. Int J Cancer 1991; 49: 551–557.
Kimberlin DW, Whitley RJ. J Neurovirol 1998; 4: 474–485.
LaCroix S, Stewart JA, Thouless ME, Black JB. Clin Diagn Lab Immunol 2000; 7: 823–827.
Levy JA, Ferro F, Greenspan D, Lennette ET. Lancet 1990; 335: 1047–1050.
Linde A, Dahl H, Wahren B, Fridell E, Salahuddin Z, Biberfeld P. J Virol Methods 1988; 21: 117–123.
Lopez C, Pellett P, Stewart J, Goldsmith C, Sanderlin K, Black J, Warfield D, Feorino P. J Infect Dis 1988; 157: 1271–1273.
Marodi CL, Csiszar A, Sierra-Vazquez B, Di Luca D, Barabas E, Nagy K, Ongradi J. Pathol Oncol Res 1998; 4: 56–61.
Neipel F, Ellinger K, Fleckenstein B. J Virol 1992; 66: 3918–3924.
Nielsen L, Vestergaard BF. J Virol Methods 1996; 56: 221–230.
Nielsen L, Vestergaard BF. J Clin Virol 2002; 25: 145–154.
Okuno T, Higashi K, Shiraki K, Yamanishi K, Takahashi M, Kokado Y, Ishibashi M, Takahara S, Sonoda T, Tanaka K, Baba K, Yabuuchi H, Kurata T. Transplantation 1990; 49: 519–522.
Okuno T, Takahashi K, Balachandra K, Shiraki K, Yamanishi K, Takahashi M, Baba K. J Clin Microbiol 1989; 27: 651–653.
Parker CA, Weber JM. J Virol Methods 1993; 41: 265–275.
Patnaik M, Komaroff AL, Conley E, Ojo-Amaize EA, Peter JB. J Infect Dis 1995a; 172: 1364–1367.
Patnaik M, Peter JB. Clin Infect Dis 1995b; 21: 715–716.
Rajcani J, Yanagihara R. Acta Virol 1994; 38: 121–123.
Ranger-Rogez S, Vidal E, Labrousse F, Riche A, Vidal J, Collineau M, Liozon F, Denis F. J Med Virol 1995; 47: 198–203.
Robert C, Agut H, Aubin JT, Collandre H, Ingrand D, Devillechabrolle A, LeHoang P, Huraux JM. Res Virol 1990; 141: 545–555.
Salahuddin SZ, Ablashi DV, Markham PD, Josephs SF, Sturzenegger S, Kaplan M, Halligan G, Biberfeld P, Wong-Staal F, Kramarsky B, Gallo RC. Science 1986; 234: 596–601.
Salonen MJ, Siimes MA, Salonen EM, Vaheri A, Koskiniemi M. Antibody status to HHV-6 in children with leukaemia. Leukemia 2002; 16: 716–719.
Saxinger C, Polesky H, Eby N, Grufferman S, Murphy R, Tegtmeir G, Parekh V, Memon S, Hung C. J Virol Methods 1988; 21: 199–208.
Sloots TP, Kapeleris JP, Mackay IM, Batham M, Devine PL. J Clin Microbiol 1996; 34: 675–679.
Sola P, Merelli E, Marasca R, Poggi M, Luppi M, Montorsi M, Torelli G. J Neurol Neurosurg Psychiatry 1993; 56: 917–919.
Soldan SS, Berti R, Salem N, Secchiero P, Flamand L, Calabresi PA, Brennan MB, Maloni HW, McFarland HF, Lin HC, Patnaik M, Jacobson S. Nat Med 1997; 3: 1394–1397.
Suga S, Yoshikawa T, Asano Y, Nakashima T, Yazaki T, Fukuda M, Kojima S, Matsuyama T, Ono Y, Oshima S. Microbiol Immunol 1992; 36: 495–506.
Suga S, Yoshikawa T, Asano Y, Yazaki T, Ozaki T. J Med Virol 1990; 30: 14–19.
Sutherland S, Christofinis G, O'Grady J, Williams R. J Med Virol 1991; 33: 172–176.
Tejada-Simon MV, Zang YC, Hong J, Rivera VM, Zhang JZ. Cross-reactivity with myelin basic protein and human herpesvirus-6 in multiple sclerosis. Ann Neurol 2003; 53: 189–197.
Tolfvenstam T, Enbom M, Ghebrekidan H, Ruden U, Linde A, Grandien M, Wahren B. J Clin Virol 2000; 16: 49–54.

Tsukazaki T, Yoshida M, Namba H, Yamada M, Shimizu N, Nii S. J Virol Methods 1998; 73: 141–149.
Ward KN, Gray JJ, Fotheringham MW, Sheldon MJ. J Med Virol 1993b; 39: 131–138.
Ward KN, Gray JJ, Joslin ME, Sheldon MJ. J Med Virol 1993a; 39: 44–49.
Ward KN, Sheldon MJ, Gray JJ. J Med Virol 1991; 34: 258–267.
Wilborn F, Schmidt CA, Brinkmann V, Jendroska K, Oettle H, Siegert W. J Neuroimmunol 1994; 49: 213–214.
Xu Y, Linde A, Dahl H, Winberg G. J Clin Microbiol 2001; 39: 1449–1455.
Xu Y, Linde A, Fredrikson S, Dahl H, Winberg G. J Med Virol 2001; 66: 394–399.
Yamamoto M, Black JB, Stewart JA, Lopez C, Pellett PE. J Clin Microbiol 1990; 28: 1957–1962.
Yamanishi K, Okuno T, Shiraki K, Takahashi M, Kondo T, Asano Y, Kurata T. Lancet 1988; 1: 1065–1067.
Yanagi K, Harada S, Ban F, Oya A, Okabe N, Tobinai K. J Infect Dis 1990; 161: 153–154.
Yoshida M, Torigoe S, Ikeue K, Yamada M. Clin Diagn Lab Immunol 2002a; 9: 388–393.
Yoshida M, Torigoe S, Yamada M. Clin Diagn Lab Immunol 2002b; 9: 394–402.
Yoshikawa T, Black JB, Ihira M, Suzuki K, Suga S, Iida K, Saito Y, Asonuma K, Tanaka K, Asano Y. Clin Diagn Lab Immunol 2001; 8: 170–173.
Yoshikawa T, Suga S, Asano Y, Nakashima T, Yazaki T, Ono Y, Fujita T, Tsuzuki K, Sugiyama S, Oshima S. Transplantation 1992; 54: 879–883.
Yoshikawa T, Suga S, Asano Y, Nakashima T, Yazaki T, Sobue R, Hirano M, Fukuda M, Kojima S, Matsuyama T. Blood 1991; 78: 1381–1384.
Zerr DM, Meier AS, Selke SS, Frenkel LM, Huang ML, Wald A, Rhoads MP, Nguy L, Bornemann R, Morrow RA, Corey L. N Engl J Med 2005; 352: 768–776.

Human Herpesvirus-6: General Virology, Epidemiology and Clinical Pathology
Perspectives in Medical Virology
Gerhard Krueger and Dharam Ablashi (Editors)

DOI 10.1016/S0168-7069(06)12008-X

Molecular Testing for HHV-6 Infection

Stephen Dewhurst, Birgit Bradel-Tretheway
Department of Microbiology and Immunology, The University of Rochester Medical Center, 575 Elmwood Avenue, Rochester, New York 14642, USA

Introduction

Since the last edition of this book (Ablashi et al., 1992), major advances have been made in the field of HHV-6 molecular diagnosis, made possible in part by advances in PCR technology and availability of genomic sequence information. As a result, it is now possible to consider molecular tests that can differentiate between the three possible states of HHV-6 infection: (1) primary infection, (2) reinfection or reactivation and (3) latency or persistence (Table 1).

Table 1

Types of HHV-6 infection, and their molecular characteristics

Infection status	Virology	Serology
Primary	Cell-associated viral DNA: + + + +	IgM: +
	Cell-free plasma viral DNA: + +	IgG[a]: –
	Viral mRNAs: + + +	
Reinfection or reactivation	Cell-associated viral DNA: + +	IgM: +
	Cell-free plasma viral DNA: +/–	IgG: + +
	Viral mRNAs: +	
Latent	Cell-associated viral DNA: + + + +	IgM: –
	Cell-free plasma viral DNA: –	IgG: + +
	Viral mRNAs: – (latent mRNA only)	

Note: All values are arbitrary.
[a]There may be maternally acquired IgG present in infants and young children.

Primary infection: characteristics and diagnostic considerations

Primary infection typically occurs in early childhood, by the age of three years (Pruksananonda et al., 1992; Zerr et al., 2005b). Infection is generally symptomatic, causing classical *Exanthem subitum* or roseola (Yamanishi et al., 1988), or febrile illness without rash, diarrhea and irritability (Pruksananonda et al., 1992; Zerr et al., 2005b). In some cases, primary infection may also be associated with an elevated incidence of seizures (Hall et al., 1994), although the overall prevalence of such seizures is uncertain (Zerr et al., 2005b). In contrast, primary infection in adulthood is rare, but may be associated with more severe symptoms (Akashi et al., 1993).

Major issues in designing diagnostic tests for primary HHV-6 infection include (1) to provide a rapid assay for useful information in a short time, and (2) to distinguish between primary virus infection versus viral reactivation or latent infection (Table 1). Of particular concern is the fact that viral DNA is likely to be present in circulating peripheral blood mononuclear cells (PBMC) from all HHV-6-infected individuals, regardless of whether that infection was recent or not. Thus, one must rely on the unique features of primary infection to distinguish it from latency/persistence or subclinical reactivation/reinfection. The most important feature of primary infection is the high levels of viremia (Asano et al., 1991) so that infectious virus can readily be cultured from patients' PBMC and viral DNA is detected in cell-free plasma samples. In addition, quantitative PCR assays can reveal high cell-associated viral DNA loads in the PBMCs, in contrast to latent or persistent infection.

Many investigators regard the ability to isolate infectious HHV-6 from PBMC of individuals as being something close to a "gold standard" for the diagnosis of primary virus infection (Yamanishi et al., 1988; Hall et al., 1994). Cell culture isolation of HHV-6 is generally performed using cord blood mononuclear cells, but these cells are in short supply due to their utility in other clinical applications, and the virus culture method is technically difficult and slow. Cell lines may also be used for virus culture, as far as susceptible to infection with both virus variants (HHV-6A and HHV-6B), with an interest in the development of rapid cell culture or antigenemia assays analogous to the shell vial assay to detect human cytomegalovirus (HCMV) infection (Ablashi et al., 1991b; Knox et al., 2000; Lautenschlager et al., 2000; Nishimura et al., 2005; Savolainen et al., 2005). There remains some uncertainty as to whether all virus strains will have equivalent growth properties in any "standard" cell culture system (Ablashi et al., 1991a; Schirmer et al., 1991; Dewhurst et al., 1992; Ablashi et al., 1993). In this context, it is noteworthy that culture-adapted strains of another human β-herpesvirus (HCMV) can change their genomic composition and cellular tropism relative to primary isolates or unpassaged strains (Cha et al., 1996; MacCormac and Grundy, 1999; Prichard et al., 2001; Murphy et al., 2003; Dolan et al., 2004). Thus, molecular assays remain important for the diagnosis of HHV-6 infections.

PCR analysis of cell-free body fluids

PCR-based assays detect HHV-6 DNA in cell-free plasma specimens, so as to identify viremic individuals (Secchiero et al., 1995; Suga et al., 1995), with successful use in many clinical applications (Chiu et al., 1998; Osiowy et al., 1998; Akhyani et al., 2000; Nitsche et al., 2001; Tomsone et al., 2001; Berti et al., 2002), including analysis of viral infection/reactivation in hematopoietic stem cell transplant (HSCT) recipients (Zerr et al., 2005a). Other acellular body fluids also have diagnostic utility—most notably, cerebrospinal fluid (CSF) specimens. Most studies have failed to detect HHV-6 DNA in cell-free CSF samples from healthy adults (Wilborn et al., 1994; Liedtke et al., 1995; Wang et al., 1999; Akhyani et al., 2000; Tomsone et al., 2001; Berti et al., 2002; Rotola et al., 2004). The one notable exception was a study by Tejada-Simon (Tejada-Simon et al., 2002), where a non-quantitative, nested PCR protocol detected HHV-6 DNA in cell-free CSF specimens from 6 out of 20 control adult subjects (20%). Thus, the presence of HHV-6 DNA in a cell-free adult CSF sample represents an unusual finding.

Detection of viral DNA in acellular CSF is of the greatest clinical /diagnostic relevance in hematopoietic stem cell transplant recipients, where it has been associated with CNS dysfunction in the absence of any other etiology (Wang et al., 1999; Singh and Paterson, 2000; Zerr et al., 2002). In other population groups with CNS disease—including persons with multiple sclerosis—the clinical significance of a positive PCR test for HHV-6 DNA in acellular CSF remain uncertain (Dewhurst, 2004). This is discussed in more detail in Chapter 16 of this book.

Other body fluids that have diagnostic potential may include saliva, which has been reported to contain high levels of viral DNA in both adults and children (Cone et al., 1993; Collot et al., 2002; Pereira et al., 2004). Shedding of virus in saliva from healthy adults may occur less frequently than in children (Di Luca et al., 1995; Tanaka-Taya et al., 1996; Fujiwara et al., 2000), possibly because for developmental reasons (Tanaka-Taya et al., 1996).

Jacobson's group reported of viral DNA in urine from persons with multiple sclerosis (MS) (Akhyani et al., 2000), and others also identified HHV-6 DNA in urine from immunocompromised persons or infants—although at a rate well below that for HCMV (Gautheret-Dejean et al., 1997; Ashshi et al., 2003). Standardized methods for the preparation and extraction of cell-free DNA from these various body fluids will be essential for diagnostic assay development. Key issues will include (i) addition of defined exogenous DNA to samples, to control for the efficiency of DNA extraction and amplification, (ii) sensitive methods to detect possible contamination of "acellular" specimens by cellular DNA and (iii) approaches to enhance assay sensitivity (such as centrifugal concentration of virus particles).

Reverse transcription PCR analysis

Reverse transcription (RT) PCR analysis for detection of lytic-phase viral transcripts distinguishes active viral replication from latent infection. Several RT-PCR assays have been developed, most of which rely upon intron-spanning primers that generate differentially sized PCR products from viral genomic DNA versus reverse-transcribed viral mRNA (Norton et al., 1999; Van den Bosch et al., 2001; Sashihara et al., 2002; Yoshikawa et al., 2003). In addition, RT-PCR assays also need to include internal controls to assure proper RNA extraction and RT—such as cellular-like glyceraldehyde 3-phosphate dehydrogenase (GAPDH), which are unaffected by HHV-6 infection (Radonic et al., 2005).

RT-PCR assays have been used to demonstrate that HHV-6 reactivation in previously infected children is most often asymptomatic, and can occur with a substantial frequency (Caserta et al., 2004). This phenomenon is an important consideration in the interpretation of results from PCR-based assays of HHV-6.

While RT-PCR assays have most commonly been used to examine lytic-phase viral gene expression (as a marker for viral reactivation or replication), it may also be useful to identify latent virus infection. Thus, Kondo et al. (2002) have described viral latency-associated transcripts derived from the immediate-early (IE)-1 and -2 genes, which are expressed *in vivo*, in freshly isolated PBMC of healthy adults.

Overall, RT-PCR assays for HHV-6 remain largely confined to research without general use in the clinical diagnostic laboratory for technical difficulties, low throughput of the assay and concerns about assay sensitivity.

Quantitative PCR analysis of HHV-6 DNA in PBMC or blood

Quantitative analysis of viral DNA load in PBMC or blood are useful for (i) identification of primary infections and reactivations, and (ii) assessment of the effectiveness of treatment. Primary virus infection can be diagnosed from high virus loads within PBMC (Clark et al., 1997; Chiu et al., 1998), with median peak virus DNA load in PBMCs from children with primary infection being on the order of $0.1 - 0.2 \times 10^5$ copies per 10^6 PBMC (Clark et al., 1997; Sashihara et al., 2002).

Peak virus loads in PBMC from persons undergoing allogeneic stem cell transplantation may predict the outcome (Wang et al., 2002b). Median peak virus loads in this population ranged between 0.1 and 0.2×10^5 copies per 10^6 PBMC (Sashihara et al., 2002; Safronetz et al., 2003), with slightly higher (Gautheret-Dejean et al., 2002), or lower (Cone et al., 1999) values. Median peak levels in other transplant recipients have been broadly similar (Griffiths et al., 1999). These values differ significantly from virus loads in PBMC of normal healthy adults with 1–10 copies per 10^6 PBMC (Clark et al., 1996; Cone et al., 1999; Gautheret-Dejean et al., 2002).

The utility of qPCR assays of PBMC virus load in post-transplant disease and encephalitis of unknown origin is less clear. There may be considerable variation in baseline values of HHV-6 virus load over time in normal subjects, with differences of up to 50-fold or more between samples collected on different dates (Cone et al., 1999). Second, high levels of virus DNA have been reported in healthy normal individuals. For example, one of nine healthy subjects had a level of roughly 1×10^7 viral genome copies per 10^6 PBMC, which remained constant over 10 months (Clark et al., 1996). In addition, the virus can become integrated in PBMC in rare individuals—leading to high levels of cell-associated viral DNA in the absence of viral replication or disease (Luppi et al., 1993; Torelli et al., 1995; Tanaka-Taya et al., 2004). Thus, the interpretation of a high virus load in an individual with a disease or disorder that has an uncertain causal relationship to HHV-6 remains problematic.

Detection of virus variants

An important consideration in molecular testing for HHV-6 is the existence of two distinct virus variants, known as HHV-6A and HHV-6B with a common genome organization, and some 90% nucleotide sequence identity (Gompels et al., 1995; Dominguez et al., 1999; Isegawa et al., 1999; Fig. 1). The two variants also have distinct biologic properties and cellular tropism (Ablashi et al., 1991a, 1993; Schirmer et al., 1991; Dewhurst et al., 1992), and differ in their pathogenicity, tissue distribution and epidemiology (Dewhurst et al., 1993; Carrigan et al., 1996; Cone et al., 1996; Kasolo et al., 1997; Hall et al., 1998; Soldan et al., 2000; Razonable et al., 2002). Discrimination between these two variants can be achieved by restriction enzyme cleavage of PCR amplimers, or by using variant-specific DNA probes (Aubin et al., 1991, 1993, 1994; Kidd et al., 1998). The former approach is particularly well suited for use in quantitative real-time PCR applications.

Other assay approaches

Fluorescent *in situ* hybridization (FISH) has been used to map HHV-6 chromosomal integration sites in those rare individuals with integrated virus DNA in their PBMC. Tanaka-Taya et al. (2004) reported of chromosomal integrated HHV-6 in the PBMC of approximately 0.2% of healthy individuals, yet confirmatory follow-up studies are still needed. Integration of HHV-6 genomic DNA has been found in the long arm of chromosome 1, short arm of chromosome 17 and in the long arm of chromosome 22 (Luppi et al., 1993; Torelli et al., 1995; Daibata et al., 1998; Morris et al., 1999). This integration appears to be latent in nature, since no lytic-phase gene expression or viremia has been detected in these subjects (Tanaka-Taya et al., 2004), and may have occurred as a result of homologous recombination between the telomer-like repeat sequences near the ends of the HHV-6 genome and their cellular counterparts (Thomson et al., 1994; Gompels and Macaulay, 1995).

Future assay methods

New approaches to the detection and quantitation of HHV-6 DNA in clinical specimens include non-PCR-based gene amplification or detection methods such as branched DNA (bDNA) hybridization (Urdea et al., 1987) and isothermal amplification (Guatelli et al., 1990; Walker et al., 1992; Ihira et al., 2004).

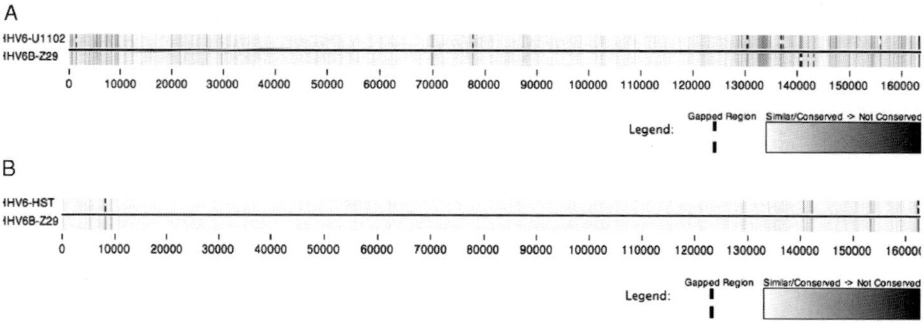

Fig. 1 Alignment of HHV-6A and HHV-6B. A: Inter-variant alignment (HHV-6A U1102 and HHV-6B Z29). B: Intra-variant alignment (HHV-6B HST and HHV-6B Z29). Shown is a visual overview of the overall genetic conservation between (panel A) and within (panel B) variants of HHV-6, based on available genomic sequences (HHV-6A U1102, HHV-6B Z29 and HHV-6B HST; respectively, GenBank Accession Numbers X83413, AF157706 and AB021506). The alignments are color-coded; dark regions represent areas of greatest sequence divergence and numbers represent the genomic location (in base pairs). Gapped regions are indicated by darkly shaded semicolons (:); these correspond to regions where there are sequence insertions or deletions in one genome, relative to another. It can be readily appreciated that sequence divergence between virus variants (panel A) is much greater than sequence divergence within a single variant (i.e., between different strains of HHV-6B, panel B; note that comparable data are not available for HHV-6A since there is to date only a single genomic sequence for HHV-6A). Global nucleotide sequence identity between variants of HHV-6 is approximately 90% but this is unevenly distributed across the viral genome (panel A). The central region of the HHV-6 genome is highly conserved between virus variants and contains sequence blocks found in other herpesviruses, whereas the terminal direct repeats (TDR) and right end of the unique region of the genome are more divergent both between (panel A) and within (panel B) variants (Dominguez et al., 1999; Isegawa et al., 1999; Mori et al., 2003). The TDR encompasses roughly 8–9 kb of largely non-coding sequence, while the segment of the unique region that is most divergent between variants is only about 72% identical between HHV-6A and HHV-6B, and spans the open reading frames (ORFs) U86–U100 (panel A) (Dominguez et al., 1999; Isegawa et al., 1999; Mori et al., 2003). Among these ORFs, the IE transactivator U89/90 (IE1) and the gQ-encoding ORF U100 are especially divergent while only the adeno-associated virus (AAV) *rep*-homolog, U94, is conserved (Dominguez et al., 1999; Isegawa et al., 1999; Mori et al., 2003). As panel B shows, the extent of intra-variant divergence is lower within this region, but some striking differences do exist—including the presence of a putative ORF (HN1) that is unique to the HST strain and absent in Z29, as well as significant divergence among viral transactivators (e.g., U89 from HST is only 92% identical to the corresponding gene product from Z29, at the amino acid level; Isegawa et al., 1999). It is important to add that two distinct groups of variant B isolates have been described, of which Z29 and HST form the prototype members; sequences derived from viruses representative of each individual group exhibit a very high level sequence conservation with one another (99% nucleotide identity or greater), which greatly exceeds the identity between viruses from different groups (Isegawa et al., 1999). Methods: the alignments shown were kindly generated by Dr. Vasily Tcherepanov and Dr. Chris Upton of the University of Victoria, BC, Canada, using the base-by-base alignment algorithm (Brodie et al., 2004). This algorithm can be downloaded freely via the WWW (http://athena.bioc.uvic.ca/). Alignments were initially generated with ClustalW 1.83.1, using a fast alignment with the following parameters: gap penalty = 3, K-tuple(word) size = 1, number of top diagonals = 5, window size = 5. The alignment from ClustalW was then loaded into Base-by-Base multiple alignment editor (BBB) and adjusted manually.

Microarray technology offers the potential to screen a single clinical specimen for many different pathogens (Wang et al., 2002a; Bryant et al., 2004; Foldes-Papp et al., 2004; Striebel et al., 2004). HHV-6-specific gene sequences will likely be included in these new arrays, yet it will take some time before this approach becomes adopted in the clinic. Practical concerns include sensitivity, cost and the complex data analysis and interpretation.

Biosensors based on quartz crystal microbalance technology (using antibodies to provide specificity) or highly sensitive nucleic acid hybridization assays, as well as living cell-based immunosensors (Cooper et al., 2001; Rider et al., 2003; Vernon et al., 2003) are at early stage of development, and its clinical utility remains unknown.

PCR-based assays: general considerations and molecular underpinnings

As the costs of quantitative PCR detection systems declined and the availability of robust and reliable assays increased, more laboratories are now in a position to perform quantitative real-time PCR assays, using a variety of probes and detection systems (e.g., TaqMan^{TM}, fluorescence resonance energy transfer (FRET) probes and molecular beacons).

TaqMan^{TM} probes are widely used system for HHV-6 detection. These probes are sequence specific oligonucleotides with a fluorophore attached to their 5′-end and a quencher to the 3′-end. When the extended primer product comes into close proximity to the probe the fluorophore is cleaved by the 5′- > 3′ exonuclease activity of the polymerase. Fluorophore and quencher then become separated, resulting in a fluorescence proportional to the amount of the amplified product (Fig. 2). Analogous technologies use oligonucleotide probes in which the fluorophore and quencher are in close proximity while the probe is folded in solution. Upon binding to its complementary target sequence within the PCR amplimer, the probe then becomes unfolded and the fluorophore physically separates from the quencher—allowing its excitation and emission at the appropriate wavelength (Fig. 2).

Another technology is FRET, where two labeled oligonucleotide probes hybridize to immediately adjacent complementary sequences in the PCR amplimer of interest. One probe harbors a donor fluorophore, while the other has an acceptor fluorophore. When these probes come into close proximity (which happens when they hybridize to their complementary sequences within the target PCR amplimer), local energy transfer occurs from the donor to the acceptor fluorophore (Fig. 2). Using appropriate excitation and emission spectra, the resulting fluoresence can be detected and quantitated.

FRET, Taqman^{TM} and other detection methods all permit the sensitive and specific quantitation of HHV-6 DNA in clinical specimens. This allowed the emergence of a diversity of commercial molecular assays for the detection of HHV-6 in clinical specimens.

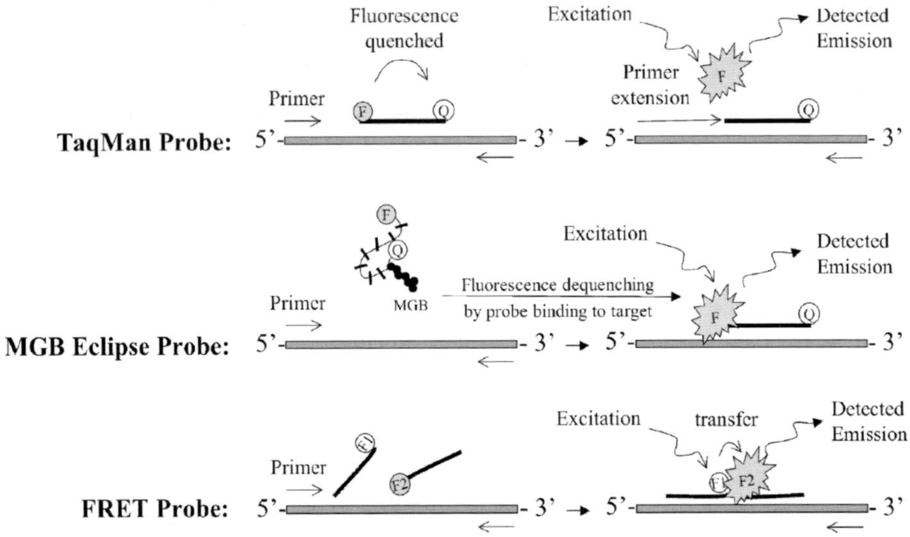

Fig. 2 Quantitative PCR assay methods. Selected PCR quantitation methods. See text for details.

Commercially available tests

Several diagnostic laboratories in the United States provide molecular testing for HHV-6 (Table 2). Many assays are purely qualitative, and their sensitivity varies considerably. From same clinical samples conflicting data were obtained from these laboratories—some labs reporting positive results, others failing to detect any viral DNA (D. Peterson, personal communication). This stresses the need of standardized assays and of rigorous external validation program.

General considerations for current molecular assays

As noted by Boivin (2004), following are key considerations for the development of molecular assays for HHV-6:

- Sample collection and preparation should be standardized.
- Detection should target conserved gene regions without cross-reaction with related viruses and should distinguish between HHV-6A and HHV-6B.
- Sensitivity (minimum of 10 copies per sample) and specificity are critical. When using acellular fluids, an initial concentration step may enhance assay sensitivity.
- Nested PCR tests are ill-suited for diagnostic use, due to the increased potential for cross-contamination, and difficulty in quantitation.
- Appropriate controls and quantitation standards must be used, including assessment of efficiency of DNA extraction and amplification.

Table 2

Selected commercially available molecular tests for HHV-6

Real-time quantitative DNA PCR assays (using CSF, serum, blood, etc.) • ARUP lab (http://www.aruplab.com): targets U67 with MGB Eclipse Probes. Nominal sensitivity: 10^3 copies/ml body fluid • ViraCor (http://www.viracor.com): TaqMan qPCR. Nominal sensitivity: 10^2 copies/ml body fluid. Can distinguish HHV-6A and HHV-6B • Commonwealth Biotechnologies, Inc. (http://www.herpes-testing.com): TaqMan PCR. Sensitivity not reported. Can distinguish HHV-6A and HHV-6B
Qualitative DNA PCR assays Focus Technologies (http://www.focusdx.com): sensitivity not reported. Can distinguish HHV-6A and HHV-6B • Medical Diagnostics Lab (http://www.mdlab.com): qualitative PCR only • Wisconsin lab (http://www.wisconsinlab.com): highly sensitive nested PCR assay, with the capacity to distinguish HHV-6A from HHV-6B. Nominal sensitivity = 2 copies

Note: All assay information is based on information provided on the public website of the corresponding company, as of May 2005.

- Exclude respiratory contamination of reactions (Kelley and McClain, 1994).
- Participation in a proficiency testing or validation program is key.

It is important to prepare guidelines and recommendations that help physicians to best utilize the available molecular tests for HHV-6. Tests need standardization, comparison and careful evaluation. It is not sufficient to simply perform assays in a Clinical Laboratory Improvement Amendments (CLIA)-certified lab, if those assays are not FDA approved or externally validated.

References

Ablashi DV, Agut H, Berneman Z, Campadelli-Fiume G, Carrigan D, Ceccherini-Nelli L, Chandran B, Chou S, Collandre H, Cone RW, Dambaugh TR, Dewhurst S, Di Luca D, Foa-Tomasi L, Fleckenstein B, Frenkel N, Gallo RC, Gompels U, Hall CB, Jones M, Lawrence G, Martin ME, Montagnier L, Neipel F, Nicholas J, Pellett P, Razzaque A, Torrelli G, Thomson B, Salahuddin S Z, Wyatt LS, Yamanishi H. Arch Virol 1993; 129: 363–366.

Ablashi DV, Balachandran N, Josephs SF, Hung CL, Krueger GR, Kramarsky B, Salahuddin SZ, Gallo RC. Virology 1991a; 184: 545–552.

Ablashi DV, Krueger GRF, Salahuddin SZ. Human Herpesvirus-6. Amsterdam: Elsevier; 1992. pp. 1–341.

Ablashi DV, Zompetta C, Lease C, Josephs SF, Balachandra N, Komaroff AL, Krueger GR, Henry B, Lukau J, Salahuddin SZ. Can Dis Wkly Rep 1991b; 17(Suppl 1E): 33–40.

Akashi K, Eizuru Y, Sumiyoshi Y, Minematsu T, Hara S, Harada M, Kikuchi M, Niho Y, Minamishima YN. Engl J Med 1993; 329: 168–171.
Akhyani N, Berti R, Brennan MB, Soldan SS, Eaton JM, McFarland HF, Jacobson S. J Infect Dis 2000; 182: 1321–1325.
Asano Y, Nakashima T, Yoshikawa T, Suga S, Yazaki T. J Pediatr 1991; 118: 891–895.
Ashshi AM, Klapper PE, Cooper RJ. J Infect 2003; 47: 59–64.
Aubin JT, Agut H, Collandre H, Yamanishi K, Chandran B, Montagnier L, Huraux JM. J Virol Methods 1993; 41: 223–234.
Aubin J, Collandre H, Candotti D, Ingrand D, Rouzioux C, Burgard M, Richard S, Huraux JM, Agut H. J Clin Microbiol 1991; 29: 367–372.
Aubin JT, Poirel L, Robert C, Huraux JM, Agut H. J Clin Microbiol 1994; 32: 2434–2440.
Berti R, Brennan MB, Soldan SS, Ohayon JM, Casareto L, McFarland HF, Jacobson S. J Neurovirol 2002; 8: 250–256.
Boivin G. Herpes 2004; 11(Suppl 2): 48A–56A.
Brodie R, Smith AJ, Roper RL, Tcherepanov V, Upton C. BMC Bioinformatics 2004; 5: 96.
Bryant PA, Venter D, Robins-Browne R, Curtis N. Lancet Infect Dis 2004; 4: 100–111.
Carrigan DR, Harrington D, Knox KK. Variant A human herpesvirus six as a cofactor in the pathogenesis of AIDS. J Acq Immun Def Synd Hum Retrovirol 1996; 13: 97–98.
Caserta MT, McDermott MP, Dewhurst S, Schnabel K, Carnahan JA, Gilbert L, Lathan G, Lofthus GK, Hall CB. J Pediatr 2004; 145: 478–484.
Cha TA, Tom E, Kemble GW, Duke GM, Mocarski ES, Spaete RR. J Virol 1996; 70: 78–83.
Chiu SS, Cheung CY, Tse CY, Peiris M. J Infect Dis 1998; 178: 1250–1256.
Clark DA, Ait-Khaled M, Wheeler AC, Kidd IM, McLaughlin JE, Johnson MA, Griffiths PD, Emery VC. J Gen Virol 1996; 77(Pt 9): 2271–2275.
Clark DA, Kidd IM, Collingham KE, Tarlow M, Ayeni T, Riordan A, Griffiths PD, Emery VC, Pillay D. Arch Dis Child 1997; 77: 42–45.
Collot S, Petit B, Bordessoule D, Alain S, Touati M, Denis F, Ranger-Rogez S. J Clin Microbiol 2002; 40: 2445–2451.
Cone RW, Huang ML, Ashley R, Corey L. J Clin Microbiol 1993; 31: 1262–1267.
Cone RW, Huang ML, Corey L, Zeh J, Ashley R, Bowden R. J Infect Dis 1999; 179: 311–318.
Cone RW, Huang ML, Hackman RC, Corey L. J Clin Microbiol 1996; 34: 877–881.
Cooper MA, Dultsev FN, Minson T, Ostanin VP, Abell C, Klenerman D. Nat Biotechnol 2001; 19: 833–837.
Daibata M, Taguchi T, Taguchi H, Miyoshi I. Br J Haematol 1998; 102: 1307–1313.
Dewhurst S. Herpes 2004; 11(Suppl 2): 105A–111A.
Dewhurst S, Chandran B, McIntyre K, Schnabel K, Hall CB. Virology 1992; 190: 490–493.
Dewhurst S, McIntyre K, Schnabel K, Hall CB. J Clin Microbiol 1993; 31: 416–418.
Di Luca D, Mirandola P, Ravaioli T, Dolcetti R, Frigatti A, Bovenzi P, Sighinolfi L, Monini P, Cassai E. J Med Virol 1995; 45: 462–468.
Dolan A, Cunningham C, Hector RD, Hassan-Walker AF, Lee L, Addison C, Dargan DJ, McGeoch DJ, Gatherer D, Emery VC, Griffiths PD, Sinzger C, McSharry BP, Wilkinson GW, Davison AJ. J Gen Virol 2004; 85: 1301–1312.
Dominguez G, Dambaugh TR, Stamey FR, Dewhurst S, Inoue N, Pellett PE. J Virol 1999; 73: 8040–8052.
Foldes-Papp Z, Egerer R, Birch-Hirschfeld E, Striebel HM, Demel U, Tilz GP, Wutzler P. Mol Diagn 2004; 8: 1–9.

Fujiwara N, Namba H, Ohuchi R, Isomura H, Uno F, Yoshida M, Nii S, Yamada M. J Med Virol 2000; 61: 208–213.
Gautheret-Dejean A, Aubin JT, Poirel L, Huraux JM, Nicolas JC, Rozenbaum W, Agut H. J Clin Microbiol 1997; 35: 1600–1603.
Gautheret-Dejean A, Manichanh C, Thien-Ah-Koon F, Fillet AM, Mangeney N, Vidaud M, Dhedin N, Vernant JP, Agut H. J Virol Methods 2002; 100: 27–35.
Gompels UA, Macaulay HA. J Gen Virol 1995; 76(Pt 2): 451 458.
Gompels UA, Nicholas J, Lawrence G, Jones M, Thomson BJ, Martin ME, Efstathiou S, Craxton M, Macaulay HA. Virology 1995; 209: 29–51.
Griffiths PD, Ait-Khaled M, Bearcroft CP, Clark DA, Quaglia A, Davies SE, Burroughs AK, Rolles K, Kidd IM, Knight SN, Noibi SM, Cope AV, Phillips AN, Emery VC. J Med Virol 1999; 59: 496–501.
Guatelli JC, Whitfield KM, Kwoh DY, Barringer KJ, Richman DD, Gingeras TR. Proc Natl Acad Sci USA 1990; 87: 1874–1878.
Hall CB, Caserta MT, Schnabel KC, Long C, Epstein LG, Insel RA, Dewhurst S. Clin Infect Dis 1998; 26: 132–137.
Hall CB, Long CE, Schnabel KC, Caserta MT, McIntyre KM, Costanzo MA, Knott A, Dewhurst S, Insel RA, Epstein LG. N Engl J Med 1994; 331: 432–438.
Ihira M, Yoshikawa T, Enomoto Y, Akimoto S, Ohashi M, Suga S, Nishimura N, Ozaki T, Nishiyama Y, Notomi T, Ohta Y, Asano Y. J Clin Microbiol 2004; 42: 140–145.
Isegawa Y, Mukai T, Nakano K, Kagawa M, Chen J, Mori Y, Sunagawa T, Kawanishi K, Sashihara J, Hata A, Zou P, Kosuge H, Yamanishi K. J Virol 1999; 73: 8053–8063.
Kasolo FC, Mpabalwani E, Gompels UA. J Gen Virol 1997; 78(Pt 4): 847–855.
Kelley PK, McClain KL. Am J Hematol 1994; 47: 325–327.
Kidd IM, Clark DA, Bremner JA, Pillay D, Griffiths PD, Emery VC. J Virol Methods 1998; 70: 29–36.
Knox KK, Brewer JH, Henry JM, Harrington DJ, Carrigan DR. Clin Infect Dis 2000; 31: 894–903.
Kondo K, Shimada K, Sashihara J, Tanaka-Taya K, Yamanishi K. J Virol 2002; 76: 4145–4151.
Lautenschlager I, Linnavuori K, Hockerstedt K. Transplantation 2000; 69: 2561–2566.
Liedtke W, Malessa R, Faustmann PM, Eis-Hubinger AM. J Neurovirol 1995; 1: 253–258.
Luppi M, Marasca R, Barozzi P, Ferrari S, Ceccherini-Nelli L, Batoni G, Merelli E, Torelli G. J Med Virol 1993; 40: 44–52.
MacCormac LP, Grundy JE. J Med Virol 1999; 57: 298–307.
Mori Y, Akkapaiboon P, Yang X, Yamanishi K. J Virol 2003; 77: 2452–2458.
Morris C, Luppi M, McDonald M, Barozzi P, Torelli G. J Med Virol 1999; 58: 69–75.
Murphy E, Yu D, Grimwood J, Schmutz J, Dickson M, Jarvis MA, Hahn G, Nelson JA, Myers RM, Shenk TE. Proc Natl Acad Sci USA 2003; 100: 14976–14981.
Nishimura N, Yoshikawa T, Ozaki T, Sun H, Goshima F, Nishiyama Y, Asano Y, Kurata T, Iwasaki T. J Med Virol 2005; 75: 86–92.
Nitsche A, Muller CW, Radonic A, Landt O, Ellerbrok H, Pauli G, Siegert W. J Infect Dis 2001; 183: 130–133.
Norton RA, Caserta MT, Hall CB, Schnabel K, Hocknell P, Dewhurst S. J Clin Microbiol 1999; 37: 3672–3675.
Osiowy C, Prud'homme I, Monette M, Zou S. J Clin Microbiol 1998; 36: 68–72.
Pereira CM, Gasparetto PF, Correa ME, Costa FF, de Almeida OP, Barjas-Castro ML. Arch Oral Biol 2004; 49: 1043–1046.

Prichard MN, Penfold ME, Duke GM, Spaete RR, Kemble GW. Rev Med Virol 2001; 11: 191–200.
Pruksananonda P, Hall CB, Insel RA, McIntyre K, Pellett PE, Long CE, Schnabel KC, Pincus PH, Stamey FR, Dambaugh TR, Stewart J. N Engl J Med 1992; 326: 1445–1450.
Radonic A, Thulke S, Bae HG, Muller MA, Siegert W, Nitsche A. J Virol 2005; 2: 7.
Razonable RR, Fanning C, Brown RA, Espy MJ, Rivero A, Wilson J, Kremers W, Smith TF, Paya CV. J Infect Dis 2002; 185: 110–113.
Rider TH, Petrovick MS, Nargi FE, Harper JD, Schwoebel ED, Mathews RH, Blanchard DJ, Bortolin LT, Young AM, Chen J, Hollis MA. Science 2003; 301: 213–215.
Rotola A, Merlotti I, Caniatti L, Caselli E, Granieri E, Tola MR, Di Luca D, Cassai E. Multiple Sclerosis 2004; 10: 348–354.
Safronetz D, Humar A, Tipples GA. J Virol Method 2003; 112: 99–105.
Sashihara J, Tanaka-Taya K, Tanaka S, Amo K, Miyagawa H, Hosoi G, Taniguchi T, Fukui T, Kasuga N, Aono T, Sako M, Hara J, Yamanishi K, Okada S. Blood 2002; 100: 2005–2011.
Savolainen H, Lautenschlager I, Piiparinen H, Saarinen-Pihkala U, Hovi L, Vettenranta K, Pediatr Blood Cancer 2005; 45: 820–825.
Schirmer EC, Wyatt LS, Yamanishi K, Rodriguez WJ, Frenkel N. Proc Natl Acad Sci USA 1991; 88: 5922–5926.
Secchiero P, Carrigan DR, Asano Y, Benedetti L, Crowley RW, Komaroff AL, Gallo RC, Lusso P. J Infect Dis 1995; 171: 273–280.
Singh N, Paterson DL. Transplantation 2000; 69: 2474–2479.
Soldan SS, Leist TP, Juhng KN, McFarland HF, Jacobson S. Ann Neurol 2000; 47: 306–313.
Striebel HM, Birch-Hirschfeld E, Egerer R, Foldes-Papp Z, Tilz GP, Stelzner A. Exp Mol Pathol 2004; 77: 89–97.
Suga S, Yazaki T, Kajita Y, Ozaki T, Asano Y. J Med Virol 1995; 46: 52–55.
Tanaka-Taya K, Kondo T, Muka T, Miyosh H, Yamamoto Y, Okada S, Yamanishi K. J Med Virol 1996; 48: 88–94.
Tanaka-Taya K, Sashihara J, Kurahashi H, Amo K, Miyagawa H, Kondo K, Okada S, Yamanishi K. J Med Virol 2004; 73: 465–473.
Tejada-Simon MV, Zang YC, Hong J, Rivera VM, Killian JM, Zhang JZ. J Virol 2002; 76: 6147–6154.
Thomson BJ, Dewhurst S, Gray D. J Virol 1994; 68: 3007–3014.
Tomsone V, Logina I, Millers A, Chapenko S, Kozireva S, Murovska M. J Neurovirol 2001; 7: 564–569.
Torelli G, Barozzi P, Marasca R, Cocconcelli P, Merelli E, Ceccherini-Nelli L, Ferrari S, Luppi M. J Med Virol 1995; 46: 178–188.
Urdea MS, Running JA, Horn T, Clyne J, Ku LL, Warner BD. Gene 1987; 61: 253–264.
Van den Bosch G, Locatelli G, Geerts L, Faga G, Ieven M, Goossens H, Bottiger D, Oberg B, Lusso P, Berneman ZN. J Clin Microbiol 2001; 39: 2308–2310.
Vernon SD, Farkas DH, Unger ER, Chan V, Miller DL, Chen YP, Blackburn GF, Reeves WC. BMC Infect Dis 2003; 3: 12.
Walker GT, Little MC, Nadeau JG, Shank DD. Proc Natl Acad Sci USA 1992; 89: 392–396.
Wang D, Coscoy L, Zylberberg M, Avila PC, Boushey HA, Ganem D, DeRisi JL. Proc Natl Acad Sci USA 2002a; 99: 15687–15692.
Wang FZ, Larsson K, Linde A, Ljungman P. Bone Marrow Transplant 2002b; 30: 521–526.

Wang FZ, Linde A, Hagglund H, Testa M, Locasciulli A, Ljungman P. Clin Infect Dis 1999; 28: 562–568.
Wilborn F, Schmidt CA, Brinkmann V, Jendroska K, Oettle H, Siegert W. J Neuroimmunol 1994; 49: 213–214.
Yamanishi K, Okuno T, Shiraki K, Takahashi M, Kondo T, Asano Y, Kurata T. Lancet 1988; 1: 1065–1067.
Yoshikawa T, Akimoto S, Nishimura N, Ozaki T, Ihira M, Ohashi M, Morooka M, Suga S, Asano Y, Takemoto M, Nishiyama Y. J Med Virol 2003; 70: 267–272.
Zerr DM, Corey L, Kim HW, Huang ML, Nguy L, Boeckh M. Clin Infect Dis 2005a; 40: 932–940.
Zerr DM, Gupta D, Huang ML, Carter R, Corey L. Clin Infect Dis 2002; 34: 309–317.
Zerr DM, Meier AS, Selke SS, Frenkel LM, Huang ML, Wald A, Rhoads MP, Nguy L, Bornemann R, Morrow RA, Corey L. N Engl J Med 2005b; 352: 768–776.

Human Herpesvirus-6: General Virology, Epidemiology and Clinical Pathology
Perspectives in Medical Virology
Gerhard Krueger and Dharam Ablashi (Editors)

DOI 10.1016/S0168-7069(06)12009-1

Epidemiology of HHV6

Caroline Breese Hall
Department of Pediatrics and Medicine, Infectious Diseases, University of Rochester School of Medicine and Dentistry, 601 Elmwood Ave, Box 689, Rochester, NY 14642, USA

The epidemiology of roseola from historical observations

The epidemiologic picture of human herpesvirus 6 (HHV6) has continually evolved over its known life span of 2 decades. Still, it remains more of a collage than a completed portrait, for HHV6 has multiple characters. It is a covert companion of life, and yet a mimicker of many maladies.

Although "HHV6" was named relatively recently, this virus probably has long been recognized as an exanthematous disease of childhood. In papers and texts from the 1800s, an acute infection of young children with rash was given such sobriquets as roseola infantilis, exanthem criticum, exanthem subitum, the rash of roses, and prophetically, the sixth exanthematous disease of childhood. These old descriptions suggest that even in ages past infections likely from HHV6 were common, geographically widespread, and occurred in young children, primarily infants.

Zahorsky's (1913) observations in the early 1900s on 44 cases of roseola and Breese's (1941) subsequent prospective study in 1941 of 111 cases confirmed the disease as an entity with distinctive characteristics, even in the pre-eruptive phase. Breese made a number of clinical and epidemiologic observations that remain true today. He noted that 16% of the infants he followed closely from the time of birth to 1 year of age developed clinical roseola, and 30% by 3 years of age. The peak occurrence (12–13%) was in the 10th month of life, and the illness was mostly limited to children between the ages of 6 months and 3 years of age. The children

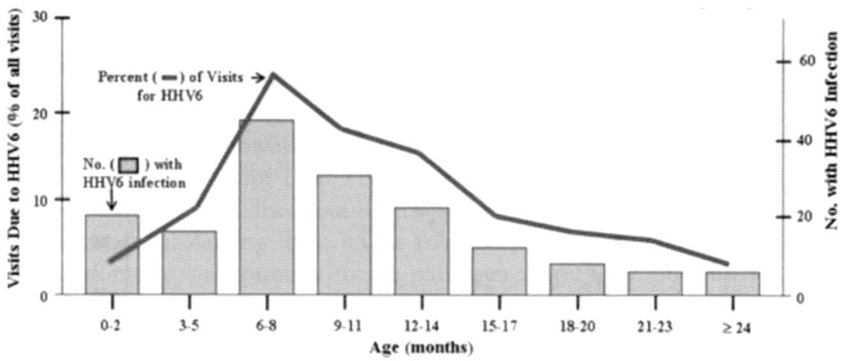

Fig. 2 Number of children presenting to the emergency department with illness from primary HHV6 infection (bars) and visits due to HHV6 infection as a percentage of all emergency department visits for acute febrile illnesses (curve), according to age. *Source*: Adapted from Hall, CB et al., N Engl J Med (1994) 331: 432–438.

documented at the time of its occurrence and, in contrast to HHV6B, no clinical illness has been conclusively associated with primary infection from variant A (Dewhurst et al., 1993; Braun et al., 1997; Ward, 2005a,b). Furthermore, because of the close genomic relatedness of the two variants, serologic studies of HHV6 cannot differentiate infection according to variant. Thus, the epidemiology of HHV6A remains mostly unknown.

HHV6A may be the initial HHV6 infection in one-third of congenitally acquired infections and when acquired after birth HHV6A may occur subsequent to HHV6B infection (DiLuca et al., 1994; Aberle et al., 1996; Tanaka-Taya et al., 1996; Kasolo et al., 1997; Hall et al., 1998; Caserta et al., 2004; Hall et al., 2004). Infection with a new variant is generally not detected, and thus the incidence of HHV6 infections may be underestimated. However, studies examining the presence of HHV6A DNA in peripheral blood mononuclear cells and secretions in normal individuals suggest that infection with HHV6A appears to be much less frequent than that with HHV6B (Aberle et al., 1996; Tanaka-Taya et al., 1996; Hall et al., 1998; Caserta et al., 2004; Ward 2005b). Alternatively, primary HHV6A infection could be silent and often transient. If this were so, the diagnosis of the initial HHV6A infection would be difficult and, without latent infection, subsequent recognition would not be possible by PCR or serology, which would not differentiate between the two variants.

In families, HHV6A DNA in saliva, and peripheral blood mononuclear cells, is infrequent in both children and their parents (Hall et al., 1998). Variant A accounted for about 3% of the DNA persisting in both saliva and the peripheral blood mononuclear cells among the members of 149 families and 2716 children, who were followed for 3–5 years after their primary infection. HHV6A DNA was detected in all age groups and did not appear to correlate with age, although young children generally tended to have HHV6 DNA detected more frequently in their saliva and peripheral blood mononuclear cells than adults.

Factors affecting acquisition of HHV6 infection

Geographical location may affect the frequency and age of occurrence of HHV6 infection, as noted above. However, some of the variation in seroprevalence among countries may be the result of differing methods for the laboratory assays utilized and for the selection of the populations surveyed. Seroprevalence studies in general have not shown that the rate of antibody positivity is appreciably affected by ethnic background, socioeconomic factors, or gender. A few reports have shown a slightly higher prevalence of HHV6 antibody in women (Briggs et al., 1988; Clark et al., 1990; Linhares et al., 1991).

Some epidemiologic studies of HHV6 infection in children have suggested that the risk of acquiring HHV6 infection or acquiring it earlier correlates with having one or more older siblings and a lower family income (Lanphear et al., 1998; Zerr et al., 2005). Breastfeeding, racial background, and daycare attendance have not been shown to be independent risk factors (Lanphear et al., 1998; Zerr et al., 2005). The occurrence of HHV6 infection does not appear to be significantly affected by season, although cases of clinical roseola have been reported as being more frequent in the spring (Breese, 1941; Asano and Grose, 1994; Lanphear et al., 1998; Zerr et al., 2005).

Mode of transmission

Clinical observations of cases of roseola from the first half of the 20th Century suggested that the disease, as noted by Breese (1941), was "only slightly, if at all, communicable," since contact with similar cases was usually lacking. Breese further commented, "however, the disease is definitely infectious" with the most likely source being "carriers among the adult attendants of these children."

A few exceptions have been noted in which contact with another similar case occurred, such as between twins (Cushing, 1927; Zahorsky, 1938; Barenberg, et al., 1939; Breese, 1941). An outbreak of 6 cases of acute illness described as roseola among 10 institutionalized infants was noted by Cushing in 1927. More recent outbreaks of exanthem subitum have been described among infants in close contact with each other in Japan (Okuno et al., 1991), United Kingdom (Reis, 1956), and Brazil (Freitas et al., 2000). However, many of these cases were diagnosed clinically and not documented as primary HHV6 infection.

Postnatal infection

Almost all children acquire HHV6 infection after birth by horizontal transmission. This suggests that the source of infection for most infants is readily and constantly present in their environment. Although not conclusively proved, the primary mode of transmission is suspected to be via the asymptomatic shedding of HHV6 in the secretions of family members and close contacts (Mukai et al., 1994). Although HHV6 DNA may be detected in saliva of adults and older children, in contrast to HHV7, it is not readily isolated from respiratory secretions (Braun et al., 1997; Black et al., 1999).

Some infants may acquire HHV6 during or around the time of birth. Perinatal HHV6 infection has been suggested by the occurrence of primary infection in the first few weeks of life (Kawaguchi et al., 1992; Hall et al., 1994; Byington et al., 2002; Zerr et al., 2002; Yoshikawa et al., 2004). However, as noted below, the incubation period for HHV6 infection has not been clearly established. Further support for possible perinatal infection comes from the detection of HHV6 DNA in genital secretions of pregnant women (Leach et al., 1994; Okuno et al., 1995; Maeda et al., 1997) and in the peripheral blood monocular cells of infants within the first month of life (Hall et al., 1994; Byington et al., 2002; Hall et al., 2006 (in press)). However, in the latter instance, some may be congenitally infected infants.

Congenital infection

Congenital infection with HHV6 could result from maternal infection acquired through transplantation or by inheritance via parental transmission of chromosomally integrated HHV6. Intrauterine transmission of HHV6 has recently been documented to occur in about 1% of children in a large and several smaller studies of normal newborns (Adams et al., 1998; Dahl et al., 1999; Daibata and Miyoshi, 1999; Hall et al., 2004). Intrauterine transmission had been suggested in a few cases by HHV6 DNA detection in fetuses and placentas, as well as cord blood (Ando et al., 1992; Aubin et al., 1992; Leach et al., 1994; Maeda et al., 1997a; Adams et al., 1998; Daibata et al., 1999; Ashshi et al., 2000; Baillargeon et al., 2000; Ohashi et al., 2002). The detection rate of HHV6 DNA in cord blood has ranged from 0 to 1.6% in studies examining 58–305 cord blood (Adams et al., 1998; Dahl et al., 1999; Daibata et al., 1999). In a study examining 5638 cord blood, the rate was 1% of live births, similar to that for cytomegalovirus (Fowler et al., 2003). The importance of this mode of transmission in terms of clinical and neurodevelopmental outcome is currently unknown. These congenitally infected infants appeared normal at birth, although the HHV6 DNA tended to persist in their peripheral blood mononuclear cells (Hall et al., 2004).

Transmission of HHV6 by integration of the virus DNA into human chromosomes has also been reported (Daibata et al., 1998; Luppi et al., 1998; Daibata et al., 1999; Hermouet and Minvielle, 2000; Tanaka-Taya et al., 2004). In Japanese subjects, the rate of detection of chromosomally integrated HHV6 has been reported as 0.2% (Tanaka-Taya et al., 2004). However, it is unclear whether the detection of the integrated HHV6 genome is truly inherited (Luppi et al., 1998; Hermouet et al., 2000).

Transmission to the infant via breast milk does not appear to occur, although studies on this are limited (Dunne and Jevon, 1993). Although HHV6 has been detected in the cervix and genital secretions of women, thus far sexual transmission has not been proven (Leach et al., 1994; Okuno et al., 1995; Maeda et al., 1997b).

Another documented mode of transmission is that from donated bone marrow containing HHV6, which caused a primary HHV6 infection in the bone marrow transplant recipient (Suga et al., 1998). Transmission of new HHV6 infection by transfusion or transplantation potentially could occur despite the recipient having had past HHV6 infection and possessing specific antibody. This raises the possibility that HHV6, which is latent in bone marrow progenitor cells, could be transmitted to recipients of blood transfusions or to those receiving other organs (Ward et al., 1989; Suga et al., 1998).

Incubation period

The incubation period for HHV6 infection has not been well defined and likely to vary according to the route, inoculum, and an individual's immune status.

In the older studies noted above in which outbreaks of roseola were observed, the incubation appeared to range from 3 to 15 days, but mostly around 10 days (Cushing, 1927; Zahorsky, 1938; Barenberg and Greenspan, 1939). However, the best information on the incubation period for primary HHV6 infection in infants comes from the studies of Kempe et al. (1950). These investigators were able to show that serum obtained on the third day of fever from an infant with roseola could transmit clinical roseola to a susceptible subject after 9 days. When the infant's serum was inoculated subcutaneously into rhesus monkeys, the animals developed, after 3 or 4 days, a highly febrile illness lasting for 3–5 days. Their further experiments showed throat washings could also transmit the disease. The washings were obtained from an infant with roseola on the second day of fever and inoculated into monkeys intranasally, resulting in a similar febrile illness in the animals. Additional experiments revealed that sera from these febrile monkeys, when subsequently inoculated into a naïve monkey, would also engender a highly febrile illness. However, if the monkeys had been previously infected, no illness occurred. Thus, these studies of Kempe et al. (1950) clearly illustrated that the infectious agent of roseola resided in both serum and nasal secretions. Second, when serum or nasal secretions were transmitted to a susceptible host, roseola would occur after approximately 9 days, or in monkeys, produce a febrile illness after 3–5 days.

Essentially, no information on the incubation period currently exists for infection documented to be from HHV6. Even for horizontally transmitted primary infection, which is the mode of infection for 99% of normal children, the incubation period remains unclear. Similarly unknown is the effect of the variant of HHV6 on the incubation period. Presumably in the studies of Kempe et al. (1950) the virus transmitted from the infant with roseola was HHV6B, since the B variant is almost exclusively associated with primary infection (Dewhurst et al., 1993). The incubation period, as well as any associated disease, with HHV6A is also unknown. Furthermore, the incubation period could vary for HHV6A or HHV6B depending on whether the infection was the initial HHV6 infection or occurred subsequent to infection with the other variant, and thus, in the presence of cross-reactive antibody.

A second infection with the same variant, but a different strain is also possible, and has been documented for HHV6B (van Loon et al., 1995).

Summary

Despite the many covert characteristics and conundrums that continue concerning HHV6, the burden that this virus imposes upon our system of healthcare is likely appreciable. This is suggested by HHV6's singular characteristics, which are first that HHV6 is an ubiquitous and worldwide infection. Second, initial infection occurs in a very young and vulnerable population and is concentrated during a relatively short interval of early life. Third, all initial infections are symptomatic and, in the young child, most require medical evaluation.

Ninety-three percent of primary HHV6 infections were symptomatic among children in Seattle who were followed during their first 2 years of life. Their illness required a visit to a physician significantly more often than occurred in age-matched controls (Zerr et al., 2005). In Rochester, NY, primary HHV6 infection accounted for 21% of emergency department visits for febrile illnesses for 6–12-months-old infants and 10% for children within the first 2 years of life (Fig. 2) (Hall et al., 1994). Thirteen percent of these young children required hospitalization.

A number of studies have also noted the association of initial HHV6 infection with seizures (Kondo et al., 1993; Suga et al., 1993; Caserta et al., 1994; (Ward and Gray, 1994; Suga et al., 2000; Yoshikawa and Asano, 2000; Dewhurst, 2004). The studies in an emergency department, primary HHV6 infection was the most frequently identified cause, accounting for one-third of all the first time febrile seizures evaluated in children 2 or less years of age (Hall et al., 1994). Of the 81 primary HHV6 infections identified among the cohort of Seattle children followed over their first 2 years of life, none of the 75 symptomatic children had a febrile seizure (Zerr et al., 2005). However, only 46 of these children had fever recorded on their symptom card. The reported differences of the findings and complications associated with HHV6 infection may relate to the design and the number of children studied.

Nevertheless, considering the number of primary HHV6 infections that occur every year among each new birth cohort, even a relatively uncommon sign or complication requiring medical attention will impose an appreciable healthcare cost. In addition, completely unknown is the potential clinical and economic impact on healthcare from reactivated infection in older children and adults. HHV6 may compromise health and medical resources across the age span of life. The epidemiologic and economic burden attributable to HHV6 currently remains unestimated and unappreciated.

References

Aberle S, Mandl C, Kunz C, Popow-Kraupp T. J Clin Microbiol 1996; 34: 3223–3225.
Adams O, Krempe C, Kogler G, Wernet P, Scheid A. J Infect Dis 1998; 178: 544–546.

Ando Y, Kakimoto K, Ekuni Y, Ichijo M. Lancet 1992; 340: 1289.
Asano Y, Grose C. Human herpesvirus type 6 infections In: Herpesvirus Infections (Glaser R, Jones J, editors). New York: Marcel Dekker Inc.; 1994; pp. 227–244.
Ashshi A, Cooper R, Klapper P, Al-Jiffri O, Moore L. Lancet 2000; 355: 1519–1520.
Aubin J-T, Poirel L, Agut H, Huraux J-M, Bignozzi C, Brossard Y, Mulliez N, Roume J, Lecuru F, Taurelle R. Lancet 1992; 340: 482–483.
Baillargeon J, Piper J, Leach C. J Clin Virol 2000; 16: 149–157.
Barenberg L, Greenspan L. Am J Dis Child 1939; 58: 983–993.
Black J, Pellett P. Rev Med Virol 1999; 9: 245–262.
Braun D, Dominguez G, Pellett P. Clin Microbiol Rev 1997; 10: 521–567.
Breese Jr. BB. New York State J Med 1941; 41: 1854–1859.
Briggs M, Fox J, Tedder R. Lancet 1988; 1: 1058–1059.
Brown NA, Sumaya CV, Liu CR, Ench Y, Kovacs A, Coronesi M, Kaplan MH. Lancet 1988; 2: 396.
Byington C, Zerr D, Taggart E, Nguy L, Hillyard D, Carroll K, Corey L. Pediatr Infect Dis J 2002; 21: 996–999.
Caserta M, Hall C, Schnabel K, McIntyre K, Long C, Costanzo M, Dewhurst S, Insel R, Epstein L. J Infect Dis 1994; 170: 1586–1589.
Caserta M, Schnabel K, Carnahan J, Gilbert L, Lathan G, Dewhurst S, Lofthus G, Boettrich C, McDermott M, Hall C. J Pediatr 2004; 145: 478–484.
Clark D, Alexander F, McKinney P, Roberts B, O'Brien C, Jarrett R, Cartwright R, Onions D. Int J Cancer 1990; 45: 829–833.
Cushing H. Can Med Assoc J 1927; 17: 905–906.
Dahl H, Fjaertoft G, Norsted T, Wang F, Mousavi-Jazi M, Linde A. J Infect Dis 1999; 180: 2035–2038.
Daibata M, Miyoshi I. J Infect Dis 1999; 178: 544–546.
Daibata M, Taguchi T, Nemoto Y, Taguchi H, Miyoshi I. Blood 1999; 94: 1545–1549.
Daibata M, Taguchi T, Sawada T, Taguchi H, Miyoshi I. Lancet 1998; 352: 543–544.
Dewhurst S. Herpes 2004; 11: 105A–110A.
Dewhurst S, McIntyre K, Schnabel K, Hall C. J Clin Microbiol 1993; 31: 416–418.
DiLuca D, Dolcetti R, Mirandola P, DeRe V, Secchiero P, Carbone A, Boiocchi M, Cassai E. J Infect Dis 1994; 170: 211–215.
Dunne W, Jevon M. J Infect Dis 1993; 168: 250.
Enders G, Biber M, Meyer G, Helftenbein E. Infection 1990; 18: 12–15.
Fowler K, Stagno S, Pass R. JAMA 2003; 289: 1008–1011.
Freitas R, Monteiro T, Linhares A. Rev Inst Med Trop Sao Paulo 2000; 42: 305–311.
Hall C, Caserta M, Schnabel K, Boettrich C, McDermott M, Lofthus G, Carnahan J, Dewhurst S. J Pediatr 2004; 145: 472–477.
Hall C, Caserta M, Schnabel K, Long C, Epstein L, Insel R, Dewhurst S. Clin Infect Dis 1998; 26: 132–137.
Hall C, Caserta M, Schnabel K, McDermott M, Lofthus G, Carnahan J, Gilbert L, Dewhurst S. J Infect Dis 2006; (in press).
Hall C, Long C, Schnabel K, Caserta M, McIntyre K, Costanzo M, Knott A, Dewhurst S, Insel R, Epstein L. N Engl J Med 1994; 331: 432–438.
Hermouet S, Minvielle S. Blood 2000; 95: 1108.
Huang L, Lee C, Chen J, Yang C, Wang J, Chang M, Hsu C, Kuo P. J Infec Dis 1992; 165: 1163–1164.

Huang L, Lee C, Lee P. Acta Paediatr 1997; 86: 604–608.
Juretic M. Helv Pediatr Acta 1963; 18: 80–95.
Kasolo F, Mpabalwani E, Gompels U. J Gen Virol 1997; 78: 847–856.
Kawaguchi S, Suga S, Kozawa T, Nakashima T, Yoshikawa T, Asano Y. Pediatrics 1992; 90: 628–630.
Kempe H, Shaw E, Jackson J, Silver H. J Pediatr 1950; 37: 561–568.
Kondo K, Nagafuji H, Hata A, Tomomori C, Yamanishi K. J Infect Dis 1993; 167: 1197–1200.
Lanphear B, Hall C, Black J, Auinger P. Pediatr Infect Dis J 1998; 17: 792–795.
Leach C, Newton E, McParlin S, Jenson H. J Infect Dis 1994; 169: 1281–1283.
Levy JA, Ferro F, Greenspan D, Lennette ET. Lancet 1990; 335: 1047–1050.
Linde A, Dahl H, Wahren B, Fridell E, Salahuddin Z, Biberfeld P. J Virol Methods 1988; 21: 117–123.
Linhares M, Eizuru Y, Tateno S, Minamishima Y. Microbiol Immunol 1991; 35: 1023–1027.
Luppi M, Barozzi P, Morris C, Merelli E, Torelli G. Lancet 1998; 352: 1707–1708.
Maeda T, Okuno T, Hayashi K, Miyamoto H, Utsunomiya A, Yamada Y, Mori T. Pediatr Infect Dis J 1997a; 16: 1176–1177.
Maeda T, Okuno T, Hayashi K, Nagata M, Ueda M, Terashima K, Kawashima T, Miyamoto H, Mori T, Yamada Y. Acta Paediatr Japonica 1997b; 39: 653–657.
Mukai T, Yamamoto T, Kondo T, Kondo K, Okuno T, Kosuge H, Yamanishi K. J Med Virol 1994; 42: 224–227.
Niederman J, Liu C, Kaplan M, Brown N. Lancet 1988; 2: 817–818.
Ohashi M, Yoshikawa T, Ihira M, Suzuki K, Suga S, Tada S, Udagawa Y, Sakui H, Iida K, Saito Y, Nisiyama Y, Asano Y. J Med Virol 2002; 67: 354–358.
Okuno T, Mukai T, Baba K, Ohsumi Y, Takahashi M, Yamanishi K. J Pediatr 1991; 119: 759–761.
Okuno T, Oishi H, Hayashi K, Nonogaki M, Tanaka K, Yamanishi K. J Clin Microbiol 1995; 33: 1968–1970.
Okuno T, Takahashi K, Balachandra K, Shiraki K, Yamanishi K, Takahashi M, Baba K. J Clin Microbiol 1989; 27: 651–653.
Pruksananonda P, Hall C, Insel R, McIntyre K, Pellett P, Long C, Schnabel K, Pincus P, Stamey F, Dambaugh T, Stewart J. N Engl J Med 1992; 326: 1445–1450.
Ranger S, Patillaud S, Denis F, Himmich A, Sangare A, M'Boup S, Itoua-N'Gaporo A, Prince-David M, Chout R, Cevallos R, Agut H. J Med Virol 1991; 34: 194–198.
Reis NLancet 1956; 270: 830.
Salahuddin S, Ablashi D, Markham P, Josephs S, Sturzenegger S, Kaplan M, Halligan G, Biberfeld P, Wong-Staal F, Kramarsky B, Gallo R. Science 1986; 234: 596–601.
Saxinger C, Polesky H, Eby N, Grufferman S, Murphy R, Tegtmeir G, Parekh V, Memon S, Hung C. J Virol Methods 1988; 21: 199–208.
Suga A, Yoshikawa T, Asano Y, Kozawa T, Nakashima T, Kobayashi I, Yazaki T, Yamamoto H, Kajita Y, Ozaki T, Nishimura Y, Yamanaka T, Yamada A, Imanishi J. Ann Neurol 1993; 33: 597–603.
Suga S, Peiris M, Chan G, Chiu D, Ha S. Bone Marrow Transplant 1998; 21: 1063–1066.
Suga S, Suzuki K, Ihira M, Yoshikawa T, Kajita Y, Ozaki T, Iida K, Saito Y, Asano Y. Arch Dis Child 2000; 82: 62–66.
Tanaka-Taya K, Kondo T, Mukai T, Miyoshi H, Yamamoto Y, Okada S, Yamanishi K. J Med Virol 1996; 48: 88–94.

Tanaka-Taya K, Sashihara J, Kurahashi H, Amo K, Miyagawa H, Kondo K, Okada S, Yamanishi.K. J Med Virol 2004; 73: 465–473.
van Loon N, Gummuluru S, Sherwood D, Marentes R, Hall C, Dewhurst S. Clin Infect Dis 1995; 21: 1017–1019.
Ward K. Curr Opin Infect Dis 2005a; 18: 247–252.
Ward K. J Clin Virol 2005b; 32: 183–193.
Ward K, Gray J. J Med Virol 1994; 42: 119–123.
Ward K, Gray J, Efstathiou S. J Med Virol 1989; 28: 69–72.
Yanagi K, Harada S, Ban F, Oya A, Okabe N, Tobinai K. J Infect Dis 1990; 161: 153–154.
Yoshikawa T, Asano Y. Brain Dev 2000; 22: 307–314.
Yoshikawa T, Suga S, Asano Y, Yazaki T, Kodama H, Ozaki T. Pediatrics 1989; 84: 675–677.
Yoshikawa T, Suzuki K, Umemura K, Akimoto S, Miyake F, Usui C, Fujita A, Suga S, Asano Y. J Med Virol 2004; 74: 463–466.
Zahorsky J. JAMA 1913; 61: 1446–1450.
Zahorsky J. Roseola infantum In: Brennemann's Practice of Pediatrics (Brennemann J, editor). vol. 2. Hagerstown: 1938; Chapter XXI.
Zerr D, Meier A, Selke S, Frenkel L, Huang M, Wald A, Rhoads M, Nguy L, Bornemann R, Morrow R, Corey L. N Eng J Med 2005; 352: 768–776.
Zerr D, Yeung L, Obrigewitch R, Huang M- L, Frenkel L, Corey L. J Med Virol 2002; 66: 384–387.

PART III: CLINICAL PATHOLOGY

Human Herpesvirus-6: General Virology, Epidemiology and Clinical Pathology
Perspectives in Medical Virology
Gerhard Krueger and Dharam Ablashi (Editors)

DOI 10.1016/S0168-7069(06)12010-8

Pathologic Features of HHV-6 Disease

Gerhard R.F. Krueger[a,b], Brigitte Schneider[c]
[a]*Department of Pathology and Laboratory Medicine, The University of Texas—Houston Medical School, 6431 Fannin St., MSB 2.246, Houston, TX 77030, USA*
[b]*Laboratory of Computational Cell Biology, Department of Anatomy II, University of Cologne, 50924 Cologne, Germany*
[c]*Immunonephrology Laboratory, Department of Research, University Hospital Basel, Hebelstrasse 20, CH-4021 Basel, Switzerland*

Introduction

HHV-6 belongs to the *Roseolavirus* genus of the β-herpesvirus subfamily (Ablashi et al., 1993). It was first discovered in 1989 at NIH (GS strain (Salahuddin et al., 1986)), with two additional isolates reported from Uganda (U1102 (Downing et al., 1987)) and Zambia (Z-29 (Lopez et al., 1988). There are two variants of HHV-6 with a different biological behavior: variant A (the original GS strain) and variant B (e.g. Z-29 strain) (Downing et al., 1987; Ablashi et al., 1991; Inoue et al., 1994). Both the HHV-6 variants are closely related to human cytomegalovirus (HCMV), the prototype of the human β-herpesviridae, as well as to HHV-7 based on serologic cross-reactivity, DNA hybridization and nucleotide sequence similarity (Braun et al., 1997; Campadelli-Fiume et al., 1999; Clark, 2000). Pathologic features of HHV-6 and HHV-7 may be indistinguishable.

HHV-6 is a ubiquitous virus with up to 90% seropositive adults (Ablashi et al., 1992; Braun et al., 1997; Campadelli-Fiume et al., 1999; Clark, 2000). Most infections occur within the first 2 years of childhood. HHV-6 DNA was detected in 87% of children aged 1 year or older, further increasing with age. HHV-6 is prevalent in salivary gland tissue and in saliva (HHV-6B being more frequent than

HHV-6A) spreading preferentially by aerogenic routes (Fox et al., 1990; Krueger et al., 1990; Campadelli-Fiume et al., 1999; Clark, 2000). Coinfections of both HHV-6A and -6B can occur (Cone et al., 1996).

A variety of clinical symptoms and diseases can be associated with HHV-6 infection (Salahuddin et al., 1993; Krueger et al., 1994a; Braun et al., 1997; Campadelli-Fiume et al., 1999; Yoshikawa and Asano, 2000), yet a causal relationship between infection and disease can be hard to prove. Primary infections must be distinguished from reactivated infections. HHV-6 type A must be separated from HHV-6 type B, and finally—similar to EBV—potential pathologic sequelae of defective replication in HHV-6 may be entertained (e.g. via NF-κB activation or defective viral antigen expression and molecular mimicry (Krueger and Ferrer Argote, 1994; Flamand et al., 1995; Deliconstantinos et al., 1998). HHV-6 lyses and thus destroys infected target cells or it may induce inflammatory and immune (or autoimmune) reactions. These are mediated by a variety of HHV-6-induced cytokines and chemokines as well as by modulation of cell membrane receptors (Deliconstantinos and Krueger, 1993; Flamand et al., 1995; Deliconstantinos et al., 1998; Hasegawa et al., 2001).

Cell and tissue tropism and strain differences

Target cells for HHV-6 infection possess CD46 as part of the virus receptor (Santoro et al., 1999; Greenstone et al., 2002), for which a HHV-6 glycoprotein H & L complex is the ligand (Mori et al., 2003). CD46 is strongly expressed on epithelial cells of salivary gland ducts and on kidney tubular cells, moderately well on lymphocytes and vascular endothelial cells, and weakly on cells of the interstitial space and on muscle cells. It may also be present on various tumor cells. CD46 participates in downregulating complement activity by binding of C′3b and C′4b complement components, thus the interaction of CD46 with HHV-6 may interfere with this function and cause inflammation via the alternate complement pathway.

In vitro, HHV-6A and HHV-6B infects and replicates in a variety of human CD4+T-cells including tissue culture lines HSB2, MoLT-3, Jihan and SupT1 (Ablashi et al., 1991; Ablashi and Hung, 1992; Braun et al., 1997; Campadelli-Fiume et al., 1999; Clark, 2000). HHV-6 can also infect monocytes and macrophages (Lusso, 1992). Occasionally, HHV-6 may apparently also infect B cells, neural cells and human fibroblasts. The propagation of HHV-6A and HHV-6B in T-cell lines varies considerably (Ablashi et al., 1991). HHV-6 easily grows in the immature T-cell line HSB2, and HHV-6B grows in the more mature MoLT-3T-cell line. Human umbilical cord mononuclear cells grow both strains to higher titers when stimulated with phytohemagglutinin. Typically, infected cells show 2–5 times enlarged multinucleated, giant cells, with ballooning degeneration (Salahuddin et al., 1986; Taniguchi et al., 2000).

In vivo, the tissue distribution of CD46 suggests as potential targets for HHV-6 infection and immediate virus-induced pathology: salivary gland epithelia, lymphoid cells, vascular endothelium, tubular epithelia of kidneys, neuroglial cells in the central nervous system and potentially muscle cells. The sole presence of CD46 in a cell or tissue does not imply, though, their infectability by HHV-6 and subsequent disease. There are also differences in the organ- and tissue pathogenicity of HHV-6A and HHV-6B.

General cellular pathology of HHV-6 infection

In vitro studies show the immediate effects of HHV-6 binding to cells, infection of susceptible cells and intracellular replication (see also Chapter 2: Ultrastructure of HHV-6). Susceptible cells (e.g. HSB2, cord blood cells) show upon exposure to HHV-6 blastic transformation with or without giant cell formation, intranuclear inclusions, eventual mitoses and production of viral particles with cellular degeneration and apoptosis (Fig. 1; Biberfeld et al., 1987; Kramarsky and Sander, 1992; Kirn et al., 1997). Blastic transformation with giant cells can occur in cells upon virus contacts even without subsequent internalization and replication of viral particles (Boehmer, 1987). Typical Reed–Sternberg-type giant cells were induced by HHV-6 in established Hodgkin cell cultures without spontaneous giant cells (L428, L540, L591, HDLM2, KMH2) and without subsequent viral replication in these cells (Fig. 1; Krueger et al., 1991, 1992, 1995). Internalization of virus particles and their replication initiate a number of cellular events, which are summarized in Table 1 and in Fig. 2. Cell membrane rigidification after HHV-6 infection and altered receptor availability, support superinfection by other viruses (Krueger et al., 1990; Schonnebeck et al., 1991; Lusso et al., 1995). NF-κB activation and other genomic effects down or upregulate other cell functions such as cell–cell cooperation and genomic transactivation of various genes (Lusso et al., 1995; Gies et al., 1998; Hasegawa et al., 2001; Arena et al., 2002), and oxidative burst and cell cycle control factors will induce cell degeneration and apoptosis (Deliconstantinos and Krueger, 1993; Kirn et al., 1997; Deliconstantinos et al., 1998).

The *in vivo* cellular effects of HHV-6 essentially correspond to those observed *in vitro*: stimulated cells may show blastic transformation and proliferate; productively infected cells may undergo degeneration and apoptosis. Prominent intranuclear inclusion bodies and giant cells are rarely seen in tissues, and when observed, these are highly suspicious for a dual infection with other herpesviruses, e.g. HHV-7.

Primary infection

Following the tissue distribution of the HHV-6 CD46 receptor, preferred target cells for virus-induced pathology are salivary glands, lymphoid tissues, vascular endothelial cells, tubular epithelia of kidneys and neuroglial cells. In addition to direct toxic effects on target cells as shown *in vitro*, damage to primarily noninfectable cells and tissues can possibly originate *in vivo* from indirect effects of viral infection such as cytokine and complement activation, immune and/or autoimmune reactions.

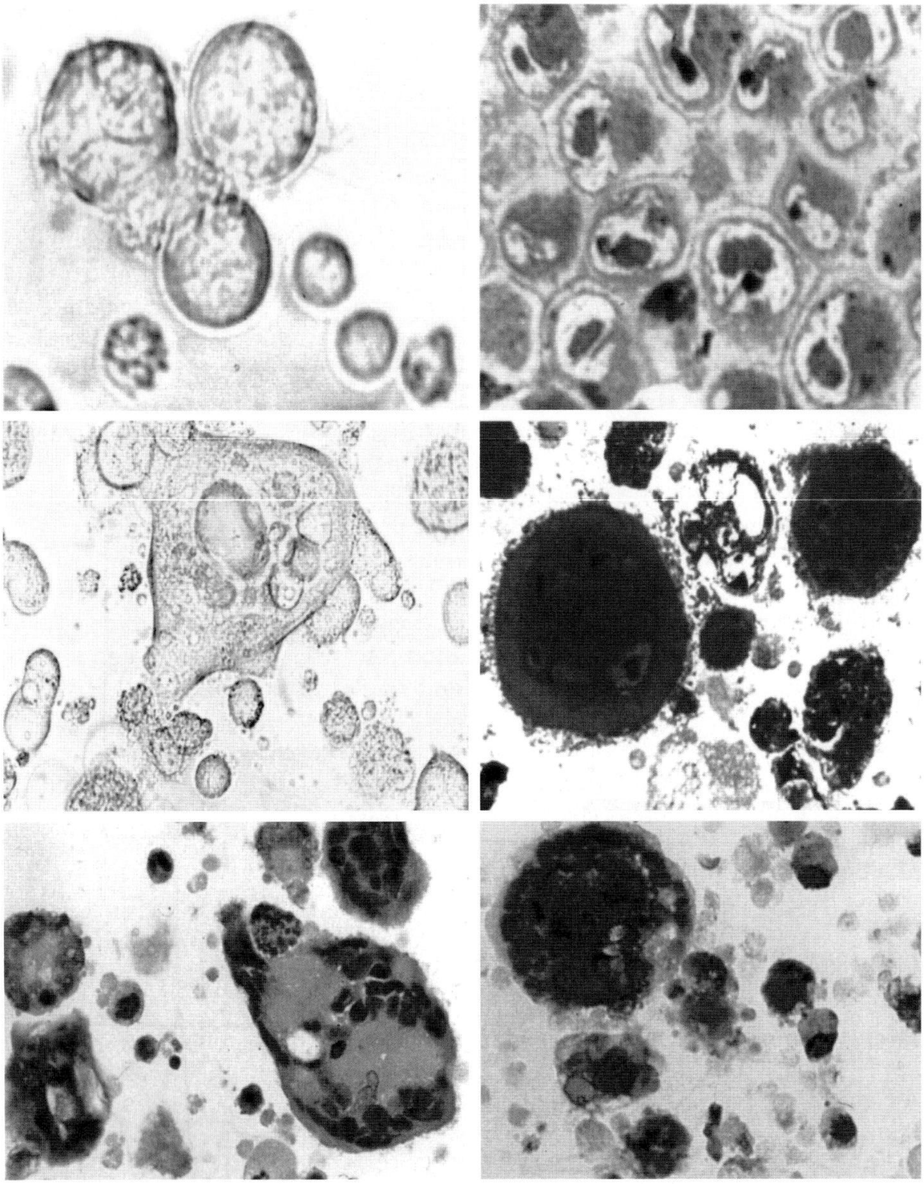

Fig. 1 Cellular reactions in HHV-6-infected cells. *Top Row*: HHV-6A-infected HSB2 cells in culture: blastic transformation of infected cells (left) and nuclear inclusions in semithin section (right). *Center Row:* Prominent giant cell formation after infection with HHV-6A: SupT1 cells in culture (left) and L-428 Hodgkin cell line (semithin section of culture; right). *Bottom Row:* Antigen expression in HHV-6A-infected SupT1 cells: p41 early antigen (left) and HAR3 (i.e. mixture of late gp antigens; right). All APAAP immunohistochemistry. (for colour version: see colour section on page 350).

Table 1

Biological effects of HHV-6

Target	Interaction	Effects	Comments
Cell Membrane	rigidification receptor expression	altred receptor availability CD4, CD5, CD21, CD57, Fc etc.	supports superinfection by other viruses interferes with immune response
Cell cycle	increase p21WAF Bax, Bcl-2	induction of apoptosis oxidative burst	kills infected cells
Genome	NK-kB activation	interference with transcription (e.g. of cytokines, cell receptors)	cell cooperation blockage
	transcript induction of CD4	increased CD4 receptor	enhanced HIV effects
	HIV-1 promoter act of EBV	increased EBV LMP protein	enhanced EBV effects
	MHC-1 downregulation	decreased cell cooperation	interferes with immune response
	U94 gene expression	late gene regulation	supports latent infection
Cytokines	increase IL-1, 2, 4, 8, 10, 12, 15, TNFa		variable according to target cell
	decrease IFNy, IL-3, 6, 18, GM-CSF	inhibits antiviral effects	Th1 to Th2 switch
Chemokines	induces beta-chemokine receptor increases MCP-1 release downregulates CXC4 reduces RANTES secretion induces adhesion molecules	all essentially interfere with the recruitment of inflammatory cells and non-specifically inflammation	
Other	CD46 binding	interferes with C3b/ C4b inactivation	alternative complement activation
		stimulation of rheumatoid factor molecular mimicry selective viral antigen expression	all contribute to immune dysregulation and to autoimmunity
		activation of myelin-directed T cells	

(from Krueger & Ablashi, 2003)

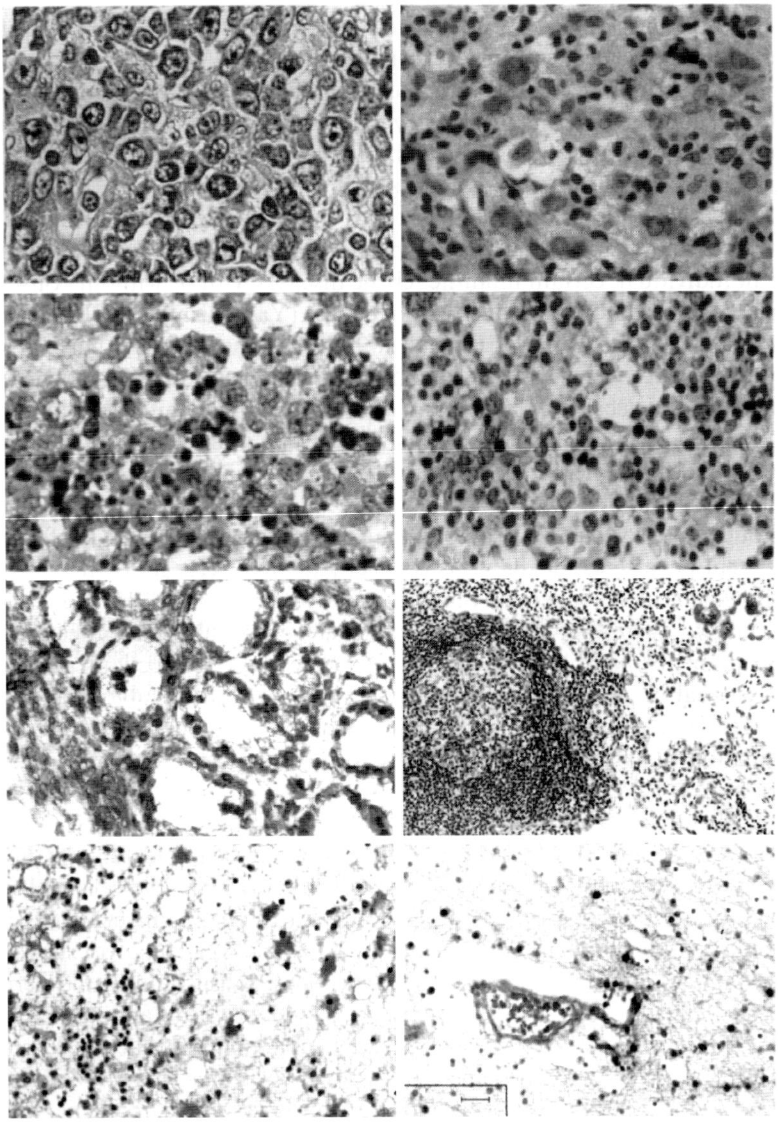

Fig. 3 Some characteristic tissue reactions in HHV-6 infections. *Top Row:* Blastic transformation of cells in tonsils with prominent nucleoli (left) with expression of HHV-6 p41 early antigen (red cells in APAAP reaction; right). *Upper Center Row:* Prominent apoptosis of HHV-6 infected cells in Kikuchi's lymphadenitis (left); hematopoietic stem cells in bone marrow expressing HHV-6 p41 antigen (red dots in APAAP reaction; right). *Lower Center Row:* Expression of HHV-6 gp135 in epithelial cells of the salivary gland (lip biopsy; left); HHV-6A-associated lymphoid interstitial pneumonitis, LIP, in an HIV infected patient (red cells carry p41 HHV-6 antigen; APAAP reaction). *Bottom Row:* Acute necrotizing encephalitis with HHV-6 p41 and DNA expression in numerous astroglial cells (red cells by APAAP left, black dots for HHV-6 DNA by *in situ* hybridization (right; Wagner et al., 1997). (for colour version: see colour section on page 351).

certain cytokines, as well as plant and food components (e.g. agglutinins, phorbol esters, aflatoxin) (Krueger et al., 1998). Reactivation *per se* does not cause disease unless the virus persists and replicates over longer periods of time secondary to immune deficiency, systemic autoimmune disorders or tumor growth. In essence, two main conditions support a pathogenic HHV-6 reactivation with reinfection of tissues: continued abnormal stimulation of cells carrying viral genomes and defective host control of virus replication and spread. Pathologic persistent activity of reactivated HHV-6 under such conditions coincides with a number of diseases in various organs.

Lymphatic and hematopoietic system (see also chapters 14 and 15)

In chronic persistent HHV-6 infection, viral DNA load and cellular changes show certain cyclic changes, suggesting some fluctuation in viral replication (Krueger et al., 2001). In about 6%, *heterophile-negative infectious mononucleosis* is caused by HHV-6A or B infection (Steeper et al., 1990; Horwitz et al., 1992; Akashi et al., 1993). More frequent is the reactivation of latent HHV-6 in patients with classical EBV-induced infectious mononucleosis resulting in a more protracted course of the disease with elevated liver enzymes (Bertram et al., 1991). Occasionally caused by HHV-6, preferentially variant B, are *angioimmunoblastic lymphadenopathy* (Luppi et al., 1993; Daibata et al., 1997), *hemophagocytic syndromes* (Sugita et al., 1995; Tanaka et al., 2000) and *Langerhans cell histiocytosis* (Leahy et al., 1993).

HHV-6 (preferentially subtype A) and/or increased viral DNA loads were found in certain malignant lymphomas, including subtypes of *Hodgkin's disease* (Krueger et al., 1989; Torelli et al., 1991; Di Luca et al., 1994; Braun et al., 1995; Hallas et al., 1996; Razzaque et al., 1996; Bandobashi et al., 1997; Luppi et al., 1998; Lin et al., 1999; Hermouet et al., 2003) as well as in a number of prelymphomatous *atypical polyclonal lymphoproliferations including Canale-Smith syndrome* (Krueger et al., 2002). Since oncogenic effects of the HHV-6 genome (Razzaque, 1990; Thompson et al., 1994) have not been proven, other mechanisms may account for a typical lymphoproliferation (Krueger and Ferrer-Argote, 1994), such as persistent antigenic stimulation or immune dysfunction (Kakimoto et al., 2002).

Both subtypes infect hematopoietic stem cells suppressing their engraftment in transplant recipients (Carrigan and Knox, 1994; Rosenfeld et al., 1995; Herbein et al., 1996; Singh and Carrigan, 1996; Isomura et al., 1997; Penchansky and Jordan, 1997; Lau et al., 1998; Wang et al., 1999). Frequent HHV-6 infections with high virus load were reported in bone marrow and stem cell transplant recipients (Clark, 2002; Ihara et al., 2002; Sashihara et al., 2002; Boutolleau et al., 2003). Besides failure of engraftment, lymphocytopenia, suppression of myelopoiesis or erythrocytopenia may ensue HHV-6 reactivation, and the virus can be even transmitted by the graft itself (Lau et al., 1998).

HHV-6 reactivates frequently in patients with renal and hepatic allografts, yet pathologic sequelae usually arise from coincident cytomegalovirus infections

(Kaden et al., 1997; Clark et al., 2003; Feldstein et al., 2003; Yoshikawa, 2003). HHV-6 antigen was found in bone marrow cells of patients with myelodysplasia and certain chronic myeloproliferative disorders (Krueger et al., 1994b; Rojo et al., 2000). Since this pertains equally well to EBV and HHV-7, a deficient clearance of virus may be responsible for this phenomenon rather than indicating some etiologic relationships.

Persistent HHV-6 may cause functional disturbances of the immune system as indicated by elevated antibody titers against HHV-6 in allergies, *drug-induced hypersensitivity reactions* and in *systemic lupus erythematosus, Sjögren's syndrome* and *progressive systemic sclerosis* (Krueger et al., 1991, 1994b; De Clerck et al., 1992; Klueppelberg, 1994; Lasch et al., 1996; Descamps et al., 1997, 2001; Tohyama et al., 1998; Conilleau et al., 1999). Virus persistence may constitute a risk factor for additional immune dysregulation and for increasing the severity of adverse reactions (Suzuki et al., 1998).

Central nervous system (see also chapter 16)

HHV-6 replicates with low efficiency in neuroglial cells (Luppi et al., 1995). Viral DNA and antigen have been successfully demonstrated in human brain tissue, both in healthy organs and in diseased tissues, with subtype A being about three times more frequent than subtype B (Fig. 2; Luppi et al., 1995; Hall et al., 1998; Cuomo et al., 2001). There are increasing reports of HHV-6-associated meningitis, encephalitis in children with febrile seizures (Caserta et al., 1994; Wilborn et al., 1994; Knox and Carrigan, 1995; Bonthius and Karacay, 2002; Eeg-Olofsson, 2003), acute necrotizing or hemorrhagic encephalitis and demyelinating brain diseases in immune-deficient patients and in persons suffering from *multiple sclerosis* (Challoner et al., 1995; Wagner et al., 1997; Solldan et al., 2000; An et al., 2002; Cirone et al., 2002; Chapenko et al., 2003; Tejada-Simon et al., 2003). CNS infections with HHV-6 appear more frequent in patients with T-cell immune deficiency (Pruitt, 2003).

Cardiovascular system (see also chapter 17)

HHV-6 infects endothelial cells of the aorta, umbilical vein and capillaries (Wu and Shanley, 1998; Rotola et al., 2000; Takatsuka et al., 2003). It may induce thrombotic microangiopathy or large vessel arteritis (Wu and Shanley, 1998; Toyabe et al., 2002; Takatsuka et al., 2003). There are single reports relating HHV-6 infection to *leukocytoclastic vasculitis* and to *Kawasaki's disease* (Hagiwara et al., 1993; Luka et al., 1995; Drago et al., 1999; Yoshikawa et al., 2003). HHV-6 genomic material was found in coronary arteries of heart allografts suggesting virus reactivation at this site. Finally, a case of *fulminant myocarditis* was described in a patient receiving steroid therapy for hepatitis. HHV-6 antibodies showed a fourfold increase in this patient, and HHV-6 DNA was demonstrated by PCR in liver and heart tissue (Fukae et al., 2000).

Other organs and tissues (see also chapters 13 and 18)

Other diseases described in association with active HHV-6 infection were *interstitial pneumonitis* (Cone et al., 1993; Know et al., 1995) and *fulminant hepatic failure* with HHV-6 present in liver tissue and in portal vein endothelium (Aita et al., 2001; Ishikawa et al., 2002). Some recipients of stem cell transplants, which suffered from severe posttransplant diarrhea showed HHV-6B DNA and antigens in their peripheral blood cells and in intestinal epithelia (Amo et al., 2003).

Finally, it is important to notice that HHV-6 can apparently activate other viral infections, e.g. those induced by EBV, CMV, HIV-1, measles, apparently also papillomavirus and parvovirus. HHV-6 may thus contribute to the pathologic effects of these viruses (Lusso et al., 1992, 1995; Flamand et al., 1993; al-Kaldi et al., 1994; Chen et al., 1994; Ablashi et al., 1995; Como et al., 1998; Singh et al., 1998). Dual active infections appear especially frequently with other herpesviruses (CMV, EBV, HHV-7) as well as with HIV-1.

Conclusions

Acute primary infection with HHV-6 and reactivation with endogenous reinfection may initiate various diseases that deserve medical attention. All cause the usual features of a viral infection without cytopathic effects specific for HHV-6 A or B. Only rarely, intranuclear inclusions (preferentially Cowdry type B) may be seen in tissues or occasional multinucleated giant cell—and in all such cases coincident or other viral infections must be excluded, such as by HHV-7, HIV or others.

The proof of HHV-6 causing a certain disease relies upon showing an active viral infection at the site of the tissue lesion. In addition, serologic testing should be done in combination with virus isolation, antigen-capture ELISA for p41 (follow-up testing in chronic persistent infections). *In situ* hybridization and polymerase chain amplification reaction (PCR) for viral DNA can indicate an increased viral load, yet not necessarily viral activity in the tissues. HHV-6 antigen indicative of viral replication (p41) should be shown by immunohistochemical techniques in tissues samples (biopsies or at autopsy) with suspected HHV-6 disease. Electron microscopy can be helpful in showing herpesvirus particles in diseased tissues; the type of herpesvirus must subsequently be confirmed by serological or molecular techniques.

References

Ablashi D, Agut H, Berneman Z, Campadelli-Fiume G, Carrigan D, Ceccerini-Nelli L, Chandran B, Chou S, Collandre H, Cone R, Dambaught T, Dewhurst S, DiLuca D, Foà-Tomas L, Fleckenstein B, Frenkel N, Gallo R, Gompels U, Hall C, Jones M, Lawrence G, Martin M, Montagnier L, Neipel F, Nicholas J, Pellett P, Razzaque A, Torrelli G, Thomson B, Salahuddin S, Wyatt L, Yamanishi K. Arch Virol 1993; 129: 363–366.

Ablashi DV, Balachandran N, Josephs SF, Hung CL, Krueger GR, Kramarsky B, Salahuddin SZ, Gallo RC. Virology 1991; 184: 545–552.
Ablashi DV, Bernbaum J, DiPaolo JA. Trends Immunobiol 1995; 3: 324–326.
Ablashi DV, Hung CL. *In vitro* propagation systems for HHV-6. In: Human Herpesvirus-6. Chapter 4 (Ablashi DV, Krueger GRF, Salahuddin SZ, editors). Amsterdam: Elsevier; 1992; pp. 37–47.
Ablashi DV, Krueger GRF, Salahuddin SZ. Human Herpesvirus-6. Amsterdam: Elsevier 1992.
Aita K, Jin Y, Irie H, Takahashi I, Kobori K, Nakasato Y, Kodama H, Yanagawa Y, Yoshikawa T, Shiga J. Hum Pathol 2001; 32: 887–889.
Akashi K, Eizuru Y, Sumiyoshi Y, Minematsu T, Hara S, Harada M, Kikuchi M, Niho Y, Minamishima Y. New Engl J Med 1993; 329: 168–171.
al-Kaldi N, Watson AR, Harris A, Irving WL. Pediatr Nephrol 1994; 8: 349–350.
Amo K, Tanaka-Taya K, Inagi R, Miyagawa H, Miyoshi H, Okusu I, Sashihara J, Hara J, Nakayama M, Yamanishi K, Okada S. Clin Infect Dis 2003; 36: 120–123.
An SF, Groves M, Martinian L, Kuo LT, Scaravilli F. J Neurovirol 2002; 8: 439–446.
Arena A, Stassi G, Speranza A, Iannello D, Mastroeni P. New Microbiol 2002; 25: 335–340.
Asano Y, Yoshikawa T, Suga S, Kobayashi I, Nakashima T, Yazaki T, Kajita Y, Ozaki T. Pediatrics 1994; 93: 104–108.
Bandobashi K, Daibata M, Kamioka M, Tanaka Y, Kubonishi I, Taguchi H, Ohtsuki Y, Miyoshi I. Blood 1997; 90: 1200–1207.
Bertram G, Dreiner N, Krueger GR, Ramon A, Ablashi DV, Salahuddin SZ, Balachandram N. In vivo 1991; 5: 271–279.
Biberfeld P, Kramarsky B, Salahuddin SZ, Gallo RC. J Natl Cancer Inst 1987; 79: 933–941.
Boehmer S. Biological effect of human herpesvirus-6 on cell populations of different phenotype. Thesis at University of Cologne Medical School, Hundt Druck, Cologne 1987: 1–219.
Bonthius DJ, Karacay B. Neurol Clin 2002; 20: 1013–1038.
Boutolleau D, Fernandez C, André E, Imbert-Marcille BM, Milpied N, Agut H, Gautheret-Dejean A. Human herpesvirus (HHV)-6 and HHV-7. J Infect Dis 2003; 187: 179–186.
Braun DK, Dominguez G, Pellett PE. Clin Microbiol Rev 1997; 10: 521–567.
Braun DK, Pellett PE, Hanson CA. J Infect Dis 1995; 171: 1351–1355.
Campadelli-Fiume G, Mirandola P, Menotti L. Emerg Infect Dis 1999; 5: 353–366.
Carrigan DR, Knox KK. Blood 1994; 84: 3307–3310.
Caserta MT, Hall CB, Schnabel K, McIntire K, Long C, Costanzo M, Dewhurst S, Insel R, Epstein LG. J Infect Dis 1994; 170: 1586–1589.
Challoner PB, Smith KT, Parker JD, Mac-Leod DL, Coulter SN, Rose TM, Schultz ER, Bennett JL, Garber RL, Chang M, Schad PA, Stewart PM, Nowinski RC, Brown JP, Burmer GC. Proc Natl Acad Sci USA 1995; 92: 7440–7444.
Chapenko S, Millers A, Nora Z, Logina I, Kukaine R, Murovska M. Correlation between HHV-6 reactivation and multiple sclerosis disease activity. J Med Virol 2003; 69: 111–117.
Chen M, Wang H, Woodworth CD, Lusso P, Berneman Z, Kingsma D, Delgado G, DiPaolo JA. Am J Pathol 1994; 145: 1509–1516.
Cirone M, Cuomo L, Zompetta C, Ruggieri S, Frati L, Faggioni A, Rago G. J Med Virol 2002; 68: 268–272.
Clark DA. Rev Med Virol 2000; 10: 155–173.
Clark DA. Int J Hematol 2002; 76(suppl 2): 246–252.
Clark DA, Emery VC, Griffith PD. Semin Hematol 2003; 40: 154–162.

Como L, Trivedi P, deGarzia U, Colagero A, D'Onofrio M, Yang W, Frati L, Faggioni A, Rymo L, Ragona G. J Med Virol 1998; 55: 219–226.
Cone RW, Hackman RC, Meei-Li WH, Bowden RA, Meyers JD, Metcalf M, Zeh J, Ashley R, Corey L. New Engl J Med 1993; 329: 156–161.
Cone RW, Huang ML, Hackman RC, Corey L. J Clin Microbiol 1996; 34: 877–881.
Conilleau V, Dompmartin A, Verneuil L, Michel M, Leroy D. Contact Dermatitis 1999; 41: 141–144.
Cuomo L, Trivedi P, Cardillo MR, Gagliardi FM, Vecchione A, Caruso R, Calogero A, Frati F, Faggioni A, Ragona G. J Med Virol 2001; 63: 45–51.
Daibata M, Ido E, Murakami K, Kuzume T, Kubonishi I, Taguchi H, Miyoshi I. Leukemia 1997; 11: 882–885.
De Clerck E, Bourgeois N, Krueger GRF, Stevens WJ. Human herpesvirus-6 in Sjoegren's syndrome. In: Human Herpesvirus-6 (Ablashi DV, Krueger GRF, Salahuddin SZ, editors). Amsterdam: Elsevier; 1992; pp. 303–315.
Deliconstantinos G, Buja LM, Rojo J, Krueger GRF. Rev Med Hosp Gen Mexico 1998; 61: 211–217.
Deliconstantinos G, Krueger GRF. J Viral Dis 1993; 1: 22–27.
Descamps V, Bouscarat F, Laglenne S, Aslangul E, Veber B, Descamps D, Saraux JL, Grange MJ, Grossin M, Navratil E, Crickx B, Belaich S. Br J Dermatol 1997; 137: 605–608.
Descamps V, Valance A, Edlinger C, Fillet AM, Grossin M, Lebrun-Vignes B, Belaich S, Crickx B. Arch Dermatol 2001; 137: 301–304.
Di Luca D, Dolcetti R, Mirandola P, De Re V, Secchiero P, Carbone A, Boiocchi M, Cassai E. J Infect Dis 1994; 170: 211–215.
Dominguez DC, Torres ML, Antony S. South Med J 2003; 96: 226–233.
Donati D, Martinelli E, Cassiani-Ingoni R, Ahlqvist J, Hou J, Major EO, Jacobson S. J Virol 2005; 79: 9439–9448.
Downing RG, Sewankambo N, Serwadda D, Honess R, Crawford D, Jarrett R, Griffin BE. Lancet 1987; ii: 390.
Drago F, Rampini P, Brusati C, Rebora A. Acta Derm Venereol (Stockh) 1999; 80: 68.
Eeg-Olofsson O. Brain Dev 2003; 25: 9–13.
Feldstein AE, Razonable RR, Boyce TG, Freese DK, El-Youssef M, Perrault J, Paya CV, Ishitani MB. Pediatr Transplant 2003; 7: 125–129.
Flamand L, Gosselin J, Stefanescu I, Ablashi D, Menezes J. Blood 1995; 85: 1263–1271.
Flamand L, Stefanescu L, Ablashi DV, Menezes J. J Virol 1993; 67: 6768–6777.
Fox JD, Briggs M, Ward PA, Tedder RS. Lancet 1990; 336: 590–593.
Fukae S, Ashizawa N, Morikawa S, Yano K. Intern Med 2000; 39: 632–636.
Gies M, Jalali Z, Wagner M, Krueger GRF. Rev Med Hosp Gen Mexico 1998; 61: 218–225.
Greenstone HL, Santoro F, Lusso P, Berger FA. J Biol Chem 2002; 277: 39112–39118.
Hagiwara K, Yoshida T, Komura H, Kishi F, Kajii T. Eur J Pediatr 1993; 152: 176.
Hall CB, Caserta MT, Schnabel KC, Long C, Epstein LG, Insel RA, Dewhurst S. Clin Infect Dis 1998; 26: 132–137.
Hall CB, Long CE, Schnabel KC, Caserta MT, McIntyre KM, Costanzo MA, Knott A, Dewhurst S, Insel RA, Epstein LG. New Engl J Med 1994; 331: 432–438.
Hallas C, Neipel F, Huettner C, Schreiner D, Fleckenstein B, Muller-Hermelink HK. Diagn Mol Pathol 1996; 5: 166–172.
Hasegawa A, Yasukawa M, Sakai I, Fujita S. J Immunol 2001; 166: 1125–1131.

Herbein G, Strasswimmer J, Altieri M, Woehl-Jaegle ML, Wolf P, Obert G. Clin Infect Dis 1996; 22: 171–173.

Hermouet S, Sutton CA, Rose TM, Greenblatt RJ, Corre I, Garand R, Neves AM, Bataille R, Casey JW. Leukemia 2003; 17: 185–195.

Horwitz CA, Krueger GRF, Steeper TA, Bertram G. HHV-6 induced mononucleosis-like illnesses. In: Human Herpesvirus-6 Chapter 13 (Ablashi DV, Krueger GRF, Salahuddin SZ, editors). Amsterdam: Elsevier; 1992; pp. 159–174.

Ihara M, Yoshikawa T, Suzuki K, Ohashi M, Suga S, Horibe K, Tanaka S, Kimura H, Kojima S, Kato K, Matsuyama T, Nishiyama Y, Asano Y. Microbiol Immunol 2002; 46: 701–705.

Inoue N, Dambaugh TR, Pellett PE. Infect Agents Dis 1994; 2: 343–360.

Ishikawa K, Hasegawa K, Naritomi T, Kanai N, Ogawa M, Kato Y, Kobayashi M, Torii N, Hayashi N. J Gastroenterol 2002; 37: 523–530.

Isomura H, Yamada M, Yoshida M, Tanaka H, Kitamura T, Oda M, Nii S, Seino Y. J Med Virol 1997; 52: 406–412.

Kaden J, May G, Wagner M, Krueger GRF. Transplantationsmedizin 1997; 9: 28–32.

Kakimoto M, Hasegawa A, Fujita S, Yasukawa M. J Virol 2002; 76: 10338–10345.

Kikuchi M, Sumiyoshe Y, Minamishima Y. Kikuchi's disease (histiocytic necrotizing lymphadenitis). In: Human Herpesvirus-6 Chapter 14 (Ablashi DV, Krueger GRF, Salahuddin SZ, editors). Amsterdam: Elsevier; 1992; pp. 175–183.

Kirn E, Krueger E, Boehmer S, Klussmann JP, Krueger GRF. Anticancer Res 1997; 17: 4623–4632.

Klueppelberg U. Der systemische Lupus erythematodes(SLE): Uebersicht und eigene Arbeiten zur Herpesvirus-6 Infektion. Hund, Cologne 1994: 1–95.

Knox KK, Carrigan DR. J Acquir Immune Defic Syndr Hum Retrovirol 1995; 9: 66–73.

Know KK, Pietryga D, Harrington DJ, Franciosi R, Carrigan DR. Clin Infect Dis 1995; 20: 406–413.

Kramarsky B, Sander C. Electron microscopy of human herpesvirus-6 (HHV-6). In: Human Herpesvirus-6 Chapter 6 (Ablashi DV, Krueger GRF, Salahuddin SZ, editors). Amsterdam: Elsevier; 1992; pp. 59–68.

Krueger GR, Ferrer Argote V. In vivo 1994a; 8: 493–499.

Krueger GR, Klueppelberg U, Hoffmann A, Ablashi DV. In vivo 1994a; 8: 457–485.

Krueger GR, Manak M, Bourgeois N, Ablashi DV, Salahuddin SZ, Josephs SS, Buchbinder A, Gallo RC, Berthold F, Tesch H. Anticancer Res 1989; 9: 1457–1476.

Krueger GR, Wassermann K, De Clerck LS, Stevens WJ, Bourgeois N, Ablashi DV, Josephs SF, Balachandran N. Lancet 1990; 336: 1255–1256.

Krueger GRF, Ablashi DV. Intervirology 2003; 46: 257–269.

Krueger GRF, Ablashi DV, Whitman JI, Luka J, Rojo J. Rev Med Hosp Gen Mexico 1998; 61: 226–240.

Krueger GRF, Brandt ME, Wang G, Berthold F, Buja LM. Anticancer Res 2002; 22: 2365–2372.

Krueger GRF, Buja LM, Rojo J, Lasch J, Koch B, Leyssens N. Pathologe 1995; 16: 120–127.

Krueger GRF, Koch B, Hoffmann A, Rojo J, Brandt ME, Wang G, Buja LM. In vivo 2001; 15: 461–466.

Krueger GRF, Koch B, Ramon A, Ablashi DV, Salahuddin SZ, Josephs SF, Styreicher HZ, Gallo RC, Habermann U. J Virol Methods 1988; 21: 125–131.

Krueger GRF, Kudlimay D, Ramon A, Klueppelberg U, Schumacher K. In vivo 1994b; 8: 533–542.

Krueger GRF, Sander C, Hoffmann A, Barth A, Koch B, Braun M. In vivo 1991; 5: 217–226.
Krueger GRF, Schonnebeck M, Braun M. FASEB J 1990; 4: A343.
Krueger GRF, Sievert J, Juecker M, Tesch H, Diehl V, Ablashi DV, Balachandran N, Luka J. J Viral Dis 1992; 1: 15–23.
Lasch JA, Klussmann JP, Krueger GR. Hautarzt 1996; 47: 341–350.
Lau YL, Peiris M, Chan GC, Chan AC, Chiu D, Ha SY. Bone Marrow Transplant 1998; 21: 1063–1066.
Leahy MA, Krejci SM, Friednash M, Stockert SS, Wilson H, Huff JC, Weston WL, Brice SL. J Invest Dermatol 1993; 101: 642–645.
Lin WC, Moore JO, Mann KP, Traweek ST, Smith C. Leuk Lymphoma 1999; 33: 377–384.
Lopez C, Pellett P, Stewart J, Goldsmith C, Sanderlin K, Black J, Warfield D, Feorino P. J Infect Dis 1988; 157: 1271–1273.
Luka J, Gubin J, Afflerbach C, Carson SD, Krueger GRF. Detection of human herpesvirus-6 (HHV-6) genome by *in situ* PCR in tissues from Kawasaki disease and in coronary arteries of transplant hearts (abstract). 3rd International Conference on Modern Methods in Analytical Morphology, Atlantic City; June 1995.
Luppi M, Barozzi P, Garber R, Maiorana A, Bonacorsi G, Artusi T, Trovato R, Marasca R, Torelli G. Am J Pathol 1998; 153: 815–823.
Luppi M, Barozzi P, Maiorana A, Marasca R, Trovato R, Fano R, Ceccherini-Nelli L, Torelli G. J Med Virol 1995; 47: 105–111.
Luppi M, Marasca R, Barozzi P, Artusi T, Torelli G. Leuk Res 1993; 17: 1003–1011.
Lusso P. Target cells for infection. In: Human Herpesvirus-6 (Ablashi DV, Krueger GRF, Salahuddin SZ, editors). Amsterdam: Elsevier; 1992; pp. 25–47.
Lusso P, Ablashi DV, Luka J. Interaction between HHV-6 and other viruses. In: Human herpesvirus-6 (Ablashi DV, Krueger GRF, Salahuddin SZ, editors). Amsterdam: Elsevier; 1992; pp. 121–133.
Lusso P, Garzino-Demo A, Crowley RW, Malnati MS. J Exp Med 1995; 181: 1303–1310.
Maric I, Bryant R, Abu-Asab M, Cohen JI, Vivero A, Jaffe ES, Raffeld M, Tsokos T, Banks PM, Pittaluga S. Mod Pathol 2004; 17: 1427–1433.
Mori Y, Yang X, Akkapaiboon P, Okuno T, Yamanishi K. J Virol 2003; 77: 4992–4999.
Penchansky L, Jordan JA. Am J Clin Pathol 1997; 108: 127–132.
Prezioso PJ, Cangiarella J, Lee M, Nuovo GJ, Borkowsky W, Orlow SJ, Greco MA. J Pediatr 1992; 120: 921–923.
Pruitt AA. Neurol Clin 2003; 21: 193–219.
Razzaque A. Oncogene 1990; 5: 1365–1370.
Razzaque A, Francillon Y, Jilly PN, Varricchio F. Cancer Lett 1996; 106: 221–226.
Rojo J, Cruz-Ortiz H, Krueger GRF. Rev Med Hosp Gen Mexico 2000; 63: 18–24.
Rosenfeld CS, Rybka WB, Weinbaum D, Carrigan DR, Knox KK, Andrews DF, Shadduck RK. Exp Hematol 1995; 23: 626–629.
Rotola A, DiLuca D, Cassai E, Giulio A, Turano A, Caruso A, Muneretto C. J Clin Microbiol 2000; 38: 3135–3136.
Salahuddin SZ, Ablashi DV, Markham PD, Josephs SF, Sturzenegger S, Kaplan M, Halligan G, Biberfeld P, Wong-Staal F, Kramarsky B, Gallo RC. Science 1986; 234: 596–601.
Salahuddin SZ, Kelley AS, Krueger GRF, Josephs SF, Gupta S, Ablashi DV. Clin Diag Virol 1993; 1: 81–100.

Santoro F, Kennedy PE, Locatelli G, Malnati MS, Berger EA, Lusso P. Cell 1999; 99: 817–827.
Sashihara J, Tanaka-Taya K, Tanaka S, Amo K, Miyagawa H, Hosoi G, Tanuguchi T, Fukui T, Kasuga N, Aono T, Sako M, Hara J, Yamanishi R, Okada S. Blood 2002; 100: 2005–2011.
Schonnebeck M, Krueger GRF, Brun M, Fischer M, Koch B, Ablashi DV, Balachandran N. In vivo 1991; 5: 255–264.
Singh N, Carrigan DR. Ann Intern Med 1996; 124: 1065–1071.
Singh VK, Lin SX, Yang VC. Clin Immunol Immunopathol 1998; 89: 105–108.
Solldan SS, Leist TP, Juhng KN, McFarland HF, Jacobson S. Ann Neurol 2000; 47: 306–313.
Steeper TA, Horwitz CA, Ablashi DV, Salahuddin SZ, Saxinger C, Saltzman R, Schwartz B. Am J Clin Pathol 1990; 93: 776–783.
Sugita K, Kurumada H, Eguchi M, Furukawa T. Acta Haematol 1995; 93: 108–109.
Sumiyoshi Y, Kikuchi M, Ohshima K, Yoneda S, Kobari S, Takeshita M, Eizuru Y, Minamishima Y. Am J Clin Pathol 1993; 99: 609–614.
Suzuki Y, Inagi R, Aono T, Yamanishi K, Shiohara T. Arch Dermatol 1998; 134: 1108–1112.
Takatsuka H, Wakae T, Mori A, Okada M, Fujimori Y, Takemoto Y, Okamoto T, Kanamaru A, Kakishita E. Bone Marrow Transplant 2003; 31: 475–479.
Tanaka H, Nishimura T, Hakui M, Sugimoto H, Tanaka-Taya K, Yamanishi K. Emerg Infect Dis 2000; 8: 87–88.
Taniguchi T, Shimamoto T, Isegawa Y, Kondo K, Yamanishi K. Virology 2000; 271: 307–320.
Tejada-Simon MV, Zang YC, Hong J, Rivera VM, Zhang JZ. Ann Neurol 2003; 53: 189–197.
Thompson J, Choudhury S, Kashanchi F, Doniger J, Berneman Z, Frenkel N, Rosenthal LJ. Oncogene 1994; 9: 1167–1175.
Tohyama M, Yahata Y, Yasukawa M, Inagi R, Urano Y, Yamanishi K, Hashimoto K. Arch Dermatol 1998; 134: 1113–1117.
Torelli G, Marasca R, Luppi M, Selleri L, Ferrari S, Narni F, Mariano MT, Federico M, Ceccherini-Nelli L, Bendinelli M, Montagnani G, Montorsi M, Artusi T. Blood 1991; 77: 2251–2258.
Toyabe S, Harada W, Suzuki H, Hirokawa T, Uchiyama M. Clin Rheumatol 2002; 21: 528–532.
Wagner M, Müller-Berghaus J, Schroeder R, Sollberg S, Luka J, Leyssens N, Schneider B, Krueger GRF. J Med Virol 1997; 53: 306–312.
Wang FZ, Linde A, Dahl H, Ljungman P. Bone Marrow Transplant 1999; 24: 1201–1206.
Wiersbitzky S, Ratzmann GW, Bruns R, Wiersbitzky H. Pädiatr Grenzgeb 1993; 31: 203–205.
Wilborn F, Schmidt CA, Brinkmann V, Jendroschka K, Ottle H, Siegert W. J Neuroimmunol 1994; 49: 213–214.
Wu CA, Shanley JD. J Gen Virol 1998; 79: 1247–1256.
Yamanishi K, Okuno T, Shiraki K, Takahashi M, Kondo T, Asano Y, Kurata T. Lancet 1988; i: 1065–1067.
Yoshikawa T. Pediatr Transplant 2003; 7: 11–17.
Yoshikawa T, Asano Y. Brain Dev 2000; 22: 307–314.
Yoshikawa T, Akimoto S, Nishimura N, Ozaki T, Ihira M, Ohashi M, Morooka M, Suga S, Asano Y, Takemoto M, Nishiyama Y. J Med Virol 2003; 70: 267–272.

Human Herpesvirus-6: General Virology, Epidemiology and Clinical Pathology
Perspectives in Medical Virology
Gerhard Krueger and Dharam Ablashi (Editors)

DOI 10.1016/S0168-7069(06)12011-X

Systemic Reactions to HHV-6 Infection

Vincent Descamps[a], Emmanuel Mahe[a], Sylvie Ranger-Rogez[b]
[a]*Department of Dermatology, Bichat Claude Bernard Hospital, Assistance Publique-Hôpitaux de Paris 46 rue Henri Huchard, 75018 Paris, France*
[b]*Department of Virology, Equipe d'accueil N3839, Dupuytren Hospital, Limoges, France*

Introduction

Drug-induced hypersensitivity syndrome illustrates the systemic reactions to human herpesvirus 6 (HHV-6). Drug induced hypersensitivity syndrome is a severe drug-induced reaction (Roujeau and Stern, 1994). Its clinical and biological manifestations are well characterized. They are associated with high fever, facial edema, erythroderma followed by an exfoliative dermatitis, diffuse lymphadenopathy, hypereosinophilia, atypical circulating lymphocytes, and abnormal results of liver function tests. But numerous other systemic manifestations may occur including pneumonitis, myocarditis, pancreatitis, nephritis, thyroiditis, and neurological symptoms. This drug adverse reaction was first described with anticonvulsant as the anticonvulsant hypersensitivity syndrome, but other drugs (including allopurinol, minocycline, dapsone, sulfonamide antibiotics, nevirapine, abacavir, etc.) may induce this reaction. These manifestations mimic viral infection. We and others have suggested that HHV-6 infection could be associated with drug-induced hypersensitivity syndrome. Accumulation of data by several groups have provided many arguments to document that HHV-6 was implicated in systemic symptoms in drug-induced hypersensitivity syndrome (Hashimoto et al., 2003). This syndrome could be considered as a paradigm of severe viral infection and systemic reactions due to HHV-6.

We discuss in this review the clinical spectrum of the drug-induced hypersensitivity syndrome, the putative role of viral infection, especially HHV-6 infection, its pathophysiology, and the clinical and therapeutic consequences of the implication of HHV-6 in this severe disease.

From the anticonvulsant hypersensitivity syndrome to the DRESS

The systemic manifestations of the anticonvulsant syndrome were first described by Chaiken (Chaiken et al., 1950) with dilantin. The authors reported a case of hepatitis with jaundice, fever, and exfoliative dermatitis. The phenytoin syndrome was later described in 1979 by Haruda (1979). Rashes were the most frequent manifestation, followed by fever, pharyngitis, lymphadenopathy, eosinophilia, hepatitis, and hematological abnormalities. It was observed that this adverse reaction was delayed in onset after drug initiation. This reaction was then reported with phenobarbital and carbamazepine (McGeachy and Bloomer, 1953; Pellock, 1987; Shear and Spielberg, 1988).

The diagnosis of this syndrome may be confused with infection, lymphoma, hypereosinophilic syndrome, and collagen vascular disease. Histological examination of skin lesions showed lymphocytic infiltration in the dermis and sometimes the epidermis (epitheliotropism). The infiltrate may resemble the infiltrate observed in the mycosis fungoides (cutaneous T-cell lymphoma): dense band-like infiltrate in the epidermis with atypical lymphocytes. Regarding this lymphocytic infiltration, the biological hematological abnormalities and the clinical lymphadenopathy, pseudolymphoma denomination was also alternatively used for the same patients. The case record of the Massachusetts general hospital reported in 1996 in the New England Journal of Medicine illustrated this discussion (Gorlin and Ferry, 1996). A 7-year-old child was admitted for the association of fever, lymphadenopathy, hepatosplenomegaly, and eosinophilia. This boy had been receiving carbamazepine and phenytoin for 2 months after a seizure. Five days after admission, atypical lymphocytes were detected on blood analysis. Many diagnoses were evoked including hematological malignancies, lupus, infection, and hypersensitivity syndrome akin to the pseudolymphoma syndrome after anticonvulsant administration.

Callot et al. (1996) helped to define in a retrospective study, the two distinct entities, e.g. drug-induced pseudolymphoma and hypersensitivity syndrome. The eruption in pseudolymphoma was composed of firm erythematous papulonodules or squamous large plaques. No systemic symptoms or visceral involvement were observed. In drug hypersensitivity syndrome, the cutaneous lesions were a widespread maculopapular rash with facial edema often with an exfoliative dermatitis. Fever, biological abnormalities (lymphocytosis, atypical lymphocytes, eosinophilia), and systemic symptoms were often observed (80%).

Some authors defined the drug hypersensitivity syndrome as the triad of fever, rash, and internal organ involvement due to drug exposure (Wong et al., 2004). Recently, the acronym for drug reaction with eosinophilia and systemic symptoms

Table 1

DRESS criteria (Begon and Roujeau, 2004)

1. Drug exposure
2. Eosinophilia > 1500 ml^{-1} and/or atypical circulating lymphocytes
3. Visceral manifestations (two or more):
a. Skin eruption
b. Lymph nodes enlargement (> 2 cm)
c. Hepatitis (alanine aminotransferase > 2 × normal values)
d. Interstitial pneumopathy
e. Nephritis
f. Pericarditis or myocarditis

(DRESS) was proposed to describe this syndrome. New criteria have been reported considering cutaneous involvement as one of the systemic symptoms (Begon and Roujeau, 2004). DRESS could occur without any skin manifestations. It needs a consensus agreement (Table 1) (Bocquet et al., 1996).

DRESS is an uncommon adverse drug reaction (1/1000 or 10,000 exposures). This syndrome is severe and is potentially life threatening. DRESS has also some special characteristics as compared to other cutaneous drug reactions. The incubation period is longer than other classical cutaneous adverse drug reactions: 2–8 weeks after initiation of drug administration. Only few drugs have been incriminated in DRESS. The syndrome is similar for each drug. No sex or age predilection has been reported. Some risk factors are suspected or identified including a personal or familial history of DRESS, ethnicity (Africans and Afro-Americans) (Begon and Roujeau, 2004). A higher prevalence of DRESS in HIV patients and immunosuppressed patients is suspected.

Clinical spectrum and systemic manifestations

Usually, high and spiking fever begins this syndrome (Knowles et al., 1999; Begon and Roujeau, 2004). Fever precedes or is concomitant to the onset of the cutaneous manifestations. Fever and skin eruption are the most common clinical manifestations. Skin manifestations are usually a maculopapular exanthema with facial edema (periorbital), which may progress to an exfoliative dermatitis (erythroderma) (Fig. 1). Sometimes pustulosis, blistering, or oral ulceration may occur. Rarely, more serious conditions mimicking Stevens–Johnson syndrome or toxic epidermal necrolysis may develop. But these other serious adverse cutaneous reactions must be considered as a differential diagnosis. Involvement of mucous membranes is uncommon in DRESS. There is no correlation between the severity of cutaneous involvement and the systemic symptoms. Hepatitis involvement is frequently asymptomatic (cytolytic hepatitis or cholestasis) but jaundice, hepatic failure, and fulminant hepatitis may occur. Other systemic involvements include nephritis, pancreatitis, pneumonitis, colitis, myocarditis, myositis, meningoencephalitis, parodititis, orchitis,

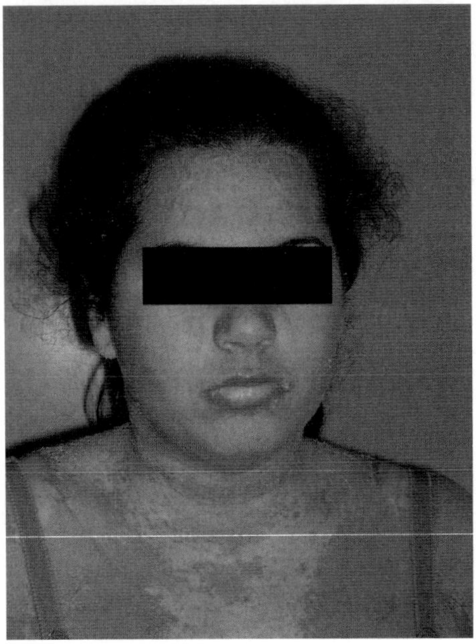

Fig. 1 Facial edema and an exfoliative dermatitis in DRESS. (for colour version: see colour section on page 352).

and thyroiditis (Hogg et al., 1981; Engel et al., 1986; Gupta et al., 1992; Knowles et al., 1999; Sekine et al., 2001; Fujino et al., 2002). Clinical hematological involvement is observed in lymphadenopathy and hepatosplenomegaly. Biological hematological abnormalities include hypereosinophilia, atypical lymphocytes, neutropenia, thrombocytopenia, hemolytic anemia, pancytopenia, and hemophagocytic syndrome.

These clinical and biological manifestations develop and may persist and progress for weeks after the offending drug is discontinued. The course is variable but it may frequently reappear with flare-ups before a complete resolution. Reexposure to the culprit drug will induce a new episode.

Fatal outcome due to fulminant hepatitis (Huang et al., 2003), myocarditis (Parneix-Spake et al., 1995), nephritis, or multiorgan failure (Mahe et al., 2004) has been reported.

DRESS and HHV-6: the first reports

DRESS mimics viral infection. Interestingly, all of the clinical and biological manifestations described in the DRESS are observed in some viral infection and especially HHV-6 infection. In 1993, Akashi et al. reported a severe infectious mononucleosis-like syndrome and primary HHV-6 infection in an adult (Akashi et al., 1993). A 43-year-old man was admitted with high fever, generalized exanthematic eruption followed by an exfoliative dermatitis, lymphadenopathy, atypical lymphocytes, hepatitis, and renal dysfunction. Lymphocyte population was mainly T-cell lymphocytes (52.6% CD8). CD19 and CD20 lymphocytes' levels were very low, 0.8 and 0.6%, respectively. The skin biopsy analysis demonstrated a diffuse infiltration with atypical lymphocytes in the dermis. HHV-6 infection was demonstrated by serial changes in titers of antibody against HHV-6 associated with an HHV-6 viremia. HHV-6 DNA was demonstrated in serum samples collected on days 10 and 13.

We had the opportunity to report the first case of DRESS associated with HHV-6 infection (Descamps et al., 1997). This DRESS was induced by phenobarbital in a 25-year-old Laotien woman. This patient fulfilled the characteristics of the DRESS (an exfoliative dermatitis with edema of the face, fever, lymphadenopathy, atypical circulating lymphocytes, eosinophilia, hepatic failure, and lymphocytic infiltrates in skin biopsy). This DRESS was complicated by a hemophagocytic syndrome. In 1997, HHV-6 has not been implicated in drug reaction. But viral infections were thought to play a role in some cutaneous drug reactions: the well-known ampicillin-induced exanthema in the Epstein–Barr virus (EBV)-mononucleosis syndrome illustrated this. Moreover, in this case, the association with a hemophagocytic syndrome was remarkable and questioning. Fulminant form of virus-associated hemophagocytic syndrome had been reported in Asia from where our patient originated with HHV-6 or EBV (Chen et al., 1995). Therefore, we investigated in this case concomitant viral infections and especially herpesvirus infections. An active HHV-6 infection was demonstrated on a rise in the anti-HHV-6 IgG antibodies titers in four consecutive sera. But PCR analysis for HHV-6 DNA in serum samples was negative. Other viral investigations (EBV, HIV) were negative. Two years later, two other groups reported three cases of DRESS caused by sulfasalazine (two cases) and allopurinol (one case) (Suzuki et al., 1998; Tohyama et al., 1998). Tohyama demonstrated an increase in the anti-HHV-6 IgG antibody titer with the detection of HHV-6 DNA from peripheral blood mononuclear cells (identified as HHV-6 variant B). No anti-HHV-6 IgM antibody was detected. Interestingly, it was not observed in the sera of four patients who developed cutaneous drug adverse reaction without any symptom of DRESS, an increase of anti-HHV-6 IgG antibodies. They concluded that this association was specific to the DRESS. A new step was reached by Suzuki et al. (1998) who detected in one patient with allopurinol-induced DRESS, HHV-6 genome in pathologic cutaneous lesions. After a rechallenge with allopurinol, the cutaneous eruption was reproduced. HHV-6 genome was found by PCR and *in situ* hybridization procedures in skin biopsies done at the initial acute phase and after a rechallenge. An increase of anti-HHV-6 IgG antibodies was also demonstrated. HHV-7 genome was also found by an *in situ* hybridization procedure at the acute phase of the initial eruption.

The conclusion after these initial reports was that HHV-6 reactivation was associated with DRESS and could participate in the pathogenic process. The sequence of the events (adverse drug reaction, virus reactivation, immune response, systemic symptoms) was not clear.

We prospectively evaluated the prevalence of HHV-6 infection in patients hospitalized in our dermatological department with DRESS (Descamps et al., 2001). Seven consecutive cases were included. All patients were seropositive for anti-HHV-6 IgG antibodies with a dramatic increase in two cases and detection of anti-HHV-6 IgM antibodies in four cases. In one patient studied, HHV-6 genome was detected in lesional skin by PCR procedure. We proposed that HHV-6 might play a role in the development of DRESS in susceptible patients. Many cases have been now reported (Conilleau et al., 1999; Sekine et al., 2001; Kunisaki et al., 2003; Masaki et al., 2003; Zeller et al., 2003; Descamps et al., 2003a,b; Enomoto et al., 2004; Ogihara et al., 2004; Michel et al., 2005).

One of the difficulties in addressing the question of the association of DRESS and HHV-6 infection is the reliability and relevance of virological tests for the diagnosis of HHV-6 active infection. HHV-6 is a ubiquitous "parasite" that is present in a vast majority of the human population. Primary infection occurs within the first 2 years of life and HHV-6 persists in peripheral blood mononuclear cells and many tissues (including salivary glands and brain) (Zerr et al., 2005). HHV-6 causes a chronic persistent or latent infection. It may be also difficult to draw any conclusion about a relationship between the presence of anti-HHV-6 IgG antibodies and a specific disease (Sullivan and Shear, 2001). While first reports of the association of HHV-6 infection and DRESS were based on serological tests with low value, recent reports used relevant real time quantitative PCR procedure in serum or from PBMC (Collot et al., 2002; Descamps et al., 2003b).

The diagnosis of reactivation, *de novo* infection or new infection may be another difficult question to answer. As the majority of adults is seropositive for HHV-6 infection, DRESS in adult seropositive patients is in its large majority clearly associated with reactivation of HHV-6. In children, it may be in some cases difficult to differentiate a primary or a reactivation. Some primary HHV-6 infections were reported in children (Mahe et al., 2004).

HHV-6 reactivation in DRESS: "innocent bystander" or causal agent of systemic symptoms

HHV-6 is perhaps the best example of an opportunistic viral infection that may reactivate in stress condition or immunosuppression. HHV-6 may reactivate in the case of immune dysregulation or deficiency. Selective reactivation of HHV-6 was demonstrated in critically ill immunocompetent patients (Desachy et al., 2001; Razonable et al., 2002). It is becoming recognized as an emerging pathogen among transplant recipients.

We and others wondered whether (1) HHV-6 played a causal role in DRESS and participated in systemic symptoms or (2) HHV-6 reactivation was only a consequence of the immune reaction. Study of the sequence of the events and the detection of HHV-6 in the visceral lesions contributed to answer this question.

Some case reports may help in the knowledge of this association. When we had the opportunity to collect serum samples early, we demonstrated that HHV-6 reactivation and viremia appeared very early in the development of DRESS. In a case of minocycline-induced DRESS we detected very high copies of HHV-6 (16 million copies/ml) on the first serum at the admission (Descamps et al., 2003b). This quantitative PCR was performed on a serum sample drawn at day 6 after the beginning of the DRESS clinical manifestations (fever, rash, lymphadenopathy). Detection of viremia was concomitant of the exanthema and mononucleosis syndrome and preceded the development of hepatitis, eosinophilia, and the increase in anti-HHV-6 IgG antibodies (Fig. 2a). The copy number of HHV-6 in the serum decreased rapidly: it was only 8400 copies/ ml 11 days later. That may explain the reason why this viremia is not always demonstrated when patients are admitted, at a hospital at distance, after the beginning of the DRESS. In another case (ethambutol-induced DRESS) HHV-6 viremia was not demonstrated at admission while an increase of anti-HHV-6 IgG antibodies was demonstrated during the first phase. HHV-6 viremia was only demonstrated at the beginning of a flare of the disease following a rechallenge with the drug (Fig. 2b). HHV-6 reactivation may progress by waves explaining that some authors demonstrated HHV-6 viremia lately in the second or third weeks of the DRESS (Aihara et al., 2003; Hashimoto et al., 2003). Aihara demonstrated HHV-6 DNA in a serum sample at day 32 in a patient with carbamazepine-induced DRESS. At this date, a flare of the DRESS began (Aihara et al., 2003). The precession of HHV-6 viremia upon systemic manifestations argues for the role of HHV-6 infection in the clinical and biological manifestations.

But the best argument of the causal role of HHV-6 in systemic symptoms is the demonstration of the presence of HHV-6 in visceral lesions. We already precised that HHV-6 genome had been detected in skin lesions (Suzuki et al., 1998, Descamps et al., 2001). Recently, we and two other groups reported cases of HHV-6 meningoencephalitis that complicated DRESS (Fujino et al., 2002; Masaki et al., 2003; Descamps et al., 2003a). Diagnosis of HHV-6 meningoencephalitis was made on the detection of HHV-6 genome in the cerebrospinal fluid. As commented by Hashimoto et al. (2003) the case reported by its group (Masaki et al., 2003) is very demonstrative of the link of HHV-6 infection and visceral involvement. An allopurinol-induced DRESS developed in a 51-year-old man, 23 days after initiation of the treatment. Corticosteroid treatment (40 mg/day of prednisolone) induced a rapid control of the symptoms. But reduction of the systemic corticosteroid was followed by the development of encephalitis and hepatitis impairment. HHV-6 DNA was detected in the whole blood sample before the development of encephalitis and in the cerebrospinal fluid. At the same time an increase of anti-HHV-6 IgG antibodies was demonstrated.

Visceral symptoms were reported in cases of DRESS associated with HHV-6 such as hepatitis, pancreatitis, and myocarditis (Sekine et al., 2001; Hashimoto et al., 2003).

But it now needs to demonstrate the presence of HHV-6 RNA or viral proteins by RT-PCR or immuno-histochemical analysis, respectively, in pathological organs (for instance, in liver in patients with hepatic failure in DRESS).

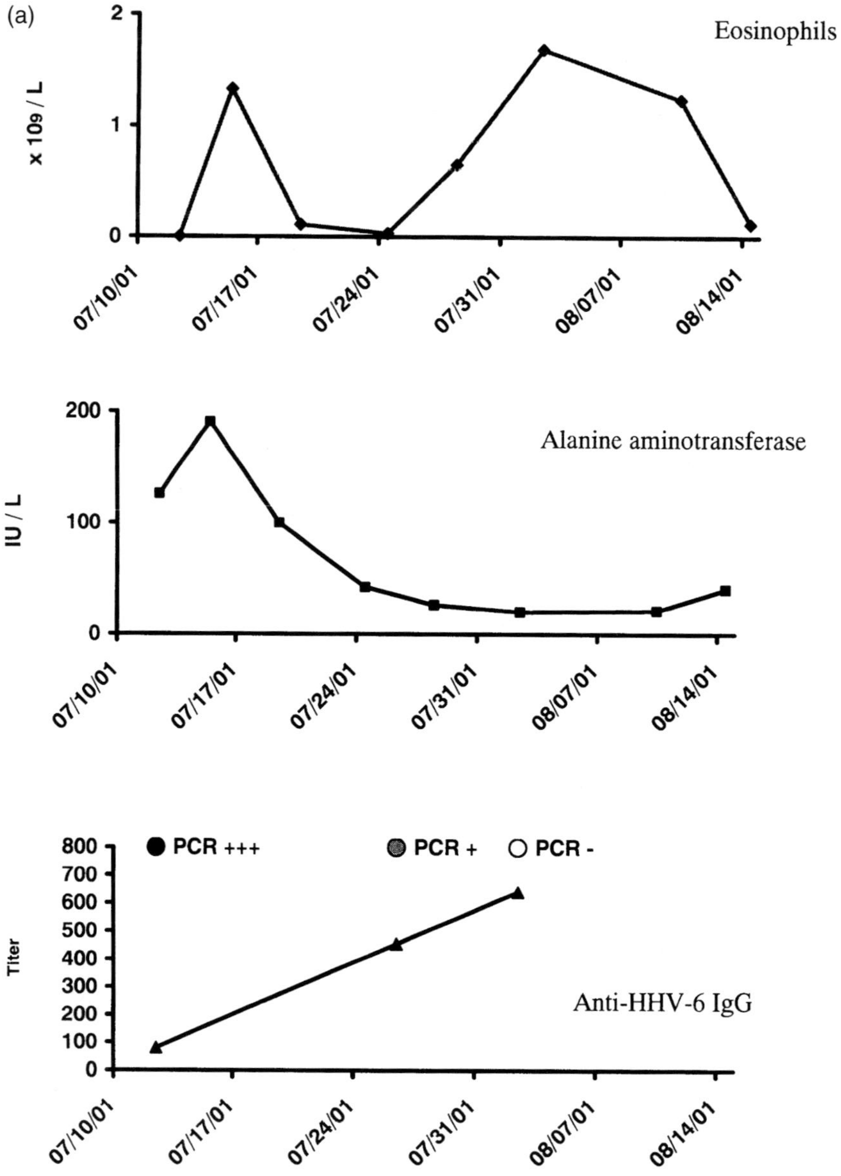

Fig. 2 Course of biological and virological parameters in two patients with DRESS.

HHV-6: a helper virus for other herpesviruses in DRESS?

HHV-6 reactivation may be associated in DRESS with other viral infections. Other herpesviruses such as Epstein–Barr virus, HHV-7, or CMV have been reported in association with HHV-6 reactivation (Suzuki et al., 1998; Aihara et al., 2001;

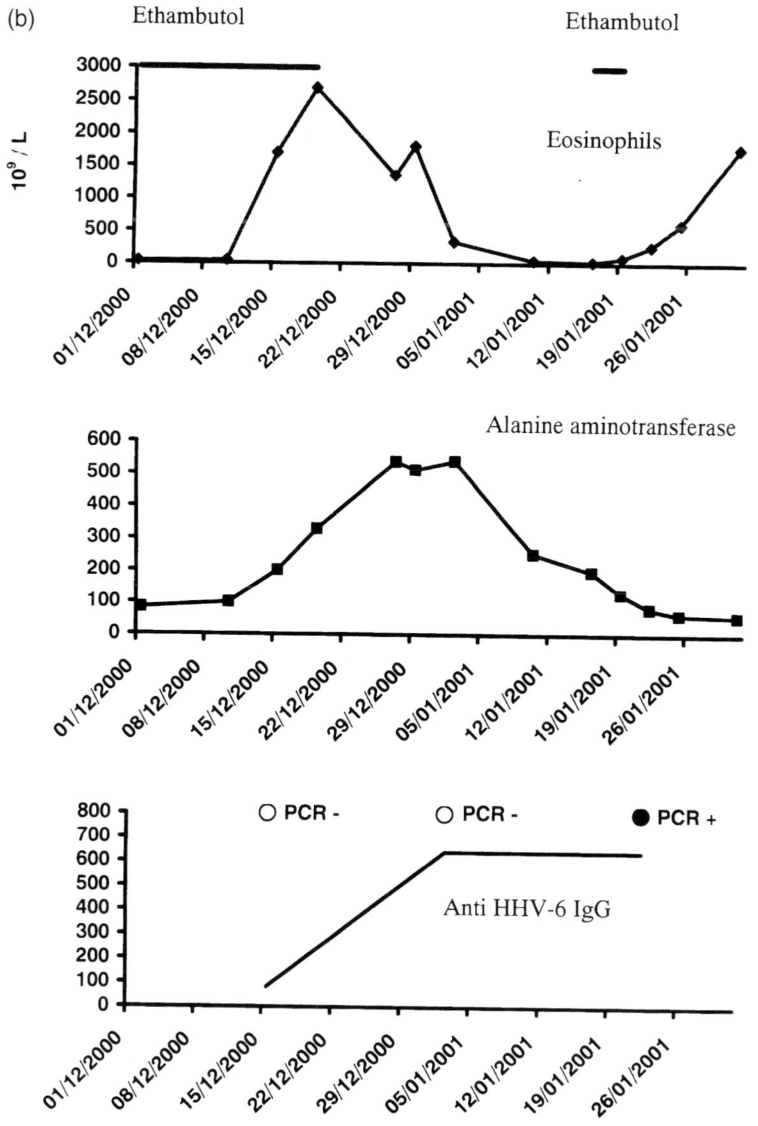

Fig. 2 (*Continued*)

Descamps et al., 2003c; Mahe et al., 2004; Sekiguchi et al., 2005). We reported a case of severe allopurinol-induced DRESS with pancreatitis associated with EBV infection (Descamps et al., 2003c). An active EBV infection was demonstrated by the detection of EBV DNA in PBMC and a seroconversion in two consecutive sera. In two recent reports this HHV-6 and EBV coinfection was fatal (Mahe et al., 2004; Sekiguchi et al., 2005). Sequential reactivation of HHV could explain the observed flare-ups of this disease (Kano and Shiohara, 2004). Independently of the DRESS, this association of viral coinfection has been observed in transplant recipients (Desjardin et al., 2001). Moreover, sequential reactivation of herpesviruses has been demonstrated after bone marrow transplantation (Maeda et al., 2000).

Pathophysiology of HHV-6 reactivation in DRESS

The mechanism by which HHV-6 is reactivated and participated in the development of DRESS is not yet understood.

Many factors are probably implicated in DRESS as recently summarized by Wong and Shear (2004): drug exposure, genetic predisposition, drug interactions, concomitant illness, host immune response with generation of drug-specific T cells, cytokines, transient hypogammaglobulinemia, reactivation of latent viral infection, viral infection, etc.

A genetic predisposition is obviously necessary. DRESS is a rare event. Patients who developed a DRESS with one drug are at risk to develop a DRESS with another. A crossreactivity among aromatic anticonvulsants is evaluated as high as 75% (Knowles et al., 1999; Allam et al., 2004). Familial clustering has been reported: three siblings of the same family developed a phenytoin-induced DRESS (Gennis et al., 1991).

Abnormalities in the drug metabolism or in the immune response are the two genetic predispositions classically considered.

Because several drugs associated with DRESS were metabolized to reactive oxidative metabolites, their role was evoked in the causation of the DRESS. Most previous studies done on lymphocytes from patients with DRESS, demonstrated a susceptibility to the toxic effects of oxidative drug metabolites (Shear et al., 1988; Green et al., 1995; Leeder, 1998). An increased risk associated with higher starting doses or concurrently administrated drugs in the case of lamotrigin or abacavir, for instance, brought clinical arguments. But genetic analysis of different genes, such as epoxide hydroxylase, implicated in detoxification did not corroborate previous studies (Gaedigk et al., 1994).

DRESS is considered to be an immune-mediated disease. The detection of drug- or drug metabolite-induced lymphocytes by stimulation test using peripheral blood mononuclear cells and the positivity of patch tests assessed that DRESS is immune-mediated (Houwerzijl et al., 1977; Sullivan and Shear, 2001). A high number of CD8 T cells are found in the skin of patients with DRESS. T-cell clonal rearrangements have been described in some patients illustrating the oligoclonal of monoclonal expansion of T lymphocytes (Choi et al., 2003; Cordel et al., 2004). Studies of cytokine profile during the course of the DRESS showed the production of TH1 (interferon gamma), and TH2 (IL-5, IL-6) cytokines (Sullivan and Shear, 2001). TNF α is also detected with elevated levels. The production of cytokines is variable during the course of the DRESS. TH2 cytokines could modulate TH1 production. The importance of TNF α is corroborated by the association of some variant of the TNF α promoter and patients with severe drug reaction to carbamazepine, mainly patients with DRESS (Pirmohamed et al., 2001). The study of TNF α promoter region gene polymorphisms in "carbamazepine-hypersensitive patients" demonstrated an association between a variant allele (TNF2) limited to the patients with DRESS. Recently, typing the major histocompatibility complex (human leukocyte antigen class I and II high-resolution typing) was performed in DNA of patients with abacavir hypersensitivity and controls (Phillips et al., 2005). A good correlation was reported between patch test and genetic testing for the presence of the haplotype HLA-B*5701 in abacavir-induced DRESS.

Recent studies demonstrated that patients with DRESS have a transient decreased level of serum gammaglobulins at onset (Aihara et al., 2003; Kano et al., 2004). We observed this hypogammaglobulinemia in three out of the seven patients in our first series (Descamps et al., 2001). In an unpublished study, we retrospectively compared gammaglobulins level in two consecutive series of DRESS and erythroderma. Hypogammaglobulinemia was statistically associated with DRESS. Interestingly it was known that anticonvulsants induced in some patients, hypogammaglobulinemia. It is not known yet whether the hypogammaglobulinemia in these patients treated by anticonvulsant is associated with a risk of reactivation of HHV-6 and could be considered as a marker of a risk of DRESS. The time necessary for the development of hypogammaglobulinemia could explain the delay for the development of DRESS.

A significant decrease of B-cell lymphocytes and NK cells is also observed (Inaoka et al., 2005). The author suggested that the low number of NK cells contributed to the absence of differentiation of B cells to immune globulin producing cells. In this regard, Inaoka recently made a link between the possible defect of NK cells observed in DRESS and the importance of these cells for the control of HHV-6 reactivation. A number of NK cells and gamma–delta-T cells and their IL-15 receptor (CD122) expression were studied as compared to B cells and CD4 and CD8 T cells. CD122 expression was significantly decreased on NK cells as compared to controls. This defect persisted after recovery. Furthermore, in NK-depletion experiments, a significant increase in HHV-6 DNA was detected in PBMC culture without expansion of gamma–delta-T cells. The author proposed that this deficiency of the innate immune system could explain reactivation of HHV-6 and other latent viruses (Gosselin et al., 1999). This immune failure may also be increased by some drugs, such as anticonvulsant (Okada et al., 2001).

Therapeutic implications

The link between HHV-6 infection and DRESS could suggest new therapeutic options with antiviral agents. Nevertheless, the most important action in the care of patients with DRESS remains to cease rapidly the causative agent or the suspected agents. In many cases, DRESS will resolve spontaneously within several days after the cessation of the culprit drug. No treatment is necessary in the mild forms. A regular follow up with blood tests, especially liver function tests is required. Topical treatment may be proposed as a symptomatic treatment. When severe systemic symptoms develop, the treatment is debated. Systemic corticosteroids may be proposed and are widely used (Roujeau and Stern, 1994). But in some cases, the decrease of corticosteroids is followed by a flare-up of the DRESS. In view of the association between DRESS and HHV-6 infection, antiviral agents must be considered in association with corticosteroids. Corticosteroids may favor virus replication. Recently, the use of intravenous gammaglobulins in association with corticosteroid was advocated considering the observed hypogammaglobulinemia at the acute phase and the antiviral action of gammaglobulins (Kano et al., 2004, 2005). A pediatric case of rapid improvement after intravenous administration of gammaglobulins had been previously reported (Scheuerman et al., 2001). Controlled studies are warranted to confirm the interest of these treatments.

Conclusion

Study of DRESS offers a very good opportunity to understand the link between viral infection, drug allergy, and systemic symptoms. Moreover, perhaps a better knowledge of the immunological factors induced by drug exposure that contribute to HHV-6 reactivation in some predisposed patients will enable us to understand the immunological factors mediating the balance between latent infection and reactivation. This model could also help to understand the factors that regulate the severity of viral infection and the reactivation in immunosuppressed patients such as organ transplants.

References

Aihara Y, Ito SI, Kobayashi Y, Yamakawa Y, Aihara M, Yokota S. Br J Dermatol 2003; 149: 165–169.

Akashi K, Eizuru Y, Sumiyoshi Y, Minematsu T, Hara S, Harada M, Kikuchi M, Niho Y, Minamishima Y. N Engl J Med 1993; 329: 168–171.

Allam JP, Paus T, Reichel C, Bieber T, Novak N. Eur J Dermatol 2004; 14: 339–342.

Begon E, Roujeau JC. Ann Dermatol Venereol 2004; 131: 293–297.

Bocquet H, Bagot M, Roujeau JC. Semin Cutan Med Surg 1996; 15: 250–257.

Callot V, Roujeau JC, Bagot M, Wechsler J, Chosidow O, Souteyrand P, Morel P, Dubertret L, Avril MF, Revuz J. Arch Dermatol 1996; 132: 1315–1321.

Chaiken BH, Goldberg BI, Segal JP. N Engl J Med 1950; 242: 897–898.

Chen RL, Lin KH, Lin DT, Su IJ, Huang LM, Lee PI, Hseih KH, Lin KS, Lee CY. Br J Haematol 1995; 89: 282–290.

Choi TS, Doh KS, Kim SH, Jang MS, Suh KS, Kim ST. Br J Dermatol 2003; 148: 730–736.

Collot S, Petit B, Bordessoule D, Alain S, Touati M, Denis F, Ranger-Rogez S. J Clin Microbiol 2002; 40: 2445–2451.

Conilleau V, Dompmartin A, Verneuil L, Michel M, Leroy D. Contact Dermatitis 1999; 41: 141–144.

Cordel N, Lenormand B, Courville P, Joly P. Ann Dermatol Venereol 2004; 131: 1059–1061.

Desachy A, Ranger-Rogez S, Francois B, Venot C, Traccard I, Gastinne H, Denis F, Vignon P. Clin Infect Dis 2001; 32: 197–203.

Descamps V, Bouscarat F, Laglenne S, Aslangul E, Veber B, Descamps D, Saraux JL, Grange MJ, Grossin M, Navratil E, Crickx B, Belaich S. Br J Dermatol 1997; 137: 605–608.
Descamps V, Collot S, Houhou N, Ranger-Rogez S. Ann Neurol 2003a; 53: 280.
Descamps V, Collot S, Mahe E, Houhou N, Crickx B, Ranger-Rogez S. J Invest Dermatol 2003b; 121: 215–216.
Descamps V, Mahe E, Houhou N, Abramowitz L, Rozenberg F, Ranger-Rogez S, Crickx B. Br J Dermatol 2003c; 148: 1032–1034.
Descamps V, Valance A, Edlinger C, Fillet AM, Grossin M, Lebrun-Vignes B, Belaich S, Crickx B. Arch Dermatol 2001; 137: 301–304.
DesJardin JA, Cho E, Supran S, Gibbons L, Werner BG, Snydman DR. Clin Infect Dis 2001; 3(3): 1358–1362.
Engel JN, Mellul VG, Goodman DB. Am J Med 1986; 81: 928–930.
Enomoto M, Ochi M, Teramae K, Kamo R, Taguchi S, Yamane T. Ann Pharmacother 2004; 38: 799–802.
Fujino Y, Nakajima M, Inoue H, Kusuhara T, Yamada T. Ann Neurol 2002; 51: 771–774.
Gaedigk A, Spielberg SP, Grant DM. Pharmacogenetics 1994; 4: 142–153.
Gennis MA, Vemuri R, Burns EA, Hill JV, Miller MA, Spielberg SP. Am J Med 1991; 91: 631–634.
Gorlin JB, Ferry JA. N Engl J Med 1996; 335: 577–584.
Gosselin J, Tomolu A, Gallo RC, Flamand L. Blood 1999; 94: 4210–4219.
Green VJ, Pirmohamed M, Kitteringham NR, Gaedigk A, Grant DM, Boxer M, Burchell B, Park BK. Biochem Pharmacol 1995; 50: 1353–1359.
Gupta A, Eggo MC, Uetrecht JP, Cribb AE, Daneman D, Rieder MJ, Shear NH, Cannon M, Spielberg SP. Clin Pharmacol Ther 1992; 51: 56–67.
Haruda F. Neurology 1979; 29: 1480–1485.
Hashimoto K, Yasukawa M, Tohyama M. Curr Opin Allergy Clin Immunol 2003; 3: 255–260.
Hogg RJ, Sawyler M, Hecox K, Eigenbrodt EJ. Pediatrics 1981; 98: 830–832.
Houwerzijl J, De Gast GC, Nater JP, Esselink MT, Nieweg HO. Clin Exp Immunol 1977; 29: 272–277.
Huang YL, Hong HS, Wang ZW, Kuo TT. J Am Acad Dermatol 2003; 49: 316–319.
Kano Y, Inaoka M, Sakuma K, Shiohara T. Toxicology 2005; 209: 165–167.
Kano Y, Inaoka M, Shiohara T. Arch Dermatol 2004; 140: 183–188.
Kano Y, Shiohara T. Acta Derm Venereol 2004; 84: 484–485.
Knowles SR, Shapiro LE, Shear NH. Drug Saf 1999; 21: 489–501.
Kunisaki Y, Goto H, Kitagawa K, Nagano M. Intern Med 2003; 42: 203–207.
Leeder JS. Epilepsia 1998; 39(Suppl 7): S8–S16.
Maeda Y, Teshima T, Yamada M, Harada M. Leuk Lymphoma 2000; 39: 229–239.
Mahe E, Bodemer C, Dupic L, Hubert P, Lacaille F, Goulet O, Leruez-Ville M, Fraitag S. Transplantation 2004; 77: 479–480.
Masaki T, Fukunaga A, Tohyama M, Koda Y, Okuda S, Maeda N, Kanda F, Yasukawa M, Hashimoto K, Horikawa T, Ueda M. Acta Derm Venereol 2003; 83: 128–131.
McGeachy TE, Bloomer WE. Am J Med 1953; 14: 600–604.
Michel F, Navellou JC, Ferraud D, Toussirot E, Wendling D. Jt Bone Spine 2005; 72: 82–85.
Ogihara T, Takahashi T, Hanihara T, Amano N, Matsumoto KJ. Clin Psychopharmacol 2004; 24: 105–106.
Okada K, Sugiura T, Kuroda E, Tsuji S, Yamashita U. Clin Exp Immunol 2001; 124: 406–413.

Parneix-Spake A, Bastuji-Garin S, Lobut JB, Erner J, Guyet-Rousset P, Revuz J, Roujeau JC. Arch Dermatol 1995; 131: 490–491.
Pellock JM. Epilepsia 1987; 28(Suppl 3): S64–S70.
Phillips EJ, Wong GA, Kaul R, Shahabi K, Nolan DA, Knowles SR, Martin AM, Mallal SA, Shear NH. AIDS 2005; 19: 979–981.
Pirmohamed M, Lin K, Chadwick D, Park BK. Neurology 2001; 56: 890–896.
Razonable RR, Fanning C, Brown RA, Espy MJ, Rivero A, Wilson J, Kremers W, Smith TF, Paya CV. J Infect Dis 2002; 185: 110–113.
Roujeau JC, Stern RS. N Engl J Med 1994; 331: 1272–1285.
Scheuerman O, Nofech-Moses Y, Rachmel A, Ashkenazi S. Pediatrics 2001; 107: e14.
Sekiguchi A, Kashiwagi T, Ishida-Yamamoto A, Takahashi H, Hashimoto Y, Kimura H, Tohyama M, Hashimoto K, Iizuka H. J Dermatol 2005; 32: 278–281.
Sekine N, Motokura T, Oki T, Umeda Y, Sasaki N, Hayashi M, Sato H, Fujita T, Kaneko T, Asano Y, Kikuchi K. JAMA 2001; 285: 1153–1154.
Shear NH, Spielberg SP. J Clin Invest 1988; 82(6): 1826–1832.
Sullivan JR, Shear NH. Arch Dermatol 2001; 137: 357–364.
Suzuki Y, Inagi R, Aono T, Yamanishi K, Shiohara T. Arch Dermatol 1998; 134: 1108–1112.
Tohyama M, Yahata Y, Yasukawa M, Inagi R, Urano Y, Yamanishi K, Hashimoto K. Arch Dermatol 1998; 134: 1113–1117.
Wong GA, Shear NH. Arch Dermatol 2004; 140: 226–230.
Zeller A, Schaub N, Steffen I, Battegay E, Hirsch HH, Bircher AJ. Infection 2003; 31: 254–256.
Zerr DM, Meier AS, Selke SS, Frenkel LM, Huang ML, Wald A, Rhoads MP, Nguy L, Bornemann R, Morrow RA, Corey L. N Engl J Med 2005; 352: 768–776.

Human Herpesvirus-6: General Virology, Epidemiology and Clinical Pathology
Perspectives in Medical Virology
Gerhard Krueger and Dharam Ablashi (Editors)

DOI 10.1016/S0168-7069(06)12012-1

HHV-6 and the Integument

Tetsushi Yoshikawa
Department of Pediatrics, Fujita Health University School of Medicine, Toyoake, Aichi 4701192, Japan

Introduction

The typical clinical course of exanthem subitum, a common disease in children and infants caused by primary human herpesvirus 6 (HHV-6) infection, begins with the abrupt onset of fever. The appearance of skin eruption coincides with waning fever on the third or fourth day. The pathogenesis of the skin eruptions in exanthem subitum remains unclear, and histological analysis of skin tissues from affected patients has been limited. Moreover, studies investigating the pathogenesis of exanthem subitum are limited by the absence of suitable animal models of primary HHV-6 infection. It is thought that, like other herpesviruses, HHV-6 establishes latency after a primary infection and gets reactivated during host immunosuppression. HHV-6 has recently been recognized as an opportunistic pathogen in transplant recipients (Yoshikawa, 2004). HHV-6 is associated with fever and skin rash resembling acute graft-versus-host-disease (GVHD) (Asano et al., 1991; Yoshikawa et al., 1991, 2001a, 2002), and may play an important role in inducing acute GVHD in hematopoietic stem cell transplant (HSCT) recipients (Appleton et al., 1994, 1995; Wilborn et al., 1994). Furthermore, several reports have recently documented HHV-6 reactivation in patients experiencing drug-induced hypersensitivity syndrome (DHS) (Suzuki et al., 1998; Tohyama et al., 1998; Drago and Rebora, 1999), characterized by fever, skin rash, lymphadenopathy, and internal organ involvement. Thus, it is likely that HHV-6 is associated with some dermatological disorders.

Exanthem subitum

Primary infection with HHV-6B causes exanthem subitum, a common febrile disease in infants (Yamanishi et al., 1988). In Japan, more than 90% of infants with primary HHV-6 infection visiting our clinic with high fever exhibited a typical clinical course of exanthem subitum (Asano et al., 1994). Additionally, about 80% of young children from several day nurseries with positive anti-HHV-6 antibody titers had a past history of exanthem subitum (unpublished data). Thus, we estimated that 70–80% of infants with primary HHV-6 infection exhibit a typical clinical course of exanthem subitum, but the remaining anti-HHV-6-positive patients probably experienced a sub clinical infection such as fever without rash (Suga et al., 1989; Yoshikawa et al., 2004a), rash without fever (Asano et al., 1989), or no clinical symptoms. In patients undergoing a typical clinical course, if the day on which the patient became afebrile was defined as day 0, rash typically appeared on day 0 ± 1.0 and persisted for 3.8 ± 1.5 days. Rashes were most typically papular (rubella-like; 54%), but macular (measles-like; 40%) and maculopapular (6%) rashes were also seen (Fig. 1). Rashes initially appear on the face, trunk, or both, and subsequently spread to other locations (Asano et al., 1994). These findings are virtually the same as the original clinical description of exanthem subitum by Zahorsky (1910). Other groups, however, have reported that only 9–17% of American children with primary HHV-6 infections visiting the emergency room develop exanthem subitum; the majority develop undefined febrile illnesses (Pruksananonda et al., 1992; Hall et al., 1994). A prospective population-based study was recently undertaken to determine the clinical features of primary HHV-6 infection outside the acute care setting (Zerr et al., 2005a). The authors prospectively studied a cohort of 277 children from birth through the first two years of life to define the pattern of acquisition of HHV-6. The children's saliva was tested every

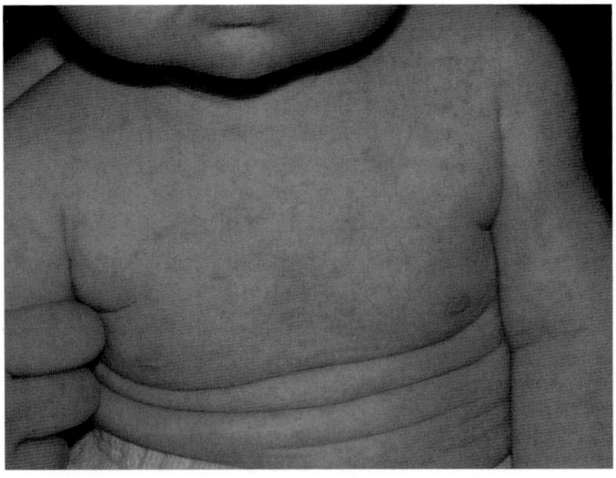

Fig. 1 Typical skin eruption observed in patient with primary HHV-6 infection. Maculopapular skin rash is observed on the trunk and extremity. (for colour version: see colour section on page 352).

week by polymerase chain reaction (PCR) for HHV-6 DNA, and parents maintained a daily log of signs and symptoms of illness in their children. Primary HHV-6 infection occurred in 130 children. Among the 81 children with a well-defined time of acquisition of HHV-6, 93% developed symptoms of disease, and 38% were seen by a physician. Compared to children with unrelated illnesses, those with primary HHV-6 infection were more likely to have rash ($P = 0.003$) and roseola ($P = 0.002$), although the actual incidences of these symptoms were low (rash; 31%, roseola; 24%). As summarized in Table 1, large differences were found in the incidence and clinical presentation of exanthem subitum in patients with primary HHV-6 infection between North America and Japan; the reasons underlying these differences remain unclear. Disease presentation and clinical course may be affected by ethnic background. Furthermore, the differences in study design shown in Table 1, the care of febrile infants differs between Japan and North America, and this may have a significant effect of the observed differences. Further international collaborative studies are needed to further clarify the differences between the incidence of exanthem subitum observed in North America and Japan.

Skin rash and HHV-6 reactivation in hematopoietic stem cell transplant recipient

Several clinical conditions such as skin rash resembling acute GVHD, bone marrow suppression, interstitial pneumonitis, and encephalitis, may be related to HHV-6 infection after HSCT (Yoshikawa, 2004b). Moreover, an association between HHV-6 infection and acute GVHD was suggested in HSCT recipients. HHV-6 and human

Table 1

Studies for evaluation of clinical features in primary HHV-6 infection (exanthem subitum)

Authors	Study design	Region	Frequency of exanthem subitum (%)
Hall et al. (1994)	Children (<3 years old) presenting to the emergency room	USA	27/160 (17)
Pruksananonda et al. (1992)	Febrile children (<2 years old) presenting to the emergency room	USA	3/34 (9)
Asano et al. (1994)	Febrile children (<1.5 years old) visiting an outpatient clinic	Japan	172/176 (98)
Zerr et al. (2005a)	Prospective cohort study from birth through the first 2 years	USA	19/80 (24)[a]

[a]Incidence of exanthem subitum was significantly higher in HHV-6 positive group than HHV-6 negative group (24% vs. 3%; $P = 0.003$).

cytomegalovirus (CMV) are highly related, and CMV is clearly an opportunistic pathogen for HSCT recipients. Although the exact frequency of HHV-6 reactivation is difficult to determine, approximately 40–50% of HSCT recipients develop HHV-6 infection, but the incidence is highly dependent on the sensitivity of the diagnostic procedure used. Our recent prospective study determined that about 40% of HSCT recipients exhibited HHV-6 viremia 2–4 weeks after HSCT (Yoshikawa et al., 1991, 2002), and this is consistent with the incidence published in other studies (Maeda et al., 1999; Imbert-Marcille et al., 2000; Ljungman et al., 2000). Although the frequency of HHV-6 infection after HSCT diagnosed by PCR is similar to that after solid organ transplantation, the frequency of HHV-6 viremia (Yoshikawa et al., 1991, 2002) is likely to be higher after HSCT than after solid organ transplantation.

We reported the isolation of HHV-6 from the blood of HSCT recipients 15 days after transplant. Two of the three patients, subsequently treated for acute GVHD, had fever and macular rash at the time of virus isolation (Asano et al., 1991). To further explore the possible association of HHV-6 infection and acute GVHD, we examined peripheral blood specimens from 25 HSCT recipients for the presence of virus by serological analysis (Yoshikawa et al., 1991). HHV-6 infection was confirmed in 12 (48%) of the 25 recipients. Four of the 12 patients developed skin rashes, and three of those four also experienced a febrile episode during the period of viral isolation; none of the remaining 13 patients developed these symptoms. These results suggest that HHV-6 infection may occur in almost half of all HSCT recipients approximately 2–3 weeks following the procedure. Viral reactivation may play an important role in acute GVHD or an acute GVHD-like illness. To clarify the association between HHV-6 and this skin rash, two institutes attempted to detect the virus genome within skin tissues obtained from patients. Collectively, these studies suggest that HHV-6 plays an important role in the development of acute GVHD (Wilborn et al., 1994; Boutolleau et al., 2003) and the development of skin rash after allogeneic HSCT (Appleton et al., 1994, 1995; Le Cleach et al., 1998; Cone et al., 1999). In a recent study, we found HHV-6 viremia in nine of fifteen (60%) patients developing skin rash within 1 month after HSCT, compared with no cases out of 10 (0%) with a skin rash occurring more than 1 month after transplantation ($P = 0.008$) (Yoshikawa et al., 2002). A similar finding was reported by Volin et al. (2004) based on an HHV-6 antigenemia assay, a good tool for the evaluation of active viral infection. In all 58 consecutive adult allogeneic HSCT recipients were prospectively monitored using the antigenemia assay, and all but one of the 11 patients who developed a rash shortly after HSCT had simultaneous HHV-6 antigenemia. Ten of the 38 (26%) HHV-6 antigenemia positive and one of the 20 (5%) negative patients had an early rash ($P = 0.077$). Given the lack of rapid diagnostic tests for active HHV-6 infection and similarities in rashes associated with HHV-6 and GVHD, many HSCT patients who develop skin rash shortly after transplant are treated with immunosuppressive therapy for presumed GVHD. HHV-6 may be directly involved in the development of skin rashes occurring within the first month after allogeneic HSCT. Fig. 2 summarizes the current hypotheses for an association between HHV-6 infection and skin rash after HSCT. It remains unclear, however,

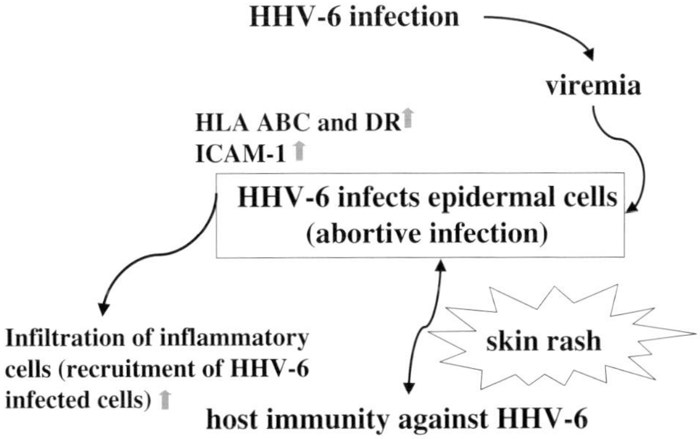

Fig. 2 The hypothesis of the mechanism for skin eruption caused by HHV-6.

whether HHV-6 causes acute GVHD or the virus causes an erythematous illness similar to acute GVHD. In addition, GVHD may induce HHV-6 reactivation, and the virus may not be associated with the observed skin eruptions. Several groups have recently used quantitative PCR to identify HHV-6 reactivation associated with acute GVHD, in particular severe GVHD (grades 3–4) (Boutolleau et al., 2003; Volin et al., 2004; Zerr et al., 2005b). The majority of patients reactivated HHV-6 after the onset of GVHD, and it is likely that immunodeficiency caused by acute GVHD and/or immunosuppressive therapy promotes HHV-6 reactivation and infection. As shown in Table 2, the appropriate strategy for patient management differs depending on the particular circumstances of HHV-6 infection. In order to improve the prognosis of patients receiving HSCT, more work is needed to clearly establish the relationship between HHV-6 infection and skin rash after HSCT.

Drug-induced hypersensitivity syndrome and HHV-6

Viral infections can be involved in the development of some drug reactions, such as ampicillin-induced eruption in mononucleosis. HHV-6 infection may be involved in the pathogenesis of DHS, defined as the clinical triad of fever, rash, and internal organ involvement due to drug exposure. The precise pathogenesis of this syndrome remains unknown, although several hypotheses have been proposed. Clearly, drug exposure is a key factor in the etiology of these reactions. Several studies have demonstrated HHV-6 reactivation near the time of onset of DHS (Suzuki et al., 1998; Tohyama et al., 1998). HHV-6 reactivation may account for a prolonged course, slow resolution, and recurrence of signs and symptoms of DHS (Suzuki et al., 1998; Tohyama et al., 1998). Various drugs including sulfasalazine, allopurinol, ibuprofen, and anticonvulsants such as phenobarbital, carbamazepine, zonisamide, phenytoin, and sodium valproate, have been proposed as drugs responsible for HHV-6 infection associated DHS (Zeller et al., 2003; Suzuki et al.,

1998, Tohyama et al., 1998; Aihara et al., 2003; Masaki et al., 2003; Sekiguchi et al., 2005). Although most patients identified with HHV-6 associated DHS are adults, one case report described a 2-year-old boy who developed DHS due to phenytoin and phenobarbital associated with HHV-6 (Criado et al., 2004). There is an association between HHV-6 infection or reactivation and encephalitis in either immunocompetent children or immunocompromised patients, and this has led to speculation that HHV-6 has some degree of neurovirulence (Yoshikawa and Asano, 2000). Several investigators have reported neurological complications in DHS patients with HHV-6 reactivation (Fujino et al., 2002; Descamps et al., 2003; Kano et al., 2004). Thus, HHV-6 reactivation in patients with DHS could lead to other clinical features such as central nervous system complications.

Many mechanisms have been proposed to explain the association between HHV-6 infection and DHS. Activation of drug-specific T-cells may induce the reactivation of latent viruses, or viral infections could modify drug metabolism and induce the production of reactive metabolites. Alternatively, viral infection could amplify the T-cell activation induced by the drug reaction. However, the precise causal role of HHV-6 reactivation in the pathogenesis of DHS remains unclear.

Pityriasis rosea and HHV-6

Pityriasis rosea is a common, acute, self-limiting papulosquamous skin disorder. The initial skin lesion is called the "herald patch," and typically appears on the trunk as a 2–3 cm oval scaly plaque with a central salmon-colored area and a darker erythematous peripheral zone. The disease normally resolves spontaneously within 4–8 weeks. The clinical and epidemiological features of this disease suggest a pathogenic role for an infectious agent. Drago et al. (1997) first suggested that reactivation of human herpesvirus 7 (HHV-7), which is the virus most similar to HHV-6, was linked to pityriasis rosea. An association between HHV-7 and the disease has been debated since then, and some investigators have suggested that HHV-6 is also linked to pityriasis rosea (Kosuge et al., 2000; Watanabe et al., 2002; Broccolo et al., 2005). HHV-6 latently infects peripheral blood mononuclear cells, and highly sensitive detection techniques such as nested PCR could falsely identify active HHV-6 infection in tissues with latently infected resident blood cells. However, HHV-6 was isolated from one patient with pityriasis rosea, and significant increases in HHV-6 antibody titers were found in some patients with this disease (Yasukawa et al., 1999). In addition, viral transcripts and antigens were detected in skin tissue obtained from patients with pityriasis rosea (Watanabe et al., 2002). These results suggest that not only HHV-7 but also HHV-6 might be involved in the pathogenesis of some cases of pityriasis rosea.

Pathogenesis of skin manifestations due to HHV-6 infection

The pathogenesis of HHV-6 related skin eruptions remains unclear, and this is largely due to the difficulty in obtaining skin tissues for histological analysis from patients with exanthem subitum as well as the absence of suitable animal models of primary HHV-6 infection. Recently, we reported the absence of rash after the resolution of fever in immunocompromised infants during primary HHV-6 infection (Yoshikawa et al., 2001a,b). Thus, the host immune response against the virus appears to play an important pathogenic role in the skin rash in patients with exanthem subitum. In addition to this *in vivo* data, we infected an epidermal cell line (A431 cells) with HHV-6 and examined the expression of several surface molecules to understand the pathogenesis of skin manifestations (Yoshikawa et al., 2003) HHV-6 can infect this epidermal cell line, and HLA-A, -B, and -C, HLA-DR, and ICAM-1 expression were upregulated in HHV-6-infected cells. Tissue culture supernatant did not affect the expression of these three surface molecules, and, therefore, direct viral infection probably causes the changes observed in these cells. When considered with a previous report (Yoshikawa et al., 2001b), our data suggest the mechanism of HHV-6-associated skin rash as shown in Fig. 2. First, direct contact between HHV-6-infected peripheral blood mononuclear cells and epidermal cells could establish viral infection in the skin. HHV-6 infection would then cause epidermal cells to increase the expression of several surface molecules including ICAM-1, causing increased inflammatory cell migration into the developing lesion. The host immune response is likely important in this step. However, additional *in vitro* studies using primary cell culture systems are needed to more clearly identify the precise mechanism responsible for the changes seen during HHV-6 infection. Additionally, these studies should be confirmed with *in vivo* histological analyses.

Lymphocytic infiltration and cytopathic changes in keratinocytes are the major features of acute GVHD. Donor T-cells reactive against recipient cells expressing different tissue antigens are thought to induce acute GVHD. MHC class I and II molecules as well as ICAM-1 are upregulated in the tissue of patients suffering from acute GVHD based on histological analysis. Viruses such as HHV-6 could trigger or enhance acute GVHD by changing the expression of these surface molecules. As discussed above, HLA-A, -B, and -C, HLA-DR, and ICAM-1 expression are increased on HHV-6 infected cells (Yoshikawa et al., 2003). Additionally, the severity of the acute GVHD was associated with HHV-6 infection (Boutolleau et al., 2003; Volin et al., 2004; Zerr et al., 2005b). The biological changes induced by HHV-6 reactivation may play an important role in promoting the trafficking of inflammatory cells into skin lesions exacerbating acute GVHD.

Although the precise mechanism of HHV-6 reactivation remains unclear, allogeneic stimulation has been hypothesized to accelerate viral reactivation in HSCT recipients. In CMV infection, the importance of allogeneic stimulation in causing viral reactivation has been demonstrated *in vitro* (Hummel et al., 2001), suggesting that pro-inflammatory cytokines induced by allogeneic reactions could promote HHV-6 reactivation, as well. Many inflammatory cytokines signal via the NF-kB pathway, and CMV immediate-early genes are induced by NF-kB signaling (Loser et al., 1998). The promoter of HHV-6 immediate-early genes also possess predicted NF-kB binding sites (Takemoto et al., 2001), and HHV-6 could be reactivated by a mechanism similar to CMV. Thus, increased cytokine levels caused by acute GVHD and DHS could trigger the reactivation of HHV-6. Further analysis is needed to clarify this hypothesis.

References

Aihara Y, Ito SI, Kobayashi Y, Yamakawa Y, Aihara M, Yokota S. Br J Dermatol 2003; 149: 165–169.

Appleton AL, Peiris JS, Taylor CE, Sviland L, Cant AJ. Lancet 1994; 344: 1361–1362.

Appleton AL, Sviland L, Peiris JS, Taylor CE, Wilkes J, Green MA, Pearson AD, Kelly PJ, Malcolm AJ, Proctor SJ, Hamilton PJ, Cant AJ. Bone Marrow Transplant 1995; 16: 777–782.

Asano Y, Suga S, Yoshikawa T, Urisu A, Yazaki T. J Pediatr 1989; 115: 264–265.

Asano Y, Yoshikawa T, Suga S, Nakashima T, Yazaki T, Fukuda M, Kojima S, Matsuyama T. N Engl J Med 1991; 324: 634–635.

Asano Y, Yoshikawa T, Suga S, Kobayashi I, Nakashima T, Yazaki T, Kajita Y, Ozaki T. Pediatrics 1994; 93: 104–108.

Boutolleau D, Fernandez C, Andre E, Imbert-Marcille BM, Milpied N, Agut H, Gautheret-Dejean A. J Infect Dis 2003; 187: 179–186.

Broccolo F, Drago F, Careddu AM, Foglieni C, Turbino L, Cocuzza CE, Gelmetti C, Lusso P, Rebora AE, Malnati MS. J Invest Dermatol 2005; 124: 1234–1240.

Cone RW, Huang ML, Corey L, Zeh J, Ashley R, Bowden R. J Infect Dis 1999; 179: 311–318.

Criado PR, Criado RF, Vasconcellos C, Pegas JR, Cera PC. J Dermatol 2004; 31: 1009–1013.

Descamps V, Collot S, Houhou N, Ranger-Rogez S. Ann Neurol 2003; 53: 280.

Drago F, Ranieri E, Malaguti F, Battifoglio ML, Losi E, Rebora A. Dermatology 1997; 195: 374–378.

Drago F, Rebora A. Arch Dermatol 1999; 135: 71–75.

Fujino Y, Nakajima M, Inoue H, Kusuhara T, Yamada T. Ann Neurol 2002; 51: 771–774.

Hall CB, Long CE, Schnabel KC, Caserta MT, McIntyre KM, Costanzo MA, Knott A, Dewhurst S, Insel RA, Epstein LG. N Engl J Med 1994; 331: 432–438.

Hummel M, Zhang Z, Yan S, DePlaen I, Golia P, Varghese T, Thomas G, Abecassis MI. J Virol 2001; 75: 4814–4822.

Imbert-Marcille BM, Tang XW, Lepelletier D, Besse B, Moreau P, Billaudel S, Milpied N. Clin Infect Dis 2000; 31: 881–886.

Kano Y, Inaoka M, Shiohara T. Arch Dermatol 2004; 140: 183–188.

Kosuge H, Tanaka-Taya K, Miyoshi H, Amo K, Harada R, Ebihara T, Kawahara Y, Yamanishi K, Nishikawa T. Br J Dermatol 2000; 143: 795–798.

Le Cleach L, Joberty C, Fillet AM, Sutton L, Cordonnier C, Frances C, Agut H, Chosidow O. Arch Dermatol 1998; 134: 759–760.

Ljungman P, Wang FZ, Clark DA, Emery VC, Remberger M, Ringden O, Linde A. Br J Haematol 2000; 111: 774–781.

Loser P, Jennings GS, Strauss M, Sandig V. J Virol 1998; 72: 180–190.

Maeda Y, Teshima T, Yamada M, Shinagawa K, Nakao S, Ohno Y, Kojima K, Hara M, Nagafuji K, Hayashi S, Fukuda S, Sawada H, Matsue K, Takenaka K, Ishimaru F, Ikeda K, Niiya K, Harada M. Br J Haematol 1999; 105: 295–302.

Masaki T, Fukunaga A, Tohyama M, Koda Y, Okuda S, Maeda N, Kanda F, Yasukawa M, Hashimoto K, Horikawa T, Ueda M. Acta Derm Venereol 2003; 83: 128–131.

Pruksananonda P, Hall CB, Insel RA, McIntyre K, Pellett PE, Long CE, Schnabel KC, Pincus PH, Stamey FR, Dambaugh TR, Stewart JA. N Engl J Med 1992; 326: 1445–1450.

Sekiguchi A, Kashiwagi T, Ishida-Yamamoto A, Takahashi H, Hashimoto Y, Kimura H, Tohyama M, Hashimoto K, Iizuka H. J Dermatol 2005; 32: 278–281.

Suga S, Yoshikawa T, Asano Y, Yazaki T, Hirata S. Pediatrics 1989; 83: 1003–1006.

Suzuki Y, Inagi R, Aono T, Yamanishi K, Shiohara T. Arch Dermatol 1998; 134: 1108–1112.

Takemoto M, Shimamoto T, Isegawa Y, Yamanishi K. J Virol 2001; 75: 10149–10160.

Tohyama M, Yahata Y, Yasukawa M, Inagi R, Urano Y, Yamanishi K, Hashimoto K. Arch Dermatol 1998; 134: 1113–1117.

Volin L, Lautenschlager I, Juvonen E, Nihtinen A, Anttila VJ, Ruutu T. Br J Haematol 2004; 126: 690–696.

Watanabe T, Kawamura T, Jacob SE, Aquilino EA, Orenstein JM, Black JB, Blauvelt A. J Invest Dermatol 2002; 119: 793–797.

Wilborn F, Brinkmann V, Schmidt CA, Neipel F, Gelderblom H, Siegert W. Blood 1994; 83: 3052–3058.

Yamanishi K, Okuno T, Shiraki K, Takahashi M, Kondo T, Asano Y, Kurata T. Lancet 1988; 8594: 1065–1067.

Yasukawa M, Sada E, MacHino H, Fujita S. Br J Dermatol 1999; 140: 169–170.

Yoshikawa T, Asano Y. Brain Dev 2000; 22: 307–314.

Yoshikawa T, Asano Y, Ihira M, Suzuki K, Ohashi M, Suga S, Kudo K, Horibe K, Kojima S, Kato K, Matsuyama T, Nishiyama Y. J Infect Dis 2002; 185: 847–853.

Yoshikawa T, Goshima F, Akimoto S, Ozaki T, Iwasaki T, Kurata T, Asano Y, Nishiyama Y. J Med Virol 2003; 71: 62–68.

Yoshikawa T, Ihira M, Ohashi M, Suga S, Asano Y, Miyazaki H, Hirano M, Suzuki K, Matsunaga K, Horibe K, Kojima S, Kudo K, Kato K, Matsuyama T, Nishiyama Y. Bone Marrow Transplant 2001a; 28: 77–81.

Yoshikawa T, Ihira M, Suzuki K, Suga S, Asano Y, Asonuma K, Tanaka K, Nishiyama Y. J Pediatr 2001b; 138: 921–925.

Yoshikawa T, Suga S, Asano Y, Nakashima T, Yazaki T, Sobue R, Hirano M, Fukuda M, Kojima S, Matsuyama T. Blood 1991; 78: 1381–1384.

Yoshikawa T, Suzuki K, Umemura K, Akimoto S, Miyake F, Usui C, Fujita A, Suga S, Asano Y. J Med Virol 2004a; 74: 463–466.

Yoshikawa T. Br J Haematol 2004b; 124: 421–432.

Zahorsky J. Pediatrics 1910; 22: 60–64.

Zeller A, Schaub N, Steffen I, Battegay E, Hirsch HH, Bircher AJ. Infection 2003; 31: 254–256.

Zerr DM, Meier AS, Selke SS, Frenkel LM, Huang ML, Wald A, Rhoads MP, Nguy L, Bornemann R, Morrow RA, Corey L. N Engl J Med 2005a; 352: 768–776.

Zerr DM, Corey L, Kim HW, Huang ML, Nguy L, Boeckh M. Clin Infect Dis 2005b; 40: 932–940.

Human Herpesvirus-6: General Virology, Epidemiology and Clinical Pathology
Perspectives in Medical Virology
Gerhard Krueger and Dharam Ablashi (Editors)

DOI 10.1016/S0168-7069(06)12013-3

HHV-6 and the Respiratory System

Sebastian M. Schmidt[a], Helga Wiersbitzky[b], Siegfried K.W. Wiersbitzky[a]
[a]*Department of General Pediatrics (incl. Infectiology, Pulmonology & Allergy), University Hospital, Ernst-Moritz-Arndt University, D-17475 Greifswald, Germany*
[b]*Institute of Diagnostic Radiology (Pediatric Radiology), University Hospital, Ernst-Moritz-Arndt University, D-17475 Greifswald, Germany*

Introduction

Primary infection with human herpesvirus-6 (HHV-6) commonly occurs in more than 90% within the first two years of life and is followed by a life-long viral persistence in the infected individual (Okuno et al., 1989; Krueger et al., 1998a). Congenital HHV-6 infections were observed in 1% of 5638 cord blood samples (Hall et al., 2004). Although the respiratory system (i.e. salivary glands and bronchial tract) belong to preferred sites for HHV-6 persistence (Fox et al., 1990; Krueger et al., 1990; Yamamoto et al., 2005), diseases caused by primary HHV-6 infections of these tissues are only rarely observed (Ablashi and Krueger, 2003). The virus may become pathogenic, if persistently active in immunodeficient patients (Krueger and Ablashi, 2003; De Bolle et al., 2005).

Making the diagnosis of a disease caused by HHV-6 relies upon the serological proof of active infection (e.g. by specific tests for IgM antibodies, virus isolation, viral p41 antigenemia or viral DNA-emia) as well as the demonstration of active virus at the site of the lesion whenever biopsy material is available (Ablashi and Krueger, 2003). One or several of these tests were done to confirm the etiological association of HHV-6 and the diseases to be discussed below.

Primary HHV-6 infection and the respiratory tract

A characteristic manifestation of primary HHV-6 infection in the baby or young child is *roseola infantum* (*exanthem subitum* (ES)) as described by Zahorsky about a century ago (Zahorsky, 1910, 1913; Yamanishi et al., 1988).

Certain infantile febrile diseases with or without exanthema, convulsions or infectious mononucleosis-like syndromes were added more recently to the list of diseases following primary HHV-6 infection (Steeper et al., 1990; Kondo et al., 1993; Barone et al., 1995).

While early descriptions of *exanthem subitum* specifically stress the absence of respiratory symptoms (Levy, 1921; Westcott, 1921; Beaven, 1924; Zahorsky, 1925), Glanzmann writes in the German *Handbuch der Inneren Medizin* that rhinopharyngitis, katarrhalic otitis and occasional bronchitis do occur. Pulmonary complications are usually not present (Glanzmann, 1952). Only after HHV-6 was identified as a causative agent for ES (Eberle et al., 1988; Takahashi et al., 1988; Yamanishi et al., 1988), and the disease was thus better classified, an occasional respiratory pathology became overt even in primary HHV-6 infections (Wiersbitzky et al., 1989a,b, 1991a,b).

Clinical features of primary HHV-6 infection in children

Clinical features as summarized in Fig. 1 and Table 1 show a phasic development (Veeder and Hempelmann, 1921).

Phase 1 (3–5 days): High fever, distinct leukocytosis with relative granulocytosis ("stress"-like blood count) and general malaise.

Phase 2 (2–3 days): Temperature falls to normal by crisis coinciding with the appearance of rubeoliform or morbilliform rash, low to normal total leukocyte counts with marked neutropenia and relative lymphocytosis.

Phase 3: Recovery with waning of rash and normalization of blood counts. The common clinical diagnosis of ES is based upon the triad, three-day fever, rubeoliform rash and characteristic blood counts (prominent neutropenia with relative lymphocytosis). The typical hematologic changes may be missing in cases without exanthema.

Several complications about primary HHV-6 infection and ES have been described. First of all, neurotropism of HHV-6 (Caserta et al., 1994) may cause febrile seizures (Levy, 1921; Willi, 1929; Eeg-Olofsson, 2003), hemiparesis (Glanzmann, 1952), serous meningitis or encephalitis (Berenberg et al., 1949; Glanzmann, 1952; Bonthius and Karacay, 2002). Gastrointestinal irritation and gastritis only occur rarely (Rauter and Mutz, 1995).

Pharynx, tonsils, eyes and ears

About 65–70% of children with ES suffer from katarrheic pharyngitis and tonsillitis with a macular enanthem and lymphofollicular hyperplasia (*Nagayama's*

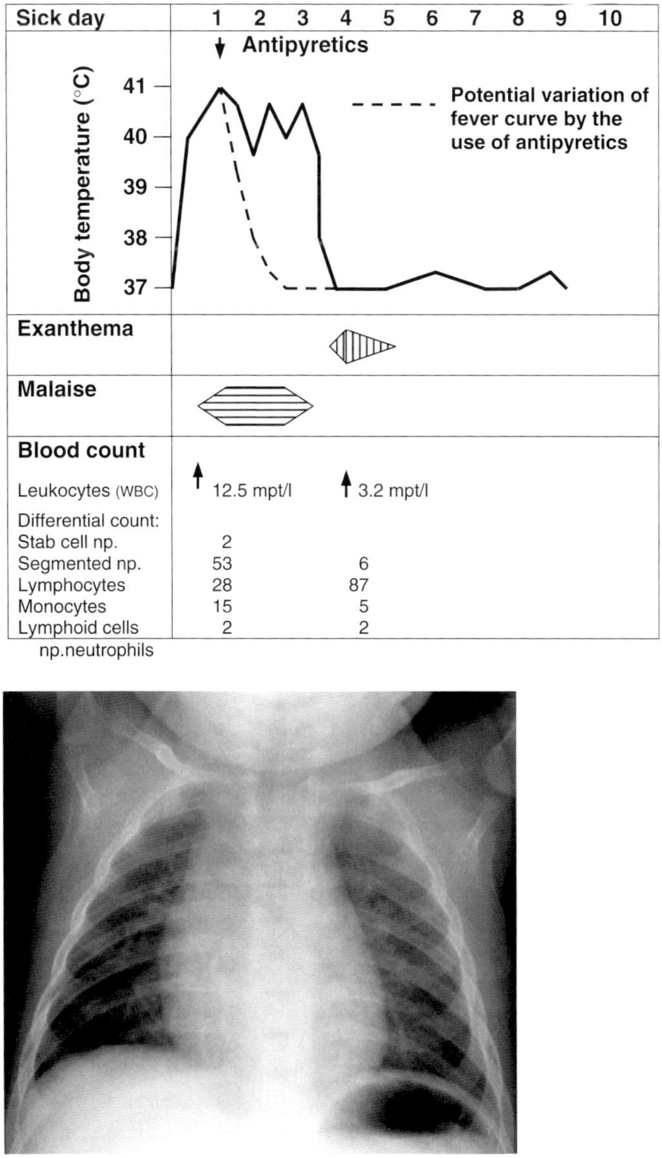

Fig. 1 Clinical course and X-ray findings in primary HHV-6 infection. Note in chest X-ray the characteristic symmetrical soft reticular pattern spreading from hilus to periphery.

papules; Fig. 2) during the highly feverish prodromal phase (Zahorsky, 1913; Berenberg et al., 1949; Nagayama et al., 1956). Nagayama regards the "*pin-point elevations on the mucous membrane in the region of the uvula and the soft palate*" as an early diagnostic sign of ES. It must be stated, however, that certain cases of ES

Table 1

Hematological findings in children with serological proof of HHV-6 infection with exanthem (Group 1, $n = 9$) as well as without exanthem (Group 2, $n = 11$). Children with febrile exanthematic diseases that cannot be clarified etiologically served as control (Group 3, $n = 13$)

	Group 1		Group 2		Group 3	
	x	SD	x	SD	x	SD
Hemoglobin (mmol/l)	7.3	0.57	6.84	1.23	7.59	0.81
Hematocrit (%)	37	3.1	39	4.4	39	7.2
Platelets (Gpt/l)	147	31.8*	211	25.4	191	35.3
Leukocytes (WBC) (Gpt/l)	5.9	2.24	7.8	2.74	8.7	4.3
Differential count (in %)						
Basophils	0		0.4	0.3	1.6	3.9
Eosinophils	0.7	1.4	2.6	3.5	2.0	2.9
Stab cell neutrophils	1.5	1.7	3.7	3.5	9.7	12.9
Segmented neutrophils	16.9	10.8*	45.2	21.0	33.0	16.0
Lymphocytes	74.8	13.0*	40.1	19.0	45.3	21.0
Monocytes	6.1	3.3	7.5	3.8	8.4	4.5
Plasma cells/Lymphoid	0		0		0	

*Mann-Whitney U-test: Polymorphonuclear granulocytes (Group 1 vs. Group 2): $p = 0.002$; Polymorphonuclear granulocytes (Group 1 vs. Group 3): $p = 0.018$; Lymphocytes (Group 1 vs. Group 2): $p = 0.001$; Lymphocytes (Group 1 vs. Group 3): $p = 0.002$; Platelets (Group 1 vs. Group 2): $p = 0.032$. *Note:* x is the mean value, and SD the standard deviation.

with similar lesions may also arise from the infection with a closely related virus, the HHV-7 (Ablashi and Krueger, 2003; Bruns et al., 2004; Wegner et al., 2004). Uvulo-palatoglossal ulcers were described in some cases of HHV-6-induced ES (Chua et al., 1999). An acute tonsillitis with lymphoid hyperplasia may cause a mononucleosis-like disease in rare cases of primary infections (Fig. 2; (Steeper et al., 1990; Horwitz et al., 1992)). It has been concluded that the tonsillar epithelium may be another site of HHV-6 persistence in latency (Roush et al., 2001). Superinfections with staphylococci or streptococci sp. are rare complications that may cause peritonsillar abscesses, purulent otitis media or purulent lymphadenitis colli (Willi, 1929; Glanzmann, 1952; Wiersbitzky, unpublished observations).

One-fourth of all ES children develop acute sero-mucoid otitis media usually secondary to serous rhinopharyngitis, also occasionally acute conjunctivitis.

Upper respiratory tract

Although negated in the original reports of ES (Zahorsky, 1910; Levy, 1921; Westcott, 1921), serous rhinitis and sinusitis maxillaris and ethmoidalis are fairly frequent during the first three days of primary HHV-6 infection (Glanzmann, 1952; Wiersbitzky, unpublished observations). Occasionally, superinfections

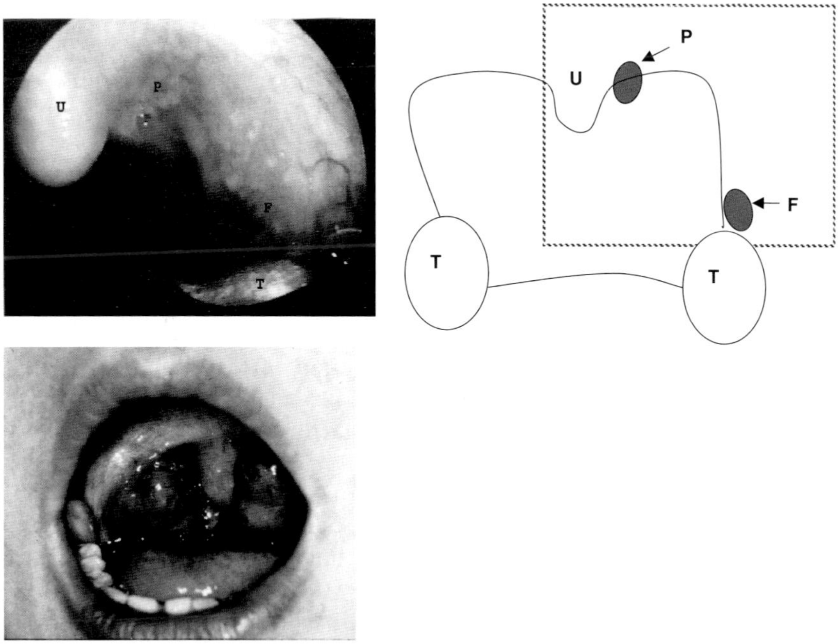

Fig. 2 Enanthema with Nagayama's papules at the parauvular region (P) and fossa supratonsillaris (F), (U) Uvula, (T) tonsils; on the right a scheme of the throat with the photo section. Tonsillar hyperplasia in primary HHV-6 infection (lower photo with courtesy of G. Bertram, University ENT Clinic Cologne, Germany). (for colour version: see colour section on page 353).

with streptococci or staphylococci sp. cause a more serious purulent sinusitis (Wiersbitzky et al., 1992) during the second or third week probably following certain local immune deficiency of the mucosa-associated lymphoid tissue (MALT).

While acute laryngitis in primary HHV-6 is rather uncommon and does not afford specific treatment, serous-katarrheic tracheitis may occur during the early phase of infection with an "irritating" cough (*hacking cough*). One-fourth to one-third of all children with primary HHV-6 infection suffer from bronchitis, which remains usually mild and is characterized by mucosal hyperemia and edematous swelling. Small children and those with an allergic predisposition may develop bronchiolitis with signs of bronchiolar obstruction and *wheezing*, requiring an occasional medication. Like tonsils and sinuses, purulent bronchitis signifies bacterial superinfection and is usually observed after the second week of HHV-6 infection.

Lower respiratory tract

Pneumonia following primary HHV-6 infection is rare, yet when occurring, it will usually require hospital admission of the seriously ill child (Wiersbitzky et al., 1992,

1993). It occurs during the initial phase of infection, may precede exanthema, enanthema and rhinopharyngitis, and is characterized by high fever, tachycardia and tachypnea. Although auscultation and percussion of lungs usually do not reveal significant alterations, thoracic radiographs are diagnostic with prominent interstitial infiltration indicating pneumonitis. The absence of preceding upper respiratory *tract* infections and the characteristic X-ray features suggest primary viral pneumonia (diagnostic biopsies are commonly not performed in these cases; for microscopy see the next section).

In the following we describe a characteristic case of primary HHV-6 infection with pulmonal involvement.

Case 1: A six-month-old male infant (B. P.-J.) was admitted with irritability, wheezing grade I, normal leukocytes, an elevation of segmented neutrophils, a decrease in lymphocytes and normal C-reactive protein (CrP). The infant suffered from high fever for three days, tachycardia and tachypnea. Thorax X-ray showed a symmetrical soft reticular pattern in medial areas of both lungs increasing toward the hilus (Fig. 1). Wheezing continued for four days. The temperature returned to normal by day 4 and the child was feeling essentially well, when a roseoliform exanthema developed on the trunk remaining visible for 48 h. The blood cell count on the first day of the exanthema revealed a low to normal total leukocyte count with neutropenia and relative lymphocytosis. HHV-6-specific IgM antibodies were positive (IFA), and the IgG antibodies were weakly positive. HHV-6—PCR was positive in the blood. Other infections were excluded such as influenza A and B, RSV, *Chlamydophila pneumoniae*, *Mycoplasma pneumoniae*, adenovirus and parainfluenzavirus type 3 (serum). Treatment was done with inhalations. The child went home after the fourth day in hospital.

Secondary bacterial bronchopneumonias (streptococcus, chlamydia and mycoplasma sp.) are fairly common to develop after two or three weeks in children with primary HHV-6 infections and ES (with or without preceding exanthema). These pneumonias essentially resemble other complicating common influenza-like illnesses. They are accompanied by purulent rhinitis, tracheobronchitis and show the typical X-ray features of bronchopneumonia. CrP values are elevated, and patients commonly require an antibiotic treatment.

Non-primary HHV-6 infection and the respiratory tract

As more than 95% of HHV-6 infections occur during the first two years of life, most HHV-6 infections observed at later ages (especially in teens and adults) are probably reactivations of latent infections or *de novo* infections with a second virus (Krueger et al., 1998b; Ablashi and Krueger, 2003). Non-primary HHV-6 infections of the respiratory tract occur preferentially in immunodeficient patients, yet may occasionally also be seen in not obviously compromised persons. Any conclusion about HHV-6 being the etiological agent for a pulmonary inflammation is complicated by the fact that there is a high incidence of HHV-6 DNA in lung tissues from not obviously sick persons (Cone et al., 1996) suggesting this organ as one site of viral persistence after primary infection. The diagnosis of HHV-6 caused pneumonitis must consequently be based upon the triad, (a) evidence of an ongoing active infection (serology and/or virus isolation), (b) demonstration of replicating virus at the site of the tissue lesion (e.g. by showing p41 antigen in biopsy specimens), and (c) effectiveness of tentative anti-HHV-6 therapy (Leland and Emmanuel, 1995; Ablashi and Krueger, 2003). Finally, in any such cases, an infection or coinfection with the closely related HHV-7 should be considered (Ross et al., 2001; Yamamoto et al., 2005).

Interstitial pneumonia in not overtly immunocompromised persons

There are only very limited case reports of HHV-6 infection causing interstitial pneumonia, and most of these patients can hardly be considered immunologically "normal" (Cone, 1995). Russler and co-workers (1991) reported a case of pneumonitis in a young healthy man caused by coinfection with HHV-6 and *Legionella pneumophila*. He suffered from severe pulmonary, renal, hepatic and CNS dysfunctions. Legionella organisms were cultured from his lungs, yet specific antibiotic therapy produced no response. HHV-6 was isolated from the patient's blood cells, and the virus was also shown in lung biopsies with pneumonitis. High-dose corticosteroid therapy suppressed the inflammatory reaction, inhibited the HHV-6 replication and supported the clearance of the Legionella infection. It was thought that HHV-6 may have worked synergistically with Legionella in causing the severe pneumonitis. Another coinfection of HHV-6 and *Pneumocystis carinii* was associated with interstitial pneumonitis in a patient with hypogamma globulinemia (Vuorinen et al., 2004). This patient, however, was obviously not fully immunocompetent. Similarly, antitumor chemotherapy supported the development of HHV-6-induced pneumonitis in a patient with malignant astrocytoma (Safdar and Brown, 2001). Of the four patients of Hammerling and co-workers (1996) with HHV-6-associated interstitial pneumonia, two were immunodeficient, one had congenital anomalies and one suffered from hepatitis of unknown etiology.

Totani and co-workers (2001) described an HHV-6-positive interstitial pneumonitis in a 47-year-old woman with Sjoegren's syndrome and Lupus erythematosus. Both autoimmune diseases were previously shown to be accompanied by higher incidences of HHV-6 reactivation (Krueger et al., 1991; De Clerck et al., 1992).

Pneumonitis in immunosuppressed patients

HHV-6 is a common latent resident in the normal lung (Cone et al., 1996), and even viral reactivation does not necessarily cause disease in immunologically competent persons. HHV-6 can be demonstrated in some 33% of bronchial lavages in healthy volunteers (Nagate et al., 2001). It may become pathogenic, though, if certain immune deficiency permits the persistent activity and replication of the virus quite similar to other "opportunistic" viruses such as Epstein–Barr virus (Purtilo et al.,

1985; Krueger and Ferrer-Argote, 1994; Singh and Carrigan, 1996; Krueger et al., 1998a).

There are large numbers of publications reporting severe and even lethal HHV-6-associated interstitial pneumonias in immunocompromised patients. These patients were preferentially various organ transplant recipients, or they suffered from HIV infection, immune deficiency or autoimmune disorders (Carrigan et al., 1991; Cone et al., 1993, 1994, 1995, 1996; Pitala et al., 1993; Knox and Carrigan, 1994; Buchbinder et al., 2000; Nagate et al., 2001; Totani et al., 2001; Michaelides et al., 2002; Taplitz and Jordan, 2002; Yata et al., 2002; Hentrich et al., 2004; Yamamoto et al., 2005). HHV-6 was usually confirmed as an etiological agent of interstitial pneumonia by showing large amounts of viral DNA in blood and lung tissue, or by finding the HHV-6 replication-associated antigen p41 at the inflammatory site in the lung. Pathologic changes are classified as non-specific interstitial pneumonitis (NIP) or lymphoid interstitial pneumonitis [LIP; (Fig. 3)]. An addition of alveolar exudates ("alveolar pneumonia") signifies secondary superinfection or coinfection. Dual or multiple infections are not uncommon in such cases with organisms

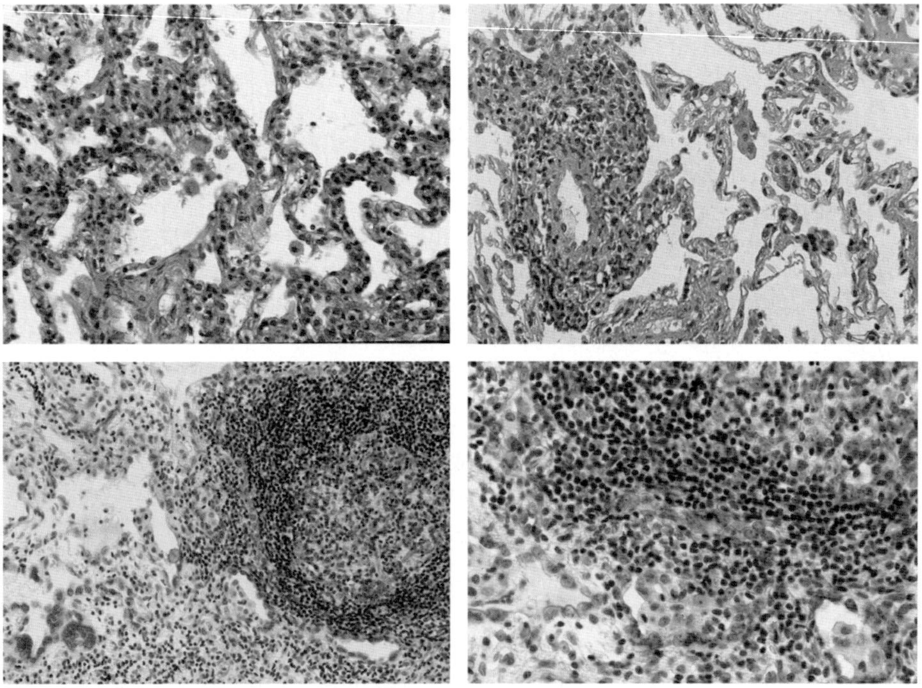

Fig. 3 Top row: non-specific interstitial pneumonitis (NIP) in patient with acute necrotizing encephalitis following primary HHV-6 infection (Wagner et al., 1997). Bottom row: lymphoid interstitial pneumonitis (LIP) in patient with HIV infection and HHV-6 reactivation red cells immunohistochemical APAAP reaction for HHV-6 p41 antigen (courtesy of G. Krueger, Immunopathology Laboratory, University of Cologne, Germany). (for colour version: see colour section on page 354).

including *P. carinii*, *L. pneumophila*, human cytomegalovirus, adenovirus or a co-infection with HHV-6 variant A and B (Russler et al., 1991; Cone et al., 1996; Hammerling et al., 1996; Vuorinen et al., 2004).

Resume

Various inflammatory reactions occur in the respiratory tract, both in primary and in non-primary infections with HHV-6. They are usually mild and rarely afford hospital admission except for the immunosuppressed patient. Similar to human cytomegalovirus and Epstein–Barr virus, HHV-6 constitutes a major risk factor in transplant recipients and in patients with acquired or inherited immune deficiency syndromes. In such patients, HHV-6 may also be a copathogen to other infectious organisms.

In addition, the observation of frequent HHV-6 reactivations in patients with certain autoimmune disorders, including in the lungs, poses the question whether this virus may be one possible copathogenic factor in the development of "idiopathic pulmonary fibrosis."

Acknowledgment

We thank Professor Renate Mentel, Friedrich Löffler—Institute for Medical Microbiology, University of Greifswald, for serological and PCR-findings.

References

Ablashi DV, Krueger GRF. The human herpesviruses HHV-6, HHV-7 and HHV-8. In: Viral Infections and Treatment (Ruebsamen-Waigmann H, editor). New York: Marcel Dekker; 2003.

Barone SR, Kaplan MH, Krilov LR. J Pediatr 1995; 127: 95–97.

Beaven PW. Arch Pediatr 1924; 41: 686–691.

Berenberg W, Wright S, Janeway CA. N Engl J Med 1949; 241: 253–259.

Bonthius DJ, Karacay B. Meningitis and encephalitis in children. An update. Neurol Clin 2002; 20: 1013–1038.

Bruns R, Müller CE, Mentel M, Wegner M, Wiersbitzky SKW. Nagayama-Papeln als Schleimhautsymptom bei einer HHV-6- und/oder HHV7-Infektion. Kinderärztl Praxis 2004; 33(Suppl 12) 33. Jahrestagung DGPI.

Buchbinder S, Elmaagacli AH, Schaefer UW, Roggendorf M. Bone Marrow Transplant 2000; 26: 639–644.

Carrigan DR, Drobyski WR, Russler SK, Tapper MA, Knox KK, Ash RC. Lancet 1991; 338: 147–149.

Caserta MT, Hall CB, Schnabel K, McIntire K, Long C, Costanzo M, Dewhurst S, Insel R, Epstein LG. J Infect Dis 1994; 170: 1586–1589.

Chua KB, Lam SK, Sazaly AB, Lim ST, Paranjothy M. Med J Malaysia 1999; 54: 32–36.

Cone RW. Semin Respir Infect 1995; 10: 254–258.

Cone RW, Hackman RC, Huang ML, Bowden RA, Meyers JD. Metcalf, M, Zeh J, Ashley R Corey L. N Engl J Med 1993; 329: 156–161.
Cone RW, Huang ML, Hackman RC. Leuk Lymphoma 1994; 15: 235–241.
Cone RW, Huang ML, Hackman RC Corey L. J Clin Microbiol 1996; 34: 877–881.
De Bolle L, Naesens L, De Clercq E. Clin Microbiol Rev 2005; 18: 217–245.
De Clerck LS, Bourgcois N, Krueger GRF, Stevens WJ. Human herpesvirus-6 in Sjoegren's syndrome. In: Human Herpesvirus-6 (Ablashi DV, Krueger GRF, Salahuddin SZ, editors). Amsterdam: Elsevier; 1992; pp. 303–315.
Eberle J, Hollinger J, Deinhardt F. Joint meeting Sektion Virologie. Dtsch Ges Hyg Mikrobiol & Soc Gen Microbiol, Hamburg (Germany) 1988; September: 15–17.
Eeg-Olofsson O. Brain Dev 2003; 25: 9–13.
Fox JD, Briggs M, Ward PA, Tedder RS. Lancet 1990; 336: 590–593.
Glanzmann E. In Handbuch der Inneren Medizin. Band I, Teil 1 (Infektionskrankheiten). Berlin-Göttingen-Heidelberg: Springer; 1952; pp. 260–268.
Hall CB, Caserta MT, Schnabel KC, Boettrich C, McDermott MP, Lofthus GK, Carnahan JA, Dewhurst S. J Pediatr 2004; 145: 472–477.
Hammerling JA, Lambrecht RS, Kehl KS Carrigan DR. J Clin Pathol 1996; 49: 802–804.
Hentrich M, Oruzio D, Jaeger G, Schlemmer M, Schleunig M, Schiel X, Hiddemann W, Kolb HJ. Br J Haematol 2004; 128: 66–72.
Horwitz CA, Krueger GRF, Steeper TA, Bertram G. HHV-6 induced mononucleosis-like illnesses. In: Human Herpesvirus-6 (Ablashi DV, Krueger GRF, Salahuddin SZ, editors). Amsterdam: Elsevier; 1992; pp. 159–174.
Knox KK, Carrigan DR. Lancet 1994; 343: 577–578.
Kondo K, Nagafuji H, Hata A, Tomomori C, Yamanishi K. J Infect Dis 1993; 167: 1197–1200.
Krueger GRF, Ablashi DV. Intervirology 2003; 46: 257–269.
Krueger GRF. Ferrer-Argote V. In Vivo 1994; 8: 493–500.
Krueger GRF, Ablashi DV, Whitman JI, Luka J, Rojo J. Rev Med Hosp Gen Mexico 1998a; 61: 226–240.
Krueger GRF, Koch B, Leyssens N, Berneman Z, Rojo J, Horwitz C, Sloots T, Margalith M, Conradie JD, Imai S, Urasinski I, Bryere M, Ferrer-Argote V, Krueger J. Vox Sang 1998b; 75: 193–197.
Krueger GRF, Sander C, Hoffmann A, Barth A, Koch B, Braun M. In Vivo 1991; 5: 217–226.
Krueger GRF, Wassermann K, De Clerck LS, Stevens WJ, Bourgeois N, Ablashi DV, Josephs SF, Balachandran N. Lancet 1990; 336: 1255–1256.
Leland DS. Emmanuel D Semin Respir Infect 1995; 10: 189–198.
Levy DJ. JAMA 1921; 77: 1785–1786.
Michaelides A, Glare EM, Spelman DW, Wesselingh SL, Hoy JF, Mijch AM Kotsimbos T. J Infect Dis 2002; 186: 173–180.
Nagate A, Ohyashiki JH, Kasuga I, Minemura K, Abe K, Yamamoto K Ohyashiki K. Int J Mol Med 2001; 8: 379–383.
Nagayama T, Inoue K, Kawasaki K. Med J Kagoshima Univ 1956; 8: 212–217.
Okuno T, Takahashi K, Balachandra K, Shiraki K, Yamanishi K, Takahashi M, Baba K. J Clin Microbiol 1989; 27: 651–653.
Pitala AK, Liu-Yin JA, Freemont AJ, Morris DJ, Fitzmaurice RJ. J Med Virol 1993; 41: 103–107.

Purtilo DT, Tatsumi E, Manolov G, Manolova Y, Harada S, Lipscomb H, Krueger G. Int Rev Exp Pathol 1985; 27: 113–183.
Rauter L, Mutz I. Mschr Kinderheilkd 1995; 143: 1217–1219.
Ross DJ, Chan RC, Kubak B, Laks H, Nichols WS. Transplant Proc 2001; 33: 2603–2606.
Roush KS, Domiati-Saad RK, Margraf LR, Krisher K, Scheuermann RH, Rogers BB, Dawson DB. Am J Clin Pathol 2001; 116: 648–654.
Russler SK, Tapper ME, Knox KK, Liepins A, Carrigan DR. Am J Pathol 1991; 138: 1405–1411.
Safdar A, Brown AE. Amer J Med 2001; 111: 329.
Singh N, Carrigan DR. Ann Intern Med 1996; 124: 1065–1071.
Steeper TA, Horwitz CA, Ablashi DV, Salahuddin SZ, Saxinger C, Saltzman R, Schwartz B. Am J Clin Pathol 1990; 93: 776–783.
Takahashi K, Sonoda S, Kawakami K, Miyata K, Oki T, Nagata T, Okuno T, Yamanishi K. Lancet 1988; 331: 1463.
Taplitz RA, Jordan MC. Semin Respir Infect 2002; 17: 121–129.
Totani Y, Demura Y, Ameshima S, Miyamori I, Ishizaki T. Nihon Kokyuki Gakkai Zasshi 2001; 39: 763–769.
Veeder BS, Hempelmann TC. J. A.M.A. 1921; 77: 1787–1789.
Vuorinen T, Kotilainen P, Lautenschlager I, Kujari H, Krogerus L Oksi J. J Clin Microbiol 2004; 42: 5415–5418.
Wagner M, Mueller-Berghaus J, Schroeder R, Sollberg S, Luka J, Leyssens N, Schneider B, Krueger GRF. J Med Virol 1997; 53: 306–312.
Wegner M, Müller CE, Mentel M, Bruns R, Wiersbitzky SKW. Klinische Erscheinungsbilder aktiver HHV7-Infektionen bei Kindern und Jugendlichen. Kinderärztl Praxis 2004; 34(Suppl 12) 34. Jahrestagung DGPI.
Westcott TS. Am J Med Sci 1921; 163: 367–372.
Wiersbitzky S, Abel E, Bruns R, Eberle J, Ladstätter L, Wiersbitzky H, Paul W, Deinhardt F. Kinderaerztl Prax 1991a; 59: 258–261.
Wiersbitzky S, Bruns R, Wiersbitzky H, Ballke E- H. Paediatr Grenzgeb 1993; 31: 195–197.
Wiersbitzky S, Eberle J, Bruns R, Weidemeier H, Bittner S, Dorn U, Frick G, Abel E, Ladstätter L, Deinhardt F. Kinderaerztl Prax 1991b; 59: 210–213.
Wiersbitzky S, Eberle J, Ladstätter L, Bruns R, Hollinger J, Deinhardt F, Abel E. Kinderaerztl Prax 1989a; 57: 155–162.
Wiersbitzky S, Eberle J, Ladstätter L, Hollinger J, Bruns R, Deinhardt F, Bittner S, Abel E. Paediatr Grenzgeb 1989b; 28: 351–363.
Wiersbitzky S, Wiersbitzky H, Bruns R, Ballke E- H. Kinderaerztl Prax 1992; 60: 159–162.
Willi H. Schweiz med Wschr 1929; 10: 953–957.
Yamamoto K, Yoshikawa T, Okamoto S, Yamaki K, Shimokata K, Nishiyama Y. J Med Virol 2005; 75: 70–75.
Yamanishi K, Okuno T, Shiraki K, Takahashi M, Kondo T, Asano Y, Kurata T. Lancet 1988; 331: 1065–1067.
Yata K, Nakajima M, Takemoto Y, Yamada O, Okada M, Takatsuka H, Okamoto T, Wada H, Otsuki T, Yawata Y, Kakishita E, Sugihara T. Kansenshogaku-Zasshi 2002; 76: 385–390.
Zahorsky J. Pediatrics 1910; 22: 60–64.
Zahorsky J. JAMA 1913; 61: 1446–1450.
Zahorsky J. Arch Pediatr 1925; 42: 610–613.

Human Herpesvirus-6: General Virology, Epidemiology and Clinical Pathology
Perspectives in Medical Virology
Gerhard Krueger and Dharam Ablashi (Editors)

DOI 10.1016/S0168-7069(06)12014-5

HHV-6 and the Lympho-Hematopoietic System

Sylvie Ranger-Rogez[a], Aurelie Lacroix[a], François Denis[a], Dominique Bordessoule[b]
[a]*Department of Virology, Equipe d'accueil EA 4021, Dupuytren University Teaching Hospital, 87042 Limoges, France*
[b]*Department of Haematology, Dupuytren Hospital, 87042 Limoges, France*

Introduction

The human herpesvirus type 6 (HHV-6) was initially discovered by Salahuddin et al. (1986). The virus was isolated from fresh peripheral blood mononuclear cells (PBMC) from six patients with various lymphoproliferative disorders (LDs): angioimmunoblastic lymphadenopathy (AIL), cutaneous T-cell lymphoma, immunoblastic lymphoma, acute lymphocytic leukaemia (ALL). The other two patients were suffering from an acquired immunodeficiency syndrome (AIDS)-related lymphoma or from a dermatopathic lymphadenopathy. This new virus was reported initially to selectively infect human B cells, but further studies revealed that it grew preferentially in $CD4^+$ T-lymphocytes. Two groups of HHV-6 were then identified, HHV-6 variants A and B: they are closely related and nucleotide sequence variations are 1–5% for the conserved genes and 10–25% for immediate-early genes.

As specific diagnostic tests became available to distinguish between active and latent infections as well as primary and non-primary virus infections (see Chapters 7 and 8), interpretation of disease associations with HHV-6 became more complicated. In addition, as with many other viral infections, the sole presence of the virus itself does not yet prove that it actually causes disease, and, unfortunately, the common species-specificity of herpesviruses limits the development of animal models for testing Koch's principles. As a consequence of this dilemma, the medical literature contains quite controversial opinions about HHV-6 and diseases, sometimes even published by the same authors. This review will attempt to give a state-of-the-art description of HHV-6 and the lympho-hematopoietic system, yet it cannot solve many of the problems as outlined above. HHV-6 infectiology remains an ongoing challenge.

In vitro data

Both HHV-6A and -B readily infect $CD34^+$ hematopoietic stem cells and cells differentiating towards the T-cell lineage, i.e. $CD34^+CD38^+$ cells, NK-cells, γ/δ T-lymphocytes and $CD4^+$ or $CD8^+$ cells. Among the differentiated T-cells, non-naive $CD4^+$ cells are equally infected by both HHV-6 strains A and B, while $CD8^+$ cells are preferentially infected by HHV-6A. In addition, HHV-6 can apparently also infect cells differentiating towards the myelo-monocytic lineage including such highly specialized cellular end-stages as dendritic reticular cells and vascular endothelial cells. *In vitro* data for HHV-6 and essential pathological effects of HHV-6 cellular infections are discussed in Chapters 5 and 10.

The reaction of lymphoid tissue components to HHV-6 infection was studied by *ex vivo* techniques. Upregulation in infected cells of certain cell membrane markers (e.g. $CD4^+CD21^+$ and others), which was in part due to virus-induced rigidification of the cell membrane, thereby providing one possible basis for superinfection of HHV-6-infected cells by other viruses such as Epstein–Barr virus (EBV) and human immunodeficiency virus (HIV), and downregulation of other markers (CD3, CD46), were demonstrated.

It may be added, though, that even under standardized tissue culture conditions, the biological effects of HHV-6 may vary somewhat. This applies not only to different variants (A or B) of the virus, but also to different isolates of the same variant.

Evolution of techniques used and evolution of findings

Given the circumstances of the first isolation of HHV-6, its lymphotropism and ability to infect and replicate in human hematopoietic cell lines and its faculty to remain in a latent state, this ubiquitous pathogen was thought to be implicated in LD. The first studies consisted of serological screening of patients' sera and the main technique used was the immunofluorescent assay (IFA). Several studies (Torelli et al., 1992) noticed a significantly higher seroprevalence of anti-HHV-6 antibodies in sera from patients suffering from Hodgkin's disease (HD) compared to controls, e.g. 95% versus 76% in India (Shanavas et al., 1992). Similar results were reported in other LD such as low-grade non-Hodgkin's lymphomas (NHL) and acute myeloid leukaemia. A statistically significant increase in the levels of anti-HHV-6 antibodies in HD patients compared to healthy controls was also reported (Clark et al., 1990), although not found by others (Shanavas et al., 1992). Such results should be interpreted prudently either as an immune activation, which is not related to the presence of HHV-6, or inversely they could be in favour of a role of HHV-6 in such diseases. Levine et al. (1992) noticed significant increase in titres during the follow-up of 37 patients with HD who relapsed, compared to the decrease observed in 39 controls who did not relapse, suggesting that HHV-6 serology may be of prognostic value in this context. Using an enzyme-linked immunosorbent assay (ELISA), Iyengar et al. (1991) detected a low percentage of sera positive for anti-p41-38 antibodies among different serum donor categories and a high percentage in the case of patients with African Burkitt's lymphoma (ABL) or HD, denoting the presence of active HHV-6 replication in these patients.

Further studies then employed molecular methods including *in situ* hybridization (ISH) or polymerase chain reaction (PCR) to detect or amplify viral sequences in tissues from patients. Numerous studies were published, with variable and sometimes discordant results; some of them are mentioned in Tables 1 and 2. Altogether, HHV-6 DNA has been found in lymph nodes from 0 to 85% of patients suffering from LD; with these figures varying according to the number of patients studied, the variety of pathologies explored and assay techniques used. Because of differences in assays employed and in sensitivity of the different assays, results reported by the numerous papers are difficult to analyze. Furthermore, the percentage of tumour cells in the tissues examined was not evaluated in most studies. However, amelioration of the employed assays, with increased sensitivity and specificity, improved the detection of HHV-6 sequences in biopsies studied.

In order to examine which role HHV-6 can play in LD, a few teams attempted to quantify viral genomes in lymph nodes of LD patient samples. Some teams used two techniques of different known sensitivities, e.g. Southern-blot analysis and PCR (Valente et al.,1996), providing an assay discriminating between low and high levels of viral DNA. More recently, quantitative PCR (qPCR) (Secchiero et al., 1995; Ohyashiki et al., 1999) and quantitative real-time PCR (qRTPCR) have been developed (Collot et al., 2002). Reported results are discussed in the following sections.

HHV-6 infection can induce atypical lymphoproliferations

Atypical lymphoproliferations (APLs) constitute a heterogenous group of lesions that clinically mimic malignant lymphomas, but are lacking the criteria of monoclonality and malignant transformation. The incidence of APL seems to be increased in patients with immune disorders, and persistently active infection by lymphotropic viruses is frequently found in APLs. They can be defined as premalignant lymphoproliferations and may finally transform to malignant lymphomas. "Type B reticulum cell sarcoma" or the Moloney and Gross virus-induced lymphoblastic lymphomas in mice represent similar polyclonal cell proliferations preceding virus-induced lymphomas. In humans, the majority of reported cases are associated with herpesvirus infections, namely EBV and/or HHV-6.

Table 1

HHV-6 prevalence in lymph nodes from HD patients

No. of samples	No. of positive (%)	Subtype of positive HD	Assays	References
25	3 (12%)	ns	Southern-blot	Torelli et al. (1991). Blood 77, 2251–2258
35	0	ND	PCR	Gledhill et al. (1991). Br J Cancer 64, 227–232
39	3 (8%)	ns	PCR	Torelli et al. (1992). Leukemia 6, 46S–48S
14	9 (64%)	ND	PCR	Sumiyoshi et al. (1993). Am J Clin Pathol 99, 609–614
45	13 (29%)	Most = ns	Nested-PCR	Di Luca et al. (1994). J Infect Dis 170, 211–215
52	38 (73%) 47 (82.4%)	ns: 73.1%, mc: 70%	PCR ISH	Valente et al. (1996). Am J Pathol 149, 1501–1510
88	11 (13%)	ns:5, mc: 4	Nested-PCR	Schmidt et al. (2000). Leuk Res 24, 865–870
47	0	ND	PCR	Shiramizu et al. (2001). J Pediatr hematol Oncol 23, 282–285
37	13 (35.1%)	Nodular sclerosis	Real-time PCR	Collot et al. (2002). J Clin Microbiol 40, 2445–2451
	39%		PCR	Tailor et al. (2004). Ann NY Acad Sci 1022, 282–285

Note: ND, not determined; ns, nodular sclerosis HD; and mc, mixed cellularity HD.

Berthold et al. (1989) described a Canal–Smith syndrome occurring in two boys from a family. Patients exhibited lymphoblasts and peripheral blood examination evoked acute lymphocytic leukaemia. Nevertheless, immunophenotyping of PBMCs revealed polyclonality and no atypical blasts were seen in the bone marrow. An investigation of an enlarged lymph node from one boy showed an overall intact structure with paracortical expansion of a polymorphic lymphoid population containing many blasts and rare giant cells that did not express CD30. ISH revealed focal accumulation of HHV-6 positive cells; HHV-6 was in a productive infectious form as demonstrated by coculture. The boys' mother, who had a past history of "acute lymphocytic leukaemia successfully treated", at the age of 3 years, still exhibited persistent lymphocytosis. One of the boys and the mother had in fact an

Table 2

Results of some of the main studies conducted on prevalence of HHV-6 in lymph node biopsies of patients suffering from non-Hodgkin's lymphoma

Pathology	No. of samples	No. of positive (%)	Assays	References
NHL (2 B cell-SS)	157	5 (3.2%)	Blot hybridization	Jarrett et al. (1988). Leukemia 2, 496–502
SS lymphomas	14	1 (7.1%)	Nucleic acid hybridization	Fox et al. (1989). J Autoimmun 2, 449–455
NHL	16	2 (12.5%)	ISH and PCR	Borisch et al. (1991). Virchows Arch B Cell Pathol Incl Mol Pathol 61, 179–187
NHL	113	3 (2.6%)	PCR	Dolcetti et al. (1996). J Med Virol 48, 344–353
NHL	106	2 (1.9%) B lineage	PCR	Torelli et al. (1992). Leukemia 6, 46S–48S
NHL	76	45 (59%) 18/29 B-cell lymph (62%)	PCR O hybridization	Sumiyoshi et al. (1993). Am J Clin Pathol 99, 609–614
AILD	8	4 (50%)	PCR	Luppi et al. (1993). Leuk Res 17, 1003–1011
AILD-L	4	3 (75%)		
NHL	45	8 (17.8%)	ISH	Yin et al. (1993). Arch Pathol Lab Med 117, 502–506
DLBCL	7	2 (28.6%)		Allory et al. (2001). Am J Surg Pathol 25, 865–874
NHL (B cell origin)	36	8 (22.2%)	qRTPCR	Collot et al. (2002). J Clin Microbiol 40, 2445–2451
NHL (T or NK cell origin)	13	3 (23.1%)		
NHL		34%	PCR	Tailor et al. (2004). Ann NY Acad Sci 1022, 282–285

Note: DLBCL, diffuse large B-cell lymphomas; AILD, angioimmunoblastic lymphadenopathy; AILD-L, angioimmunoblastic lymphadenopathy-like lymphoma; NHL, non-Hodgkin's lymphomas; and SS, Sjögren's syndrome.

IgG2 deficiency. Interpretation of these data concluded as unusual lymphostimulation by persistently active HHV-6 infection.

Krueger et al. (1989) conducted a large study focusing on LD and APL associated with persistent infection by EBV and/or HHV-6. They underlined the improved detection of herpesviruses-associated APL in immunocompromised subjects such as AIDS patients or allograft recipients; two cases were also associated with Gougerot–Sjögren's syndrome (SS). APL was reported to be a histological reminiscence of excessive infectious mononucleosis or mononucleosis syndrome, and although EBV and HHV-6 infect different lymphocyte populations, positive APL exhibited no histologic differences according to the virus found.

Borisch et al. (1991), in a study conducted to identify forms of lymphoproliferation linked to HHV-6, detected two positive lymphadenopathies by PCR and ISH occurring in young adults. Because the diagnosis of HHV-6 is not always undertaken in cases of APL, its association with this pathology is probably underestimated. Interestingly, reported cases of APL mentioned higher amounts of HHV-6 DNA as shown with PCR and ISH, compared subjects without specific histologic abnormalities where it could rarely be detected. HHV-6 is therefore suspected to contribute to lymphomagenesis by inducing progressive lymphoproliferation.

A case of Kikuchi–Fujimoto disease, concomitant with an acute HHV-6 infection, was diagnosed in a 34-year-old man. Its involvement in Rosai–Dorfman disease, a rare benign, paediatric lymphadenopathy has also been envisaged.

The main LD associated with HHV-6, in immunocompetent subjects, is probably HD

HD is characterized by the expansion of Reed–Sternberg (RS) cells, which are now postulated to be of B-cell lineage. Numerous teams have worked on the association of HHV-6 with HD testing; first, the seroprevalence, then prevalence of viral DNA in lymph nodes from patients and results reported are variable and sometimes contradictory, probably due to the difference in assays employed and also in patients tested as discussed previously. Table 1 summarizes some of the results reported by molecular methods. Some teams detected no HHV-6 in lymph nodes of HD patients, for example, Gledhill et al. (1991) or Shiramizu et al. (2001). In this latter case, the patients studied were children suffering from paediatric HD. In contrast, a prevalence as high as 77% or 82% was found by others (Table 1). In our experience, the same lymph node population examined by two different assays revealed a positivity of 35.1% or 78%. Although it was difficult to conclude definitively on this subject, it seems that HHV-6 can be detected in more than one-third of HD biopsies suggesting that it might be associated with at least a subset of this disorder.

Characteristics of HHV-6 found in lymph nodes

Viral quantification

As mentioned previously, a few papers attempted to quantify viral genomes in samples of HD patients. Ohyashiki et al. (1999), using a qPCR–ELISA, observed significant higher levels of HHV-6 DNA for all LD patients combined (21 blood specimens and 19 lymph nodes), compared to controls (23 peripheral blood leukocytes from healthy volunteers, four lymph nodes with reactive hyperplasia). Valente et al. (1996) obtained <10,000 HHV-6 copies/μg DNA for positive lymph node samples, while Secchiero et al. (1995) reported <10–1000 HHV-6 genome equivalents per μg DNA among AIDS–HD samples. A qRTPCR developed in our laboratory revealed from 100 to 864,640 copies/μg DNA (Collot et al., 2002).

All the reports underline the fact that HHV-6 may be present in variable quantities in lymph nodes according to patients studied, but, however, in amounts higher than those found in healthy subjects.

HHV-6B is more often detected than HHV-6A

The detection of HHV-6 variants in HD samples was done by several authors and in most cases, the HHV-6 variant B was found almost exclusively (Valente et al., 1996; Collot et al., 2002). This determination was often done by restriction polymorphism analysis of amplified fragments and electrophoresis, taking advantage of the presence of a well-characterized *Hind* III site polymorphism within the viral genome. This fact is not surprising considering the greater lymphotropism of HHV-6B compared to HHV-6A. This raises the additional question of variant-specific HHV-6 pathogenicity.

HHV-6 is found essentially in scleronodular HD in young adults

EBV, another herpesvirus frequently associated with HD, is not distributed equally among the different subtypes of EBV-positive HD; 70% of mixed cellularity, >95% of lymphocyte depleted, 10–40% of nodular sclerosis, and almost absent from lymphocyte-predominant HD subtypes. In addition, HD cases associated with EBV occurred most frequently in children or in elderly people. Previous epidemiological studies suggested multiple etiologies for HD and led to the hypothesis of an infectious viral non-EBV etiology for cases occurring in young adults. It was therefore interesting to look for the presence of HHV-6 sequences according to HD subtype.

Torelli et al. (1992) described three HHV-6-positive HD, belonging to the nodular-sclerosis-lymphocyte-depletion subgroup, which occurred in young women (27-, 28- and 31-years old). Similarly, different authors obtained the highest prevalence and highest mean copy for the scleronodular subtype of HD (Table 1). In a large study conducted on 86 adult HD patients, we obtained a high prevalence (83.6%) of HHV-6 in scleronodular HD (unpublished data). However, scleronodular subtype is the most frequent in HD patients examined. In the same study, the mean age of scleronodular HD patients positive for HHV-6 and negative for EBV, was 29.5 years, whereas 45 years if HHV-6 negative and EBV positive. Therefore, HHV-6 seems to be more predominant in scleronodular HD and much more in young adults, arguing a potential role for this virus in the etiology of HD in this context. Clark et al. (1990) found an association of increased HHV-6 seropositivity and geometric mean titre ratio with HD among young adults lacking social contact in the family group, suggesting that those patients were exposed late to HHV-6.

HHV-6 may be associated with EBV

Some cases of HD are associated with EBV, especially cases of mixed cellularity occurring in children and in elderly people. Although HHV-6 and EBV infect different cells, they can be present in the same HD tissue. Using two different real-time PCRs (one for HHV-6 and one for EBV), we found 68.2% of HD patients positive for both viruses in their lymph nodes. In this case, EBV was often weakly present compared to HHV-6 (unpublished data). Identical findings have also been reported by others (Torelli et al., 1992; Valente et al., 1996).

HHV-6 is present in different cell types and also in RS cells

Cells harbouring HHV-6 were studied in positive lymphoid lesions by ISH and/or immunohistochemistry (IHC) (Valente et al., 1996; Luppi et al., 1998): elevated number of cells carrying HHV-6 was found in cases of HD not yet in reactive lymphoid hyperplasia. Virus-infected cells were identified primarily as lympho-histiocytic cells, less frequently as Hodgkin's and RS cells. In some studies, HHV-6 positive-cells appeared small and lymphoid in appearance (Krueger et al., 1989). In other studies, RS or HD cells expressing HHV-6 non-structural and/or structural antigens were detected with monoclonal antibodies (MAbs) in up to 20% of HD patients. In our experience, not all, but at least 30–50% of lymph nodes from HHV-6 positive HD exhibit RS cells expressing HHV-6 structural antigens (HHV-6A or -B gp116/54/64) as determined by the use of MAbs. In these specimens and also in other HHV-6-positive lymph nodes, HHV-6 is also detected in lymphocytes–plasmocytes and in histiocytic cells (unpublished data). This phenomenon is also described for EBV, which, when present, may be localized in RS cells and also in lymphocytes.

The objective of these studies was to determine if HHV-6 can be detected in uninvolved tumour tissues and therefore be considered as a simple passenger, or if it is restricted to tumour cells and can be proposed as an etiological agent of HD. Unfortunately, biology is often much more complex and questions are still open. Does HHV-6 present in RS cells play a role in the cell transformation? Does HHV-6 present in the "cell environment" act as a transactivator, for example? Or is HHV-6 just a passenger?

HHV-6 can also be detected in other lymphoproliferative diseases

Some teams looked for the presence of HHV-6 in lymph nodes from NHL patients and reported no viral sequences in these tissues. In contrast, HHV-6 presence has been detected by Southern blot or by ISH in some cases of NHL (Table 2). In some papers, values recorded were close to those obtained for HD. In a cohort of NHL patients, we found HHV-6 DNA in 22.2% of B-cell lymphomas and 23.1% of T-cell lymphomas.

As for HD, and probably more, results reported by different teams vary largely. This could be in relation to the great diversity of LD examined in each study, with often just a few cases representative of each haematological entity. Histologically, HHV-6-positive LD cases were not distinguishable from negative ones.

NHL in immunocompetent subjects

B-cell lymphomas

On the basis of the detection of active replication, HHV-6 was suspected to be associated with some ABLs (Iyengar et al., 1991). HHV-6 sequences were also detected in lymph nodes of European Burkitt's lymphoma patients at a high copy number.

Other B-cell lymphomas are not frequently reported to be associated with HHV-6 and the virus, probably, has no direct role in the development of B-cell NHL. Borisch et al. (1991) reported two positive B-cell lymphomas of high-grade malignancy with clonal proliferations, originating from elderly patients. Using a qPCR–ELISA, Ohyashiki et al. (1999) obtained a low amount of HHV-6 (6.4 copies/μg of DNA) for a B-cell lymphoma, although in other B-cell lymphomas they studied, number of HHV-6 genomes was <5 copies/μg of DNA in blood and lymph node specimens.

More recently, Daibata et al. (2000) studied a subset of low-grade primary ocular lymphomas, which arise from mucosa-associated lymphoid tissue (MALT), and found four HHV-6-positive cases among 14 tested (28.5%). Detection of HHV-6 DNA was rarely found in tumour cells and mostly in healthy tissues.

T-cell lymphomas

One of the initial isolates described by Salahuddin et al. (1986) was from a patient suffering from angioimmunoblastic lymphadenopathy disease (AILD). At the time, different reports underlined the high percentage of AILD associated with HHV-6 and Ohyashiki et al. (1999) found very high levels of HHV-6 DNA in lymph nodes obtained from an AILD and from a T-cell lymphoma. A case of AILD also preceded a T-cell lymphoma found positive for HHV-6 (Jarrett et al., 1988).

AILD is a peripheral T-cell lymphoma characterized by the effacement of lymph node architecture and replacement by a proliferation in arborescence of small vessels and a hypocellular polymorphous infiltrate. Progression to lymphoma occurs when lymphoid cells are prominent and form clusters: expanded B-cell clones are often present beside the T-cell clones. A fatal issue is rapidly observed in most cases. The disease is thought to be due to excessive immune activity triggered by an antigen, which leads to a clonal expansion of B- and T-lymphocytes. A viral infection is suspected to directly or indirectly trigger this disorder: HHV-6 could represent a good candidate for this role, either as an etiologic agent or as a modulating element.

HHV-6 involvement in S100-positive T-cell chronic LD has been reported. Because HHV-6 was rarely found in cutaneous T-cell lymphomas or in mycosis fungoide tissues (1.1–3.3%), it seems improbable that it could play a role in these pathologies.

Cerebral lymphomas and lymphomas with a predilection for the nervous system

A few papers looked for the presence of HHV-6 in cerebral lymphomas and reported an absence or a very low number of positive samples. Interestingly, Viali et al. (2000) related the case of a 53-year-old man who developed an angiotropic large-cell lymphoma, which manifested as neurological symptoms resulting from intravascular lymphomatosis.

Lymphomas in SS patients

A few cases of patients suffering from SS followed, during the multi-step course of the disease, by an NHL associated with HHV-6, were described (Fox et al., 1989; Krueger et al., 1989). Because antibody titres were markedly elevated in several SS patients with lymphoma, a role for HHV-6 in the neoplastic transformation occurring in SS patients has been postulated.

HHV-6-associated lymphomas in immunocompromised individuals

In the context of immunosuppression, HHV-6 seems to be more rarely found in lymphoproliferative tissues.

Post-transplant LD

Lin et al. (1999) described the case of a gammadelta T-cell lymphoma associated with HHV-6, occurring in a 48-year-old man who had kidney transplantation 4 years ago. A case of large-cell-immunoblastic lymphoma succeeding HHV-6 viraemia and infection of the bone marrow, after allogeneic bone marrow transplant, was also reported. Both cases were EBV negative.

AIDS-related LD

Some studies, conducted either in paediatric or in adult patients, demonstrated no positivity for HHV-6 in AIDS-related LD, while the others reported about 30–40% AIDS-lymphomas harbouring HHV-6 sequences (Fillet et al., 1995; Asou et al., 2000). When the HHV-6 variant was determined, HHV-6B was largely predominant. Although it is quite difficult to have a general consensus on this question, most authors, citing the low number of DNA copies and the lack of significantly higher prevalence of HHV-6 in AIDS-associated LD, conclude against a strong association between this virus and the genesis of AIDS-lymphomas.

Acute lymphoblastic leukaemia

A high copy number of HHV-6 genome has been demonstrated in one case of a Philadelphia chromosome-positive early pre-B-cell ALL (Jarrett et al., 1988).

Different mechanisms have been proposed for HHV-6 in lymphoproliferative diseases

During the course of LD, conditions could be favourable for the reactivation of an opportunistic virus, such as HHV-6, when an immunosuppression, which often concerns T lymphocytes, occurs. Nevertheless, the fact that less HHV-6-associated LDs are found in post-transplant or AIDS patients, argues against this hypothesis.

The presence of HHV-6 could inversely be considered in favour of a direct role or as a cofactor in cell transformation. Therefore, different mechanisms have been put forth.

HHV-6 exhibits transforming capacities

The 3.9 kbp *SalI*-L fragment located within the direct repeat region of HHV-6 variant A was shown to transform the murine NIH3T3 cell line, human epidermal keratinocytes RHEK-1, and both primary and established rodent cells (Razzaque et al., 1993; Thompson et al., 1994). This transforming activity was localized to the DR7 gene and cells expressing DR7 protein-induced tumours when injected into immunodeficient nude mice, while cells expressing truncated DR7 protein did not (Kashanchi et al.,1997). Moreover, DR7 protein from HHV-6A strain U1102 binds to the human tumour-suppressor protein p53, which is a major control point of the cell cycle, involving the inhibition of p53-activated transcription (Collot et al., 2004).

Because DR7 from variants A and B share only 42.2% homology, it seemed interesting to us to look for the same oncogenic properties for DR7B. Consequently, using a two-hybrid system, we demonstrated that DR7B, like DR7A, was able to bind to human p53, indicating that the p53-binding site on the DR7 protein could be on the common portion of both viral proteins (unpublished data).

Recently, we looked for the expression of DR7 protein in tissue samples from HD previously detected as positive by HHV-6 structural MAbs. DR7 was strongly positive and it was principally found in RS cells, and to a lesser extent in other lymphoid cells (Fig. 1), signalling the expression of this oncogenic protein in transformed cells.

As a result of these transforming capacities, a latently HHV-6-infected cell line, named "Katata" cell line, was established from pathological tissues derived from

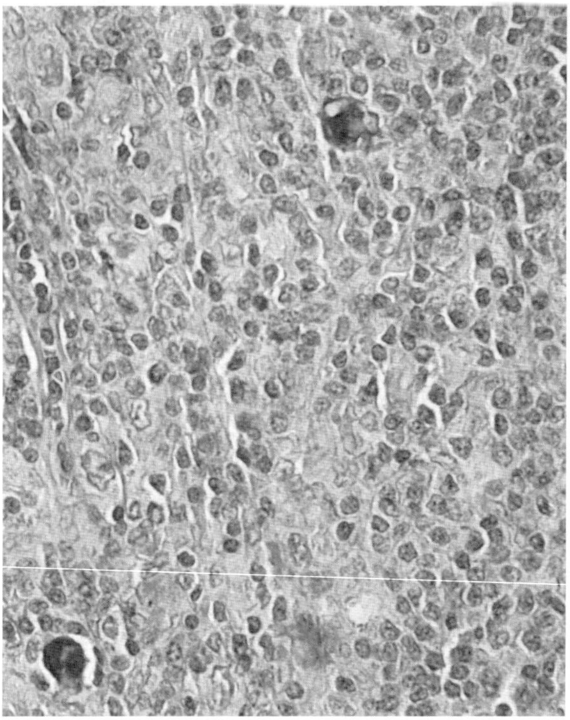

Fig. 1 Expression of DR7 protein in tissue samples from patients suffering from HD, previously detected positive by HHV-6 structural monoclonal antibodies. DR7 was strongly positive and was principally found in RS cells, and to a lesser extent in other lymphoid cells. (for colour version: see colour section on page 355).

an HHV-6-positive and EBV-negative Burkitt's lymphoma (Bandobashi et al., 1997). HHV-6 was also present in two leukaemia cell lines.

HHV-6 retains p53 within the cytoplasm

Recently, it was shown that p53 increases in the cytoplasm of HHV-6-infected cells, this stability of p53 being dependent on deubiquitination. The infected cells revealed resistance to apoptosis (Takemoto et al., 2004).

HHV-6 can transactivate genes

Transactivation of human genes

An interesting feature associated with HD is the abundant constitutive activation of the p50/p65 nuclear factor kappa B (NFkB) complex in cultured RS cells, this factor being normally observed only for limited time intervals after stimulation with diverse inducers. Moreover, constitutive NFkB activation has been shown to be required for survival and proliferation of RS cells, preventing them from undergoing apoptosis under stress conditions (Bargou et al.,1997). These findings tended to identify NFkB as an important component for understanding the HD pathogenesis. We found enhanced expression of p50 and p65 mRNAs and NFkB activation, as demonstrated by the use of reporter genes, in HHV-6B-infected cells compared to mock-infected cells. The increase in the NFkB levels in infected cells could be considered to be the result of transcriptional upregulation of the p50 and p65 genes (unpublished data).

Transactivation of viral genes

EBV transactivation. The EBV, considered as an oncogenic virus, is found in 50–60% of HHV-6-positive LD tissues. In this case, the intensity of signals obtained for EBV is weaker than for HHV-6, suggesting that EBV sequences are markedly less represented (Torelli et al.,1992). Techniques used do not indicate if both viruses were present in the same cells.

More recently, by superinfection of EBV latently infected cells with HHV-6A, an HHV-6 transactivation effect on EBV replication cycle has been demonstrated. This activation, mediated via a cyclic AMP-responsive element located within the EBV Zebra promoter requires infectious virus. This promoter controls the EBV gene product ZEBRA, which is responsible for disrupting EBV latency and initiating the lytic replication cascade (Flamand et al.,1996).

Activation of LMP1 and EBNA2 expression has also been reported, after HHV-6A infection of EBV-positive Burkitt's lymphoma cell lines. The authors have demonstrated the presence of positive and negative regulatory elements responsive to HHV-6A infection in LMP1 regulatory sequences (Cuomo et al., 1998).

HHV-8 transactivation. HHV-6, which was reported to be rarely present in HHV-8-related LD (Asou et al., 2000), is able to induce HHV-8 replication as demonstrated by coculture of HHV-6-infected T cells with HHV-8. HHV-8 replication results from activation of the first promoter activated during HHV-8 replication (ORF-50 promoter) (Lu et al., 2005).

HHV-6 has the ability to integrate into host cell chromosomal DNA

Luppi et al. (1993) demonstrated the pulsed-field gel electrophoresis HHV-6 integration to samples from two LD patients. Then, fluorescent ISH (FISH) on PBMCs from the patients in complete remission, showed integration of the HHV-6 genome in the distal part of the short arm of chromosome 17 (17p13) (Torelli et al., 1995). A possible deregulation of two human-known oncogenes, CRK and ABR could subsequently occur. FISH was also used to demonstrate integration of HHV-6 in the long arm of chromosome 1 (1q44) of leukaemic cells from an ALL (Daibata et al., 1998). The authors reported a surprising observation: the chromosomally integrated HHV-6 genome was transmitted in the same location (1q44) to one son and one granddaughter of that ALL patient, who are otherwise healthy. Another case localized HHV-6 to 22q13.

Thus, the high-viral load observed in some LD may be due to a chromosomal integration in some or even in all cells of those patients. However, viral integration seems to be rare, without any specific integration site. On the other hand, the detection of HHV-6 sequences integrated into host DNA of lymphoma cells does not prove the presence of a complete viral genome and does not indicate whether this chromosomally integrated HHV-6 genome has any role in the pathogenesis of HHV-6-positive LD.

HHV-6 may exert a role in cell proliferation by dysregulation of cytokine network

The hypothesis that HHV-6 may contribute to the development of LD by deregulation of cytokine control rather than by direct oncogenic involvement has been suggested. Recently, Takaku et al. (2005) studied the network of dynamic gene and protein interactions occurring during the infection of an adult T-cell leukaemia cell line by HHV-6B using a microarray and analysing the data by the Bayesian statistical framework. They reported the possible association between chemokine genes regulating the Th1/Th2 balance and genes regulating T-cell proliferation. Moreover, a gene encoding a TEC-family kinase, ITK, might be a putative modulator of the host immune response against HHV-6B infection.

An interesting approach consisted in the design of a computer model in order to simulate cell changes happening in LD and disturbances of the T-cell immune system (Krueger et al., 2003). The model uses the concept that these disturbances, identified as proliferation, differentiation and inhibition factors, may lead to hyperplastic, aplastic or neoplastic diseases. This computer model simulated acute and chronic persistent HHV-6 infections to study the influence of cytokines or chemokines in the Canale–Smith syndrome.

Conclusion

The presence of HHV-6 sequences in a variable proportion of LD, and more frequently in HD, raises the question of its role in these contexts. It could be considered just as a passenger, being so an opportunistic agent, or more probably, it could be thought as implicated in the pathogenesis involvement of some cases, such as scleronodular HD of young adults. Different potentially complementary mechanisms are envisaged.

References

Allory Y, Challine D, Haioun C, Copie-Bergman C, Delfau-Larue MH, Boucher E, Charlotte F, Fabre M, Michel M, Gaulard P. Am J Surg Pathol 2001; 25: 865–874.

Asou H, Tasaka T, Said JW, Daibata M, Kamada N, Koeffler HP. Leuk Res 2000; 24: 59–61.
Bandobashi K, Daibata M, Kamioka M, Tanaka Y, Kubonishi I, Taguchi H, Ohtsuki Y, Miyoshi I. Blood 1997; 90: 1200–1207.
Bargou RC, Emmerich F, Krappmann D, Bommert K, Mapara MY, Arnold W, Royer HD, Grinstein E, Greiner A, Scheidereit C, Dorken B. J Clin Invest 1997; 100: 2961–2969.
Berthold F, Krueger GR, Tesch H, Hiddemann W. Anticancer Res 1989; 9: 1511–1518.
Borisch B, Ellinger K, Neipel F, Fleckenstein B, Kirchner T, Ott MM, Muller-Hermelink HK. Virchows Arch B Cell Pathol Incl Mol Pathol 1991; 61: 179–187.
Clark DA, Alexander FE, McKinney PA, Roberts BE, O'Brien C, Jarrett RF, Cartwright RA, Onions DE. Int J Cancer 1990; 45: 829–833.
Collot S, Bass J, Denis F, Ranger-Rogez S. Rev Med Virol 2004; 14: 301–319.
Collot S, Petit B, Bordessoule D, Alain S, Touati M, Denis F, Ranger-Rogez S. J Clin Microbiol 2002; 40: 2445–2451.
Cuomo L, Trivedi P, de Grazia U, Calogero A, D'Onofrio M, Yang W, Frati L, Faggioni A, Rymo L, Ragona G. J Med Virol 1998; 55: 219–226.
Daibata M, Komatsu T, Taguchi H. Leuk Lymphoma 2000; 37: 361–365.
Daibata M, Taguchi T, Sawada T, Taguchi H, Miyoshi I. Lancet 1998; 352: 543–544.
Di Luca D, Dolcetti R, Mirandola P, De Re V, Secchiero P, Carbone A, Boiocchi M, Cassai E. J Infect Dis 1994; 170: 211–215.
Dolcetti R, Di Luca D, Carbone A, Mirandola P, De Vita S, Vaccher E, Sighinolfi L, Gloghini A, Tirelli U, Cassai E, Boiocchi M. J Med Virol 1996; 48: 344–353.
Fillet AM, Raphael M, Visse B, Audouin J, Poirel L, Agut H. J Med Virol 1995; 45: 106–112.
Flamand L, Menezes J. J Virol 1996; 70: 1784–1791.
Fox RI, Saito I, Chan EK, Josephs S, Salahuddin SZ, Ablashi DV, Staal FW, Gallo R, Pei-Ping H, Le CS. J Autoimmun 1989; 2: 449–455.
Gledhill S, Gallagher A, Jones DB, Krajewski AS, Alexander FE, Klee E, Wright DH, O'Brien C, Onions DE, Jarrett RF. Br J Cancer 1991; 64: 227–232.
Iyengar S, Levine PH, Ablashi D, Neequaye J, Pearson GR. Int J Cancer 1991; 49: 551–557.
Jarrett RF, Gledhill S, Qureshi F, Crae SH, Madhok R, Brown I, Evans I, Krajewski A, O'Brien CJ, Cartwright RA, Venables P, Onions DE. Leukemia 1988; 2: 496–502.
Kashanchi F, Araujo J, Doniger J, Muralidhar S, Hoch R, Khleif S, Mendelson E, Thompson J, Azumi N, Brady JN, Luppi M, Torelli G, Rosenthal LJ. Oncogene 1997; 14: 359–367.
Krueger GR, Brandt ME, Wang G, Buja LM. Anticancer Res 2003; 23: 123–135.
Krueger GR, Manak M, Bourgeois N, Ablashi DV, Salahuddin SZ, Josephs SS, Buchbinder A, Gallo RC, Berthold F, Tesch H. Anticancer Res 1989; 9: 1457–1476.
Levine PH, Ebbesen P, Ablashi DV, Saxinger WC, Nordentoft A, Connelly RR. Int J Cancer 1992; 51: 53–57.
Lin WC, Moore JO, Mann KP, Traweek ST, Smith C. Leuk Lymphoma 1999; 33: 377–384.
Lu C, Zeng Y, Huang Z, Huang L, Qian C, Tang G, Qin D. Am J Pathol 2005; 166: 173–183.
Luppi M, Barozzi P, Garber R, Maiorana A, Bonacorsi G, Artusi T, Trovato R, Marasca R, Torelli G. Am J Pathol 1998; 153: 815–823.
Luppi M, Marasca R, Barozzi P, Ferrari S, Ceccherini-Nelli L, Batoni G, Merelli E, Torelli G. J Med Virol 1993; 40: 44–52.
Ohyashiki JH, Abe K, Ojima T, Wang P, Zhou CF, Suzuki A, Ohyashiki K, Yamamoto K. Leuk Res 1999; 23: 625–630.

Razzaque A, Williams O, Wang J, Rhim JS. Virology 1993; 195: 113–120.
Salahuddin SZ, Ablashi DV, Markham PD, Josephs SF, Sturzenegger S, Kaplan M, Halligan G, Biberfeld P, Wong-Staal F, Kramarsky B, Gallo RC. Science 1986; 234: 596–601.
Schmidt CA, Oettle H, Peng R, Binder T, Wilborn F, Huhn D, Siegert W, Herbst H. Leuk Res 2000; 24: 865–870.
Secchiero P, Zella D, Crowley RW, Gallo RC, Lusso P. J Clin Microbiol 1995; 33: 2124–2130.
Shanavas KR, Kala V, Vasudevan DM, Vijayakumar T, Yadav M. J Exp Pathol 1992; 6: 95–105.
Shiramizu B, Chang CW, Cairo MS. J Pediatr Hematol Oncol 2001; 23: 282–285.
Sumiyoshi Y, Kikuchi M, Ohshima K, Yoneda S, Kobari S, Takeshita M, Eizuru Y, Minamishima Y. Am J Clin Pathol 1993; 99: 609–614.
Tailor PB, Saikia TK, Advani SH, Mukhopadhyaya R. Ann N Y Acad Sci 2004; 1022: 282–285.
Takaku T, Ohyashiki JH, Zhang Y, Ohyashiki K. Biochem Biophys Res Commun 2005; 336: 469–477.
Takemoto M, Mori Y, Ueda K, Kondo K, Yamanishi K. J Gen Virol 2004; 85: 869–879.
Thompson J, Choudhury S, Kashanchi F, Doniger J, Berneman Z, Frenkel N, Rosenthal LJ. Oncogene 1994; 9: 1167–1175.
Torelli G, Barozzi P, Marasca R, Cocconcelli P, Merelli E, Ceccherini-Nelli L, Ferrari S, Luppi M. J Med Virol 1995; 46: 178–188.
Torelli G, Marasca R, Luppi M, Selleri L, Ferrari S, Narni F, Mariano MT, Federico M, Ceccherini-Nelli L, Bendinelli M, et al. Blood 1991; 77: 2251–2258.
Torelli G, Marasca R, Montorsi M, Luppi M, Barozzi P, Ceccherini L, Batoni G, Bendinelli M, Muyombano A. Leukemia 1992; 6(Suppl 3): 46S–48S.
Valente G, Secchiero P, Lusso P, Abete MC, Jemma C, Reato G, Kerim S, Gallo RC, Palestro G. Am J Pathol 1996; 149: 1501–1510.
Viali S, Hutchinson DO, Hawkins TE, Croxson MC, Thomas M, Allen JP, Thomas SM, Powell KH. Muscle Nerve 2000; 23: 1295–1300.
Yin SY, Ming HA, Jahan N, Manak M, Jaffe ES, Levine PH. Arch Pathol Lab Med 1993; 117: 502–506.

Human Herpesvirus-6: General Virology, Epidemiology and Clinical Pathology
Perspectives in Medical Virology
Gerhard Krueger and Dharam Ablashi (Editors)

DOI 10.1016/S0168-7069(06)12015-7

Human Herpesvirus-6 and the Cellular Immune System

Judith A. Britz
Cylex, Inc., 8980–I Old Annapolis Road, Columbia, MD 21045, USA

Abstract

The emergence of human immunodeficiency virus (HIV) in the 1980s heightened our awareness that immunodeficiency could be acquired through viral infection. Historically, measles and hepatitis B and C viruses were implicated as immunosuppressive viruses; however, there is growing evidence that a number of the herpesviruses have properties similar to HIV with serious clinical sequelae. This chapter is devoted to a discussion of the impact of human herpesvirus-6 (HHV-6) infection on the immune system, focusing on the cell-mediated immune response. It is unclear whether preexisting immunodeficiency is a prerequisite for infection with HHV-6 or whether these viruses use stealth strategies for initial entry, establish latency and then exert their immunosuppressive effects. Given the complexity of the immune system, the mechanism(s) by which this virus family exerts its effect is not well understood. This chapter will review the range of evidence of immune system involvement from *in vitro* studies, clinical complications, and the number of immune-mediated disease associations that continue to mount as well as identify the need for the tools that will be required to research this highly prevalent, commensal organism. Even when causality cannot be clearly established and the coexistence of viruses in the same individual presents enormous scientific and epidemiological challenges to proving the etiological agent, there is a growing body of data that HHV-6 is an emerging pathogen with serious consequences ranging from its first clinical manifestation as exanthem subitum in infants to more debilitating fibromyalgia, chronic fatigue syndrome (CFS), and potentially multiple sclerosis. As molecular, serological, and cellular methodologies for HHV-6's detection improve, insight into causality will provide avenues for developing prophylactic and treatment strategies for this ubiquitous virus that affects millions of individuals.

Introduction

Human herpesvirus-6 (HHV-6) is the recognized etiological agent of roseola, a self-limited benign infection in infants characterized by rash and often fever (Yamanishi et al., 1988). In the Western world, it is transmitted to over 90% of children before the age of two (Okuno et al., 1989). However, even in immunocompetent children, latency is established in the brain and lymphoid system where HHV-6 exists as a commensal organism. Only when immune balance is seriously perturbed by superinfection with another virus (e.g., HIV) or immunosuppressive drugs (e.g., transplant) does reactivation occur. This chapter will review the current literature on HHV-6 and the immune system. Antibody responses to HHV-6 are detectable in 95% of most populations in the world after the age of three (Enders et al., 1990; Hall et al., 1994), and therefore the percentage of serologically positive individuals does not increase substantially with increasing age. There are significant differences in the prevalence of antibodies to the two variants, A and B, which has been thoroughly reviewed by Romagnoli et al. (2000) and will not be repeated here. The focus of this paper is on the cellular immune response evoked (or more appropriately suppressed) by HHV-6, a betaherpesvirus characterized by its tropism for lymphocytes (Yasukawa et al., 1998) with a recognized impact on the nervous system (Riel-Romero, 2005).

Isolation history: association with immunodeficiency disorders

Consistent with its reputation as an opportunistic infection, HHV-6 was originally isolated in 1986 by Salahuddin et al. (1986) from patients with AIDS where it is believed to be a cofactor in the natural history of HIV infection. Two biologically and genomically variant forms have been discovered, Types A and B (Clark, 2000). Variant A is of lower prevalence (16%) and can be isolated from the peripheral blood mononuclear cells (PBMC) of drug-induced immunosuppressed patients as well as those with lymphoproliferative disorders. Variant B is the more prevalent form (98%) and in contrast to Type A is frequently isolatable from saliva (Krueger et al., 1990; Levy et al., 1990; Aberle et al., 1996). The viability of the virus in saliva probably accounts for the ease of transmission in children under three.

Subversion of the immune system

It is unclear whether preexisting immunodeficiency is necessary for infection with HHV-6 or whether this virus uses stealth strategies for initial entry and then exerts its immunosuppressive effects. However, it is clear that once infection has occurred, HHV-6 is able to establish latency and remain quiescent for prolonged periods of time unless the immune system is otherwise immunocompromised.

One of the mechanisms by which viruses establish long-lived infections and subvert the immune system is by directly infecting immune cells thereby impairing surveillance. The original report of the isolation of HHV-6 by Salahuddin et al. (1986) described an isolate from patients with a B-cell lymphoproliferative disorder, although these lymphomas frequently contain T lymphocytes that were the likely source of the virus (G. Krueger, personal communication). Horvat et al. (1993) reported that HHV-6 inhibited the lymphoproliferative responses of human peripheral blood monocytes *in vitro*. Kondo et al. (2002) identified monocytes/macrophages as target cells during acute infection. According to Lusso et al. (1991, 1993, 1995), HHV-6 is able to productively infect CD4 and CD8 cells, although CD4 is most often selected and T-memory cells are preferentially targeted over naïve lymphocytes. Productive infections with HHV-6 are also possible in thymocytes, γ-δ T cells, and NK cells. In fact, since HHV-6 can induce *de novo* expression of CD4 antigens in CD8 cells, NK cells, and γ-δ T cells, the patient coinfected with HIV hypothetically has more cells bearing the receptor for HIV (Lusso et al., 1991, 1993, 1995; Schonnebeck et al., 1991). In contrast, dendritic cells and monocytes are non-productively infected, but their functionality is impacted by viral infection.

In vitro *studies: Th1 to Th2 switch*

HHV-6 replicates especially efficiently in CD4+ cells, the primary cells that orchestrate the immune response through complex direct cell to cell interactions as well as the secretion of multiple cytokines with both autocrine and paracrine effects (Gosselin et al., 1992). *In vitro*, HHV-6 has been shown to influence the immune balance between pro-inflammatory (Th1) and anti-inflammatory (Th2) cytokines. Interleukin 12 (IL-12) produced by macrophages is the predominant stimulator of Th1 activity, whereas IL-4 stimulates Th2. Human measles virus has been described to selectively suppress IL-12 (Karp, 1999), which drives the immune balance toward a Th2 response. Similarly, Arena et al. (1999) have described *in vitro* downregulation of IL-12 in peripheral blood monocytes as well as upregulation by HHV-6 of IL-10, a known suppressor of Th1 cells. These studies are consistent with those of Smith et al. (2003, 2004) in which HHV-6-induced suppression of IL-12 in macrophages following stimulation with interferon α and LPS. In addition, Smith et al. (2003, 2004) observed that even UV-inactivated HHV-6 downregulated IL-12, similar to observations with UV-irradiated measles virus. The fact that inactivated virus can inhibit IL-12 suggests that receptor blockade was made by the mechanism of inhibition, rather than a requirement for active replication. Over 10 years before, Flamand et al. (1991) had demonstrated that HHV-6 induced the secretion in peripheral blood cells of IL-1-beta, interferon-α, and tumor necrosis factor-α, also a known suppressor of IL-12. Flamand et al. (1995) later reported that HHV-6 decreased the production of IL-2, a product of Th1 cells, when using CD3 or phytohemagglutinin (PHA) as a stimulant. All of these *in vitro* observations confirm that HHV-6 is pivotal in shifting immune activity from a Th1 to a Th2 response, which is less likely to neutralize viral activity.

While specific immunity is undoubtedly involved in protection against the clinical consequences of HHV-6 infection, innate immunity, including NK cells may provide the first line of defense. Flamand et al. (1996) showed that HHV-6 actually stimulated PBMC *in vitro* to produce IL-15 that had an enhancing effect on NK-cell activity. This is also consistent with increases in NK-cell activity during the febrile phase of exanthema subitum. On the other hand, Lusso et al. (1993) had shown earlier than HHV-6 could induce CD4 expression on NK cells, postulating that it increased their susceptibility to infection with HIV-1. Collectively, these observations have led others to postulate that NK defect increases in susceptibility of individuals to the clinical sequelae associated with HHV-6 infection.

Another mechanism for susceptibility may involve CD46, widely represented on the surface of most nucleated cells, which has been suggested as a possible receptor for the HHV-6. This cell surface molecule has been implicated as a regulator of complement activation and may also be a receptor for the measles virus (Greenstone et al., 2002), Neisseria, and Group Strep A. It has been suggested by Lusso et al. (1991, 1993, 1995) that CD46 can act as a bridge between the innate and adaptive immune responses via regulation of IL-12. Fusion induced by CD46 may be the mechanism by which the cytopathic effect of the HHV-6 is expressed in heart transplant recipients as pancreatitis and gastroduodenitis (Mori et al., 2002).

Although *in vitro* studies can be valuable in forming hypotheses about the interactions of viruses with immune cells, the most compelling evidence of immunosuppression induced by HHV-6 is derived from clinical observations in pharmacologically suppressed patients undergoing transplant.

Reactivation: transplantation

Solid organ transplant

In 1992, Yoshikawa first published that 14% of kidney transplant recipients developed HHV-6 viremia in the first 2–4 weeks posttransplant and 55% showed an increase in anti-HHV-6 antibody titer in the first 3 months (Yoshikawa et al., 1992). Since then, Singh and Carrigan (1996), Singh and Patterson (2000), Ljungman (2002), and Lautenschlager et al. (2000) have reported HHV-6 as an emerging pathogen in solid organ transplantation. HHV-6 is expressed in the early weeks posttransplant, often exacerbating the severity of other diseases in the transplant recipient (Des Jardin et al., 1998, 2001; Dockrell et al., 1997, 1999), including cytomegalovirus (CMV) (Humar et al., 2000; Boeckh and Garret, 2003).

Until the routine availability and use of prophylactic anti-viral medications like ganciclovir in solid organ transplant recipients, CMV was one of the leading causes of morbidity and mortality (Tolkoff-Rubin and Rubin, 1998). Serologically CMV-negative patients receiving an organ from a CMV-positive donor are at greatest risk, although use of potent immunosuppressives like anti-lymphocyte globulin, steroids, and mycophenolate mofetil as well as a history of rejection increases the risk (Abbott et al., 2002). CMV seropositivity therefore confers some protection against CMV disease. Most recipients of a solid organ transplant have previously been infected with HHV-6 and the organ donors are also highly likely to have been infected by the age of two, therefore the probability of a D+/R− combination for HHV-6 is extremely low. If so, what is the explanation for the fact that infections with HHV-6 are estimated to recur in 14–82% of transplanted individuals (Des Jardin et al., 1998, 2001; Dockrell et al., 1997, 1999)? Do these HHV-6 antibodies fail to confer protection as in CMV? In fact, Dockrell et al. (1999) reported that HHV-6 seronegativity was a risk factor for fungal infection in liver transplant recipients. Since HHV-6 seroprevalence is so high, it was hypothesized that these individuals may be "false" negative reflecting that fact that they are already severely suppressed prior to transplant and therefore antibody negative (despite prior infection with HHV-6).

Management of the immunosuppressive regimen of transplant recipients has always required a balancing act to avoid the risks of infection and rejection. HHV-6 clearly poses heightened risks for the transplant recipient that is thoroughly reviewed by Ljungman in Chapter 22 and shown by others (Morris et al., 1989; Robert et al., 1994; and Snydman et al., 1993). In fact, Tong et al. (2002) published that HHV-6 was associated with chronic allograft nephropathy after renal transplantation. In addition, Acott et al. (1996) have reported that failure to use anti-viral therapy at the time of pulsed steroid therapy for HHV-6 associated rejection episodes in febrile pediatric kidney transplant recipients resulted in chronic allograft rejection. Fortuitously, ganciclovir, which shows some efficacy against HHV-6, may have protected many transplant recipients since CMV anti-viral prophylaxis is now a standard of care in the first months following transplant.

In the transplant community, neurological side effects are common including tremors, cognitive disorders, psychiatric complications, and depression. These sequelae have been described in most of the clinical trials of calcineurin inhibitor clinical trials in transplant recipients. Is it possible that the drugs are not themselves the source of these side effects but the immunosuppressive environment that they are designed to create? If so, reactivation of the herpesviruses, particularly those known to have neurotropisms like HHV-6 may actually be implicated.

Bone marrow transplant

The immunosuppressive regimens used in bone marrow transplant (BMT) tend to be even more severe than those in solid organ transplant. Most BMT recipients are immunocompromised initially by their underlying disease that is further exacerbated by total body irradiation, lymphodepleting antibodies, steroids, and the anti-proliferative drugs associated with chemotherapy. Although the immunosuppressive protocol for allogeneic transplants is considerably more aggressive than autologous transplants, there is no apparent significant difference in the incidence of HHV-6 recurrence between these transplants nor sibling versus unrelated donor grafts in a pediatric population (Yoshikawa et al., 2002; Savolainen et al., 2005). HHV-6 reactivation is common in BMT transplant patients (Wang et al., 2002) and associated with skin rash and fever, the same symptoms often manifested at the time of primary infection in children. Idiotypic myelosuppression is characterized by delayed neutrophil and platelet engraftment and is highly associated with the reactivation of HHV-6 (Dobryski et al., 1993; Carrigan and Knox, 1994). Bethge et al. (1999) has reported two cases of BMT patients with HHV-6 PCR-positive spinal fluid who display neurological symptoms, including disorientation, sleepiness, and short-term memory loss and showed improvement following treatment with foscarnet. Appleton et al. (1995) conducted a study in BMT recipients to determine the potential role of HHV-6 in graft-versus-host disease (GVHD). They established a significant relationship between the detection of HHV-6 in biopsy tissue and the severity of GVHD suggesting a causal role for HHV-6 in exacerbation of GVHD rather than as a consequence of GVHD therapy.

However, despite the severity of immunosuppression in BMT patients, a number of researchers report that life-threatening complications with HHV-6 are rare (Cone et al., 1999; Savolainen et al., 2005). It is possible that the extended prophylactic use of ganciclovir for CMV in BMT recipients has had the unintended benefit of limiting HHV-6 disease in these patients.

Current immunological tools

Although lymphoproliferation (LPA) has remained largely an immunological tool for the research laboratory, there are now a variety of new methodologies that do not require use of radioactive materials or the prolonged incubation times of LPA. These include flow cytometry assays to quantify subsets of circulating cells based on cell surface characteristics, ELISPOT to assess expression of cytokines, and newer cytokine genotype microarray methodologies. Genotype microarray technologies offer tremendous potential for identifying the activation of a wide array of genes throughout the course of disease, but so far these methods have not translated easily into the clinical lab setting and have been hampered by high cost and intense laboratory processing within hours of sample collection. While all of these methods offer insight into mechanisms of action of the immune system, there is a concomitant pressing need for clinical assays that can be standardized for the routine monitoring of patients over time. In addition, such methods must be amenable to use in the reference laboratory by providing results that are comparable from laboratory to laboratory and can withstand shipping conditions. In this context, a general biomarker for the immunosuppressive state would be a useful adjunct in the monitoring of HHV-6 susceptible and recently infected individuals. The FDA has recently cleared an assay for the detection of cell-mediated immunity in immunosuppressed populations, ImmuKnow™, which measures global T-cell function by stimulating whole blood cells with the plant mitogen, PHA, and then quantifying lymphocyte activation by the intracellular content of ATP in the selected CD4+ cell subset.

Disease associations between HHV-6 infection and multiple sclerosis, fibromyalgia, and chronic fatigue syndrome

Ablashi et al. (2005) have performed a meta analysis of over 85 published papers in which the relationship between HHV-6 and patients with multiple sclerosis or fibromyalgia, or chronic fatigue syndrome (CFS) was examined. Although a clear causal relationship has not been established between HHV-6 and these chronic illnesses, his study discerned that there is evidence of correlation in more than 75% of those papers that distinguish between active and latent virus. Since both *in vitro* studies and *in vivo* clinical experience provide evidence that HHV-6 is immunosuppressive, patients with a history of CFS (data provided through the courtesy of Daniel Peterson, MD) were tested for the functionality of their global T-cell response using an FDA-cleared test for cell-mediated immunity assessment (Cylex Immune Function Assay, ImmuKnow™) and these cellular immune responses were compared with individuals infected with HIV and stable transplant patients on immunosuppressive regimens.

Method: rapid immune function assay

Whole blood was collected into an 8 ml sodium heparin vacutainer tube during inpatient or routine clinical visits, stored at room temperature, and tested within 30 h of draw at Viracor (Kansas City, MO). The ImmuKnow™ assay (Catalog No. 4400, Cylex, Inc. Columbia, MD), which is FDA-cleared, was performed according to manufacturer's package insert. Briefly, 250 μl of whole blood was diluted with sample diluent, added to wells of a 96-well microtiter plate and incubated for 15–18 h with PHA in a 37°C, 5% CO_2 incubator. CD4+ cells were then positively selected within the microwells using magnetic particles coated with anti-human CD4 monoclonal antibodies (Dynabeads®, Dynal, Oslo, Norway) and a strong magnet (Cylex® Magnet Tray 1050, Cylex Inc®. Columbia, MD). Microwells were washed to remove residual cells, and the isolated CD4+ cells were lysed to release intracellular ATP. ATP was measured using luciferin/luciferase and a luminometer (Turner Biosystems, Sunnyvale, CA). A patient's level of immune response was assessed based on the amount of ATP expressed in ng/ml.

Results: comparing healthy adults, stable transplant patients, HIV-infected patients, and CFS patients

In these preliminary studies, the immune function characteristics of healthy adults, stable transplant recipients, HIV-infected individuals and CFS patients were compared on the basis of their distribution into low, moderate, or strong immune response zones as assessed by the intracellular ATP (ng/ml) of their T cells shown in Fig. 1.

Remarkably, all three immunosuppressed populations had median immune function values that were not statistically different. Transplant patients ran the lowest at 259 ng/ml ATP, followed by HIV patients at 262 ng/ml ATP, and CFS patients at 281 ng/ml ATP. In addition, the relative distributions of patients in the low, moderate, and strong zones were statistically equivalent. Distribution of transplant recipients into three zones of immune response was initially defined by a three center clinical trial of stable patients (Kowalski et al., 2003). The majority (52%) of stable transplant patients ran in the moderate zone with a significant percentage in the low zone (40%) and only 8% in the strong zone (Table 1). By comparison, HIV patients distributed nearly identically to solid organ transplant with the majority in the moderate zone (55%), followed by the low zone (37%), and a minor proportion in the strong zone (8%). The CFS patients showed remarkable consistency with both the transplant patients and the HIV patients with the

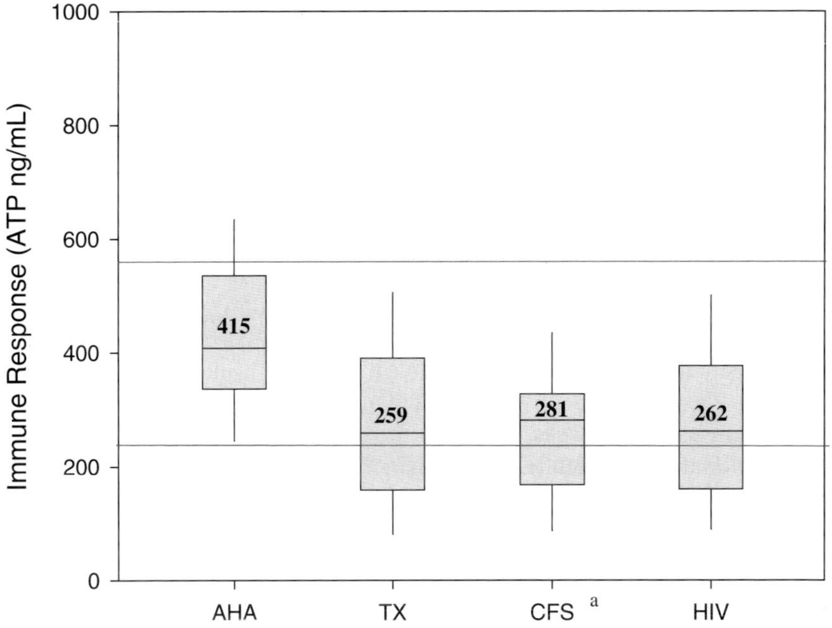

Fig. 1 Comparison of immunosuppressed populations: HIV-infected, transplant, and CFS patients. AHA = apparently healthy adults; TX = transplant recipients; CFS = chronic fatigue syndrome patients; and HIV = HIV-infected individuals.

[a]Data for CFS patients provided through the courtesy of Daniel Peterson, MD.

The median values among transplant, CFS, and HIV-infected patients are not statistically significant. All disease state medians are statistically different from apparently healthy adults.

Note: The box and whiskers plot defines the range of individual values between 25 and 75% within the box, and 10–90% outside the box. The value inside the box is the median ng/ml ATP.

Table 1

Percent distribution of transplant recipients, CFS patients, and HIV-infected individuals in three immune response zones: low, moderate, and strong compared with apparently healthy adults

Immune function (ng/ml ATP)	Healthy adults	Transplant recipients	HIV-infected	CFS[a] patients
Low	5	40	39	37
Moderate	67	52	57	55
Strong	28	8	4	8
P values	<0.001	>0.5	>0.5	>0.5

Note: The percent distributions of each disease state (transplant, CFS, HIV) are statistically equivalent at all three levels of immune function: low, moderate, and strong and statistically different from healthy adults. The *p* value between the AHA (Apparently Healthy Adults) and each of the other three groups (Transplant, HIV, and CFS) is <.001 and therefore significant. The *p* value between any two groups (TX vs CFS, CFS vs HIV, or TX vs HIV) is >0.5 and therefore not significant.
[a]Data provided through the courtesy of Daniel Peterson, MD.

majority of individuals in the moderate zone (57%), followed by the low zone (39%), and the smallest proportion in the strong zone (4%). Therefore, despite the dramatic differences in etiology, these three disease diagnosis categories were statistically equivalent on the basis of their T-cell immune function distributions. Additional studies are being conducted in which HHV-6 serology and viral load will be assessed prospectively in parallel with immune function testing to more definitively establish the link between the immunological behavior and virus expression.

Although understanding global immunosuppression is a useful indicator in the management of the HHV-6-infected patient, ultimately tests for HHV-6 specific memory responses will be needed to identify what cells are involved in protection and can reflect the efficacy of drugs under development. Wang et al. (1999) demonstrated LPA to glycine extracts of HHV-6 variants A and B in healthy subjects. Future assays would most likely address CD4 and CD8 functions and be used to delineate the epitopes of the HHV-6 antigens that are involved in neutralizing the virus.

Discussion and conclusions

Twenty years after its discovery, HHV-6 is an emerging pathogen with an increasing body of data to support disease associations spanning the self-limiting rash of childhood to the ravages of multiple sclerosis. Applying scientific rigor to assessing these associations will be necessary to verify the etiology of these diseases; particularly those like fibromyalgia, which remains very difficult to diagnose. The development of additional immunological research tools (both cellular and humoral) as well as molecular, which can also distinguish variants A and B will be synergistic to scientific discovery while providing future tools for monitoring the efficacy of therapy, whatever is the eventual target of treatment. By virtue of its dual tropisms for lymphocytes and nervous tissue, HHV-6 is one of a growing list of viruses, which present both the research challenge and excitement of bridging two important disciplines: immunology and neurobiology. This critical connection has enormous implications for understanding the relationships between behavior (stress) and the immune system in emerging the field of psychoneuroimmunology. Although this chapter fails to address the conundrum of which came first, immunosuppression or the virus? It is clear that with one in four individuals in the United States is potentially immunosuppressed at any point in time. Therefore, there is a developing need to understand our own immune status as a routine measure of immunological health. While we can generally control our exposure to agents like HIV, most of us are already infected with a number of herpesviruses, one of which is HHV-6, which remains latent as long as the immune system is healthy. The risks of therapeutic immunosuppression are clear for recipients of transplants. Unintended immunosuppression generated by prolonged stress and increasingly casual use of immunosuppressants pose risks that can be avoided. The increasing prevalence of fibromyalgia and multiple sclerosis may be the manifestations of a population under immunological stress allowing the expression of latency from otherwise harmless organisms.

References

Abbott KC, Hypolite IO, Viola R, Poropatich RK, Hshieh P, Cruess D, Hawkes CA, Agodoa LY. AEP 2002; 12(6): 402–409.

Aberle S, Mandl CW, Kunz C, Popow-Kraupp T. J Clin Microbiol 1996; 34(12): 3223–3225.

Acott PD, Lee SHS, Bitter-Suerman H, Lawen JG, Crocker JFS. Transplantation 1996; 62(5): 689–691.

Appleton AL, Sviland L, Peiris JS, Taylor CE, Wilkes J, Green MA, Perason AD, Kelly PJ, Malcolm AJ, Proctor SJ. Bone Marrow Transplant 1995; 169(6): 777–782.

Arena A, Liberto MC, Iannello D, Capozza AB, Foca A. New Microbiol 1999; 22(4): 293–300.

Bethge W, Beck R, Jahn G, Mundinger P, Kanz L, Einsele H. Bone Marrow Transplant 1999; 24(11): 1245–1248.

Boeckh M, Garrett NW. Herpes 2003; 10(1): 12–16.

Carrigan DR, Knox KK. Blood 1994; 84(10): 3307–3310.

Clark DA. Rev Med Virol 2000; 10: 155–173.

Cone RW, Meei-Li WH, Corey L, Zek J, Ashley R, Bowden R. J Infect Dis 1999; 179: 311–318.

Des Jardin JA, Cho E, Supran S, Gibbons L, Werner BG, Snydman DR. Clin Infect Dis 2001; 33: 1358–1362.

Des Jardin JA, Gibbons L, Cho E, Supran SE, Falagas ME, Werner BG, Syndman DR. J Infect Dis 1998; 178: 1783–1786.

Dobryski WR, Dunne WM, Burd EM, Knox KK, Ash RC, Horowitz MM, Fomemberg M, Carrigan DR. J Infect Dis 1993; 167: 735–739.

Dockrell DH, Mendez JC, Jones M, Harmsen WS, Patel F, Ilstrup DM, Smith TF, Wiesner RH, Krom RAF, Paya CV. Transplantation 1999; 67(3): 399–403.

Dockrell DH, Prada J, Jones MF, Patel R, Badley AD, Harmsen WS, Ilstrup DM, Wiesner RH, Krom RAF, Smith TF, Paya CV. J Infect Dis 1997; 176: 1135–1140.

Enders G, Biber M, Meyer G, Helfentein E. Infection 1990; 18: 12–15.
Flamand L, Gosselin J, D'Addario M, Hiscott J, Ablashi DV, Gallo RC, Menezes J. J Virol 1991; 65: 5105.
Flamand L, Gosselin J, Stefanescu I, Ablashi D, Menezes J. Blood 1995; 85(5): 1263–1271.
Flamand L, Stefanescu I, Menezes J. The Am Soc Clin Invest Inc 1996; 97(6): 1373–1381.
Gosselin J, Flamand L, D'Addario M, Hiscott J, Stefanescu I, Ablashi D, Gallo RC, Menezes J. J Immunol 1992; 149(1): 181–187.
Greenstone HL, Santoro F, Lusso P, Berger E. J Biol Bhem 2002; 277(42): 39112–39118.
Hall CB, Long CE, Schnabel KC, Casenta MT, McIntyre KM, Costanzo MA, Knott A, Dewhurst S, Insel RA, Epstein CG. New Eng J Med 1994; 331: 432–438.
Horvat RT, Parmely MJ, Chandran B. J Infect Dis 1993; 167: 1274–1280.
Humar A, Malkan G, Moussa G, Greig P, Levy G, Mazzulli T. J Infect Dis 2000; 181: 1450–1453.
Karp CL. Immunol Rev 1999; 168: 91–101.
Kondo K, Kondo T, Shimada K, Amo K, Miyagawa H, Yamanishi K. J Med Virol 2002; 67(3): 364–369.
Kowalski RK, Post D, Schneider MC, Deierhoi TJ, Lobashevsky A, Redfield R, Schweizter E, Heredia A, Reardon E, Davis C, Bentlejewski C, Fung J, Shapiro R, Zeevi A. Clin Transplant 2003; 17: 77–78.
Krueger GRF, Wassermann K, Declerck LS, Stevens WJ, Bouroeois N, Ablashi DV, Josephs SF, Balachandran N. Lancet 1990; 336: 1255–1256.
Lautenchlager I, Linnavuori K, Hocherstedt K. Transplantation 2000; 69: 2561–2566.
Levy JA, Ferro F, Greenspan D, Lennette ET. Lancet 1990; 335: 1047–1050.
Ljungman P. J Infect Dis 2002; 186(Suppl 1): S99–S109.
Lusso P, Garzino-Demo A, Crowley RW, Mlanati M. J Exp Med 1995; 18: 1303–1310.
Lusso P, Malnati M, Garzino-Demo A, Crowley RW, Long EO, Gallo RC. Nature 1993; 362: 458–462.
Lusso P, Maria AD, Malnati M, Lori F, Rocco SE, Baseler D, Gallo RC. Nature 1991; 349: 533–535.
Mori Y, Seya T, Huang HL, Akkapaiboon P, Dhepakson P, Yamanishi K. J Virol 2002; 76(13): 6750–6761.
Morris DJ, Littler E, Arrand JR, Jordan D, Mallick NJ, Johnson RWG. New Eng J Med 1989; 320: 1560.
Okuno T, Takahashi K, Balanchandra K, Shiraki K, Yamanishi K, Takahashi M, Baba K. J Clin Microbiol 1989; 27: 651–653.
Riel-Romero RMS. Infect Dis Clin Pract 2005; 13(2): 48–53.
Robert C, Agut H, Lunel-Fabian F, Leger PH. La Press Med 1994; 23: 1209.
Romagnoli PA, Nates SV, Pavau JV, Serra HM. Trans R Soc Trop Med Hyg 2000; 94(6): 669–672.
Salahuddin SZ, Ablashi DV, Markham PD, Josephs SF, Sturzenegger S, Kaplan M, Halligan G, Biverfeld P, Wong-Staal F, Kramarsky B, Gallo RC. Science 1986; 234: 596–601.
Savolainen H, Lautenschlager I, Piiparinen H, Saarinen-Pihkala U, Hovi L, Vettenranta K. Pediatr Blood Cancer 2005; 44: 1–6.
Schonnebeck M, Krueger GRF, Braun M, Fischer M, Koch B, Ablashi DV, Balachandran N. In Vivo 1991; 5(3): 255–264.
Singh N, Carrigan R. Ann Intern Med 1996; 124: 1065–1071.

Singh N, Patterson DL. Transplantation 2000; 69: 2474–2479.
Smith AP, Paolucci C, Di Lullo G, Burastero SE, Santoro F, Lusso P. J Virol 2004; 79(5): 2807–2813.
Smith AP, Santoro F, Di Lullo G, Dagna L, Verani A, Lusso P. Blood 2003; 102(8): 2877–2884.
Snydman DR, Werner BG, Dougherty NN, Griffith J, Rubin RH, Dienstag JL, Rohrer RH, Freeman R, Jenkins R, Lewis WD, Hammer S, O'Rourke E, Grady GF, Fawaz K, Kaplan MM, Hoffman MA, Katz AT, Doran M. Ann Intern Med 1993; 119: 984–991.
Tolkoff-Rubin NE, Rubin RH. Transplant Proc 1998; 30: 1060–1063.
Tong CYW, Bakran A, Peirus JSM, Muir P, Simon Herrington C. Transplantation 2002; 74(4): 576–578.
Wang FZ, Dahl H, Ljungman P, Linde A. J Med Virol 1999; 57(2): 134–139.
Wang FZ, Larsson K, Linde A, Ljungman P. MMT 2002; 30: 521–526.
Yamanishi K, Okuno T, Shiraki K, Tahahashi M, Kando T, Asano Y, Kurata T. Lancet 1988; 1: 1065–1067.
Yasukawa M, Ihorne Y, Ohminami H, Terada K, Fujita S. J Gen Virol 1998; 79(1): 143–147.
Yoshikawa T, Asano Y, Ihira M, Suzuki K, Ohashi M, Suga S, Kudo K, Horibe K, Kojima S, Kato K, Matsuyama T, Nishiyama Y. J Infect Dis 2002; 185: 847–853.
Yoshikawa T, Suga S, Asano Y, Nakashima T, Yazaki T, Ono Y, Fujita T, Tsuzuki K, Sugikawa S, Oshima S. Transplantation 1992; 54: 879–883.

Human Herpesvirus-6: General Virology, Epidemiology and Clinical Pathology
Perspectives in Medical Virology
Gerhard Krueger and Dharam Ablashi (Editors)

DOI 10.1016/S0168-7069(06)12016-9

HHV-6 and the Central Nervous System

Samantha S. Soldan[a], Andrew D. Goodman[b], Steven Jacobson[c]
[a]*Department of Neurology, University of Pennsylvania, Philadelphia, PA 19104, USA*
[b]*Department of Neurology, University of Rochester, Rochester, NY 14642, USA*
[c]*Viral Immunology Section, NINDS/NIH, Building 10, Room 5B-16, Bethesda, MD 20892, USA*

HHV-6 infection of the central nervous system

HHV-6 seroprevalence rates vary from 72 to 100% in healthy adults worldwide (Yamanishi, 1992). Moreover, HHV-6 DNA has been detected by PCR in brain tissue from normal specimens and a variety of diseased material (Lusso, 1996; Braun et al., 1997). Therefore, detection of HHV-6 genome in brain *per se* does not necessarily indicate productive infection, and establishing a primary pathogenic role for HHV-6 in central nervous system (CNS) disorders is challenging. Nevertheless, HHV-6 has been associated with aseptic meningoencephalitis, encephalitis (Asano et al., 1992), transverse myelitis (Hill et al., 1994), Guilain-Barré syndrome (Yoshikawa and Asano, 2000), and cerebellar astrocytoma (Rantala et al., 2000). Although the HHV-6A variant has increased neurotropism and is associated with viral persistence and reactivation in the CNS (Hall et al., 1998), both HHV-6 variants are clearly neurotropic. The HHV-6B variant is a causative agent of *exanthem subitum* and accounts for the majority of symptomatic HHV-6 infections in infants (Dewhurst et al., 1993). In addition, HHV-6B has been isolated from patients with limbic encephalitis (Wainwright et al., 2001) and seizure disorders (Zerr et al., 2002; Donati et al., 2003; Lanari et al., 2003). The HHV-6A variant has yet to be definitively implicated in a particular disease. Of interest, HHV-6A has been isolated from the CNS of acquired immuno deficiency syndrome

(AIDS) patients with areas of demyelination (Knox and Carrigan, 1995) and associated with multiple sclerosis (MS) (Akhyani et al., 2000; Kim et al., 2000; Soldan et al., 2000; Rotola et al., 2004). This chapter will focus on evidence suggesting an association of HHV-6 with four major disorders of the CNS: childhood febrile seizures, mesial temporal-lobe epilepsy (MTLE), MS, and progressive multifocal leukoencephalopathy (PML).

HHV-6 in seizure disorders

Febrile seizures are the most common type of seizures observed in children. Primary HHV-6 and HHV-7 infections are the causative agents of the common childhood illness *exanthem subitum* (Yamanishi et al., 1988), and febrile seizures as a complication of *exanthem subitum* have been reported in 1–50% of patients (Yoshikawa and Asano, 2000). Owing to the high incidence of febrile seizures complicating *exanthem subitum*, HHV-6 may play a causative role in febrile seizures. It has been suggested that HHV-6 enters the brain during the acute phase of *exanthem subitum* with recurrent seizures manifesting upon virus reactivation (Caserta et al., 1994; Hall et al., 1998). Several reports have detected HHV-6 by PCR or viral culture from the CSF of children with febrile seizures, supporting a role for the virus in this disorder (Kondo et al., 1993; Hall et al., 1994; Eeg-Olofsson, 2003). A survey of children under 2 years of age hospitalized in Britain and Ireland with suspected encephalitis and/or febrile seizures indicates that HHV-6 and HHV-7 infections account for 17% of these admissions (Ward et al., 2005), suggesting that HHV-6 and 7 are significant causes of neurological disease and hospitalization of young children. Most recently, a multisite prospective study found evidence of primary HHV-6 infection in 22.9% of children presenting with febrile status epilepticus (FSE). In contrast, no evidence of primary infection with HHV-7 was found (Epstein et al., 2005). These data suggest that HHV-6 may be a common cause of FSE and associated with an increased risk of both hippocampal injury and subsequent temporal-lobe seizures (Epstein et al., 2005).

Presumably related to the magnitude and/or rapidity of the increase in body temperature, the pathophysiology of febrile seizures is largely undefined. It remains unclear whether HHV-6 plays a role in the pathogenesis of febrile seizures, or if these seizures are caused by the febrile illness itself rather than the neurotropic properties of the virus. In addition, it is possible that immune-mediated responses to HHV-6 infection, including the release of IL-6, may contribute to the neurological sequelae that follow primary infection (Go and Nakamura, 2002). Recently, evidence for primary infection with the HHV-6B variant was demonstrated in a 3-week-old infant and a newborn with afebrile seizures (Zerr et al., 2002; Lanari et al., 2003), providing further evidence in support of a direct, neuropathogenic role of HHV-6 independent of fever in childhood seizures and suggesting that HHV-6 may be a more common cause of afebrile seizures than previously appreciated. Of interest, the frequency of atypical, more severe seizures in patients experiencing their first febrile seizure and primary HHV-6 infection was significantly higher than that in patients with non-primary HHV-6 infection (Suga et al., 2000; Yoshikawa and Asano, 2000). These data indicate that febrile seizures caused by primary HHV-6 infection may be the result of viral invasion of the CNS and not simply fever. Moreover, febrile seizure associated with primary HHV-6 infection is associated with the development of focal, prolonged, and repeated seizures and could be a risk factor for the subsequent development of epilepsy (Suga et al., 2000; Yoshikawa and Asano, 2000; Eeg-Olofsson, 2003).

Donati et al. (2003) examined brain samples from patients with MTLE and neocortical epilepsy (NE) undergoing therapeutic, surgical resection for pharmacologically untreatable seizures. These samples were analyzed quantitatively for the presence of HHV-6 DNA. In addition, HHV-6 expression was assessed by Western blot analysis for HHV-6 p41 antigen and HHV-6 reactive cells were identified by double immunofluorescence for HHV-6 gp116/54/64 and glial fibrillary acidic protein (GFAP). While HHV-6 was not amplified in any of the brain samples from NE patients undergoing surgery, significantly elevated HHV-6 viral loads were demonstrated in brain resections of four of the eight MTLE patients. Higher levels of HHV-6 were obtained in hippocampal sections compared to lateral temporal-lobe material from the same MTLE patients. Notably, viral DNA amplified from brain tissue of MTLE patients was subtyped as HHV-6B; the viral variant also identified in children with both febrile and afebrile seizures (Dewhurst et al., 1993; Chua et al., 1997; Zerr et al., 2002; Lanari et al., 2003). In addition, HHV-6 gp116/54/64 colocalized with GFAP in cells that morphologically appeared to be astrocytes. Astrogliosis is frequently observed in MTLE lesions and may play a role in disease pathogenesis by releasing epileptogenic factors (Represa et al., 1995). Importantly, no evidence of inflammation was observed in the MTLE samples examined (Donati et al., 2003). Collectively, these data suggest that HHV-6B is localized to astrocytes in a subset of patients with MTLE and reactivation of HHV-6B may play a role in the development of MTLE (Donati et al., 2003). The possible association between HHV-6 and MTLE is supported by the increased detection of HHV-6 DNA in Japanese patients with temporal-lobe epilepsy (Uesugi et al., 2000).

The association of HHV-6 with MS

MS is the most prevalent demyelinating disease of the CNS (Pugliatti et al., 2002). The variability of the diseases suggests that many factors, including genetic and immune components, are involved in the spectrum of clinical syndromes that are defined as MS. In addition, epidemiological, laboratory, and clinical findings suggest that an infectious agent(s) (Marie, 1895; Johnson, 1994) may be involved in the induction and progression of this disorder (reviewed in Soldan and Jacobson, 2001). Over the years, several viruses and bacteria have been associated with MS based primarily on elevated antibody titers or the isolation of a particular virus/bacterium from MS tissue or CSF. However, most of these reported agents have been eliminated from consideration in the pathogenesis of MS (reviewed in Soldan and Jacobson, 2001).

HHV-6 is one of the latest infectious agents postulated to play a role in the pathogenesis of MS. The putative role of HHV-6 in MS was first suggested by Challoner and colleagues based on an unbiased search for non-human DNA by representational difference analysis (Challoner et al., 1995). In this study, over 70 DNA fragments were analyzed and one was found to be homologous to the MDBP gene of the HHV-6B variant Z29. HHV-6 DNA was found in 78% of MS brains and 74% of control brains. However, monoclonal antibodies against the HHV-6 101 K protein and the DNA binding protein/early antigen p41 were detected in the brain tissues of MS patients and not in controls (Challoner et al., 1995).

Further investigations into the possible association of HHV-6 with MS have focused on immunological and molecular studies. To distinguish between latent and active HHV-6 infection, early antibody responses to the HHV-6 p41/38 early antigen and the presence of HHV-6 serum DNA by nested PCR were examined (Soldan et al., 1997). A significant increase in IgM response to the p41/38 early antigen was demonstrated in patients with the relapsing remitting form of MS and other inflammatory disease (Soldan et al., 1997). Additional studies have confirmed increased antibody responses to HHV-6 in patients with MS, although other reports have not (reviewed in Fotheringham and Jacobson, 2005). The detection of viral sequences in cell-free compartments such as serum has been considered a measure of productive virus. HHV-6 serum DNA was detected in 30% of MS patients and in 0% of controls consisting of healthy individuals, patients with other inflammatory diseases and other neurologic diseases (Soldan et al., 1997). Subsequent studies by a number of other groups have reexamined the presence of HHV-6 serum DNA in MS patients. Overall, the results from these studies have been equivocal (reviewed in Soldan and Jacobson, 2001). These discrepancies may be attributable to differences in patient selection, techniques, and reagents used.

To extend the observation of cell-free serum HHV-6 DNA in MS patients, a longitudinal study consisting of 215 samples obtained from 59 MS patients followed prospectively for a 5-month period of time was conducted (Berti et al., 2002). Serum HHV-6 DNA was detected from significantly fewer sera obtained during periods of clinical remission. These data suggest a statistically greater likelihood of detecting HHV-6 DNA in the serum of an MS patient during an exacerbation (Berti et al., 2002). This report supports a role for HHV-6 in the pathogenesis of MS by suggesting that the presence of serum HHV-6 DNA, similar to the presence of gadolinium-enhancing lesions, coincides with clinical worsening in a subset of patients (Berti et al., 2002). Additional studies have demonstrated that HHV-6A active replication is associated with a subset of relapsing-remitting multiple sclerosis (RRMS) patients and that HHV-6A active infection increases the risk of exacerbations in these patients (Alvarez-Lafuente et al., 2004).

Cellular immune response to HHV-6 has been compared in MS patients and controls (Soldan et al., 2000; Cirone et al., 2002; Tejada-Simon et al., 2002, 2003). In a report examining the T-cell lymphoproliferative responses of healthy controls and patients with MS to both variants of HHV-6 and HHV-7 using whole virus lysates (Soldan et al., 2000), it was demonstrated that there was no difference in either the frequency or magnitude of proliferative responses between healthy controls and patients with MS to either the HHV-6B variant or HHV-7. However, a significantly higher percentage of patients with MS had proliferative responses to the HHV-6A variant compared with healthy controls. It is unknown whether the increased frequency of lymphoproliferative response to HHV-6A lysate in patients with MS is the result of a higher seroprevalence of the HHV-6A variant in MS patients or an altered host immune response (Soldan et al., 2000). Subsequent studies have demonstrated the amplification of the HHV-6A variant in peripheral blood mononuclear cells (PBMC), serum, and urine of MS patients, but not in controls (Akhyani et al., 2000; Kim et al., 2000). Collectively, the description of an increased lymphoproliferative response to the HHV-6A variant and the unique amplification of HHV-6A DNA in the PBMC, serum, and urine of MS patients further supports the association of HHV-6 with MS. These studies suggest that the highly neurotropic A variant rather than the B variant may play a role in this disease (Soldan et al., 2000).

In a recent study, a lower precursor frequency of HHV-6B101 K specific T-cells was observed in MS patients compared with controls (Tejada-Simon et al., 2002). The impaired T-cell response to the 101 K protein of HHV-6 was associated with increased HHV-6-specific IgM responses. Of interest, HHV-6-specific T-cell lines derived from MS patients demonstrated Th-1 biased cytokine profiles marked by the inability to produce IL-4 and IL-10 (Tejada-Simon et al., 2002). In addition, cross-reactivity between MBP (residues 96–102) and HHV-6 U24 (residues 4–10) and increased precursor frequencies of MBP/HHV-6 cross-reactive T-cells were found in MS patients compared with healthy controls (Tejada-Simon et al., 2003). Importantly, this study suggests a potential role for HHV-6-specific T-cells in the pathogenesis of MS via molecular mimicry (Tejada-Simon et al., 2003).

Recent studies have specifically addressed the frequency of HHV-6 in control and MS brains and identified infected cell phenotypes in pathologically defined tissue (Cermelli et al., 2003; Goodman et al., 2003). Goodman et al. identified individual cells containing the HHV-6 genome from surgical biopsy specimens from MS patients presenting with acute disease who had not received immunomodulatory therapy. This study demonstrated that high frequencies of HHV-6 genome-positive neuroglial and inflammatory cells are present in acute-phase tissue and oligodendrocytes are the predominant cell infected in the acute MS lesion (Goodman et al., 2003). In an accompanying study, pathologically defined material from brain autopsies were isolated by LASER-assisted microdissection prior to HHV-6 DNA amplification (Cermelli et al., 2003). HHV-6 DNA was detected at a significantly higher frequency in MS plaques compared to brain tissue from non-MS neurologic disorders, non-MS inflammatory, and normal appearing white matter (NAWM) from MS brains. In a subsequent study Virtanen et al. (2005) detected HHV-6 antigen in 67% of MS brains compared to 30% of controls. All MS and control brains were negative for CMV antigen and Epstein-Barr virus (EBV) EBV-encoded small RNA (EBER) RNA by *in situ* hybridization. In addition, HHV-6 mRNA expression has been demonstrated to be significantly elevated in NAWM and lesion material in MS patients compared with controls (Opsahl and Kennedy, 2005). Collectively, these studies suggest that HHV-6 is present early in the evolution of the MS lesion and may play a significant role in the demyelinative pathogenesis of MS.

Over the years, researchers have sought to fulfill Koch's postulates and identify a virus found in MS patients, but not in unaffected individuals. To date, this search has failed. The involvement of multiple infectious agents in MS, as suggested originally by Dr. Pierre Marie over 100 years ago (Marie, 1895), may explain the difficulty in identifying a single agent responsible for this highly variable and chronic disease. An alternative paradigm of the disease process, which suggests that a common or ubiquitous virus may act as a trigger in genetically or immunologically predisposed individuals, has emerged (Soldan and Jacobson, 2001). It is also possible that multiple viruses may be involved in the etiology of MS and that particular viruses trigger disease in different subsets of individuals through a common mechanism such as use of a common receptor.

In 1999, CD46, a molecule expressed on all human nucleated cells was identified as the cellular receptor of HHV-6A (Santoro et al., 1999). CD46 is a member of a family of glycoproteins that regulate complement activation (RCA) and prevent spontaneous activation of complement on autologous cells. The use of a virtually ubiquitous human molecule as a surface receptor helps to explain the pleiotropism of HHV-6. Notably, CD46 is also the primate-specific receptor for measles virus (Dorig et al., 1993), another virus tentatively associated with MS, and other viruses use various members of the RCA family as cellular receptors (Bergelson et al., 1995).

Increased levels of the complement regulatory protein CD46 have been demonstrated in the serum and CSF of MS patients compared with healthy controls and other neurologic disease patients (Soldan et al., 2001; Fogdell-Hahn et al., 2005). Elevated levels of soluble serum CD46 were also found in other inflammatory disease cohorts, indicating that an increase in soluble serum CD46 may be a common phenomenon in autoimmune disorders (Cuida et al., 1997; Kawano et al., 1999). Therefore, the increased levels of CD46 demonstrated in the serum and CSF of MS patients may indicate increased activation of the complement system in MS, both peripherally and intrathecally. A significant correlation was observed between elevated levels of serum soluble CD46 and the detection of serum HHV-6 DNA in serum from MS patients, while no serum HHV-6 DNA was detected in other inflammatory disease controls (Soldan et al., 2001). HHV-6 DNA was subsequently co-purified with CD46 from the eluates of a CD46 immunoaffinity column in 4 of 42 MS samples tested and subtyped as variant A in three of the four positive samples (Fogdell-Hahn et al., 2005). The co-purification of soluble CD46 (sCD46) and HHV-6A DNA from MS sera indicates that HHV-6 is tightly connected to its receptor, CD46, in the serum of MS patients and lends further support to the association of HHV-6A with MS (Fogdell-Hahn et al., 2005).

HHV-6 and PML

PML is an often fatal disorder of the CNS that primarily affects individuals with impaired immune systems including patients treated for cancers such as leukemia or lymphoma, allograft recipients, and up to 10% of patients with AIDS (Padgett et al., 1971; Berger and Major, 1999). The etiologic agent of PML is JC virus (JCV), a human neurotropic member of the polyomaviruses; a family of non-enveloped tumor viruses with small, circular, double-stranded DNA genomes. JCV is widespread throughout the human population with a worldwide distribution of more than 80% seropositivity (Major et al., 1992).

The pathology of PML is distinctive and characterized by multiple foci of demyelinating lesions of variable size caused by lytic infection of oligodendrocytes with JCV (Gordon and Khalili, 1998; Post et al., 1999; Safak and Khalili, 2003). This lytic infection represents reactivation of JCV in the immunocompromised host (Major et al., 1992). Oligodendrocytes in the peripheral zone surrounding an area of demyelination are grossly abnormal, exhibiting nuclei engorged with JC virions. The death of infected oligodendrocytes leads to the development of focal areas of progressive demyelination (White et al., 2005). In addition, recent evidence suggests that necrotic death of JCV-infected astrocytes may contribute to the pathogenesis of PML (Seth et al., 2004). The disease course of PML progresses rapidly with death typically occurring within 1 year (Hou and Major, 2000). However, prolonged survival of AIDS patients with PML has been achieved with the advent of highly active antiretroviral therapy (HAART) (Albrecht et al., 1998).

After several reports had suggested an association of HHV-6 with MS, Mock et al. (1999) examined the possible association of HHV-6 with the demyelinative lesions of PML. In this study, a highly sensitive, two-step *in situ* PCR (ISPCR) procedure was used to amplify HHV-6 DNA from formalin-fixed paraffin-embedded archival brain tissues from normal, AIDS, and other neurological disease controls. A significantly higher frequency of infected cells was found in PML white matter compared to control tissues. Of interest, the HHV-6 genome was detected primarily within oligodendrocytes (Mock et al., 1999). Immunocytochemistry for HHV-6 antigens revealed active HHV-6 infection of the abnormal oligodendrocytes within demyelinative PML lesions, but not in adjacent, unaffected tissue or control tissues, including brains from individuals with HIV-1 encephalopathy. The detection of active HHV-6 infection coupled with the collocation of JCV large T antigen and HHV-6 in swollen intralesional oligodendrocytes suggested an association of HHV-6 activation in the context of JCV infection with PML lesions.

ISPCR was again employed to amplify HHV-6 genome from normal, PML, AIDS (no PML), MS, and other neurologic disease control brains (Blumberg et al., 2000). In support of the previous report, a high frequency of HHV-6 DNA was detected in demyelinative white matter lesions from both PML and MS compared with controls. Remarkably, a greater frequency of HHV-6 than JCV genomes was detected in oligodendrocytes from PML lesion areas. HHV-6 p41 and gp101 antigens were detectable by immunocytochemistry in PML lesions. Subsequently,

HHV-6 was detected in demyelinating PML lesions of a Japanese patient with follicular lymphoma (Daibata et al., 2001) and an HIV-infected patient with both PML and meningoencephalitis (Ito et al., 2000) further supporting the possible association of HHV-6 with PML. Of interest, in cultured human fetal astrocytes, both HHV-6 and the related β-herpesvirus CMV transactivate the HIV-1 long terminal repeat (LTR) (McCarthy et al., 1998). Co-infection of oligodendrocytes with HHV-6 and JCV presents an opportunity for these ordinarily commensal viruses to stimulate or activate one another in a synergistic fashion, thereby participating as co-factors in the pathogenesis of PML (Blumberg et al., 2000).

In recent years, RRMS patients have participated in a clinical trial investigating the efficacy of natalizumab (Tysabri; Biogen Idec and Elan) in combination with interferon beta-1a (Avonex; Biogen Idec). Tragically, two MS patients on combination therapy have developed the seemingly unrelated, demyelinating disorder, PML (Langer-Gould et al., 2005; Kleinschmidt-DeMasters and Tyler, 2005). Natalizumab is a humanized monoclonal antibody against α_4 integrins that is thought to prevent migration of autoreactive T-cells into the CNS. The development of PML in these patients raises several questions. Does the inhibition of normal immune cell trafficking into the brain allow for opportunistic reactivation of JCV? Is HHV-6, a virus speculated to be involved in the pathogenesis of MS, implicated in the reactivation of JCV or development of PML in these RRMS patients? It remains to be determined whether there will be an increased frequency in the detection of HHV-6 in the CSF of MS patients who received natalizumab compared with control patients.

HHV-6 is a highly neurotropic virus that has been both tentatively and firmly associated with a wide array of CNS disorders. Further understanding of the biology of this virus is paramount in deciphering the neuropathology of several diseases of the CNS including encephalitis, seizure disorders, MS, and PML. Further, the development of antiviral agents for HHV-6A and B may result in new treatment modalities for many neurological diseases.

Acknowledgments

S.S.S. is supported by NIH grant NS-007180.

References

Akhyani N, Berti R, Brennan MB, Soldan SS, Eaton JM, McFarland HF, Jacobson S. J Infect Dis 2000; 182: 1321–1325.

Albrecht H, Hoffmann C, Degen O, Stoehr A, Plettenberg A, Mertenskotter T, Eggers C, Stellbrink HJ. AIDS 1998; 12: 1149–1154.

Alvarez-Lafuente R, De las Heras V, Bartolome M, Picazo JJ, Arroyo R. Arch Neurol 2004; 61: 1523–1527.

Asano Y, Yoshikawa T, Kajita Y, Ogura R, Suga S, Yazaki T, Nakashima T, Yamada A, Kurata T. Arch Dis Child 1992; 67: 1484–1485.

Bergelson JM, Mohanty JG, Crowell RL, St John NF, Lublin DM, Finberg RW. J Virol 1995; 69: 1903–1906.
Berger JR, Major EO. Semin Neurol 1999; 19: 193–200.
Berti R, Brennan MB, Soldan SS, Ohayon JM, Casareto L, McFarland HF, Jacobson S. J Neurovirol 2002; 8: 250–256.
Blumberg BM, Mock DJ, Powers JM, Ito M, Assouline JG, Baker JV, Chen B, Goodman AD. J Clin Virol 2000; 16: 159–178.
Braun DK, Dominguez G, Pellett PE. Clin Microbiol Rev 1997; 10: 521–567.
Caserta MT, Hall CB, Schnabel K, McIntyre K, Long C, Costanzo M, Dewhurst S, Insel R, Epstein LG. J Infect Dis 1994; 170: 1586–1589.
Cermelli C, Berti R, Soldan SS, Mayne M, D'Ambrosia JM, Ludwin SK, Jacobson S. J Infect Dis 2003; 187: 1377–1387.
Challoner PB, Kristen ST, Parker JD, MacLeoad DL, Coulter SN, Rose TM, Schultz ER, Bennett L, Garber RL, Chang M, Schad PA, Stewart PM, Nowinski RC, Brown JP, Burmer GC. Proc Natl Acad Sci USA 1995; 92: 7440–7444.
Chua KB, Lam SK, AbuBakar S, Koh MT, Lee WS. Med J Malaysia 1997; 52: 335–341.
Cirone M, Cuomo L, Zompetta C, Ruggieri S, Frati L, Faggioni A, Ragona G. J Med Virol 2002; 68: 268–272.
Cuida M, Legler DW, Eidsheim M, Jonsson R. Clin Exp Rheumatol 1997; 15: 615–623.
Daibata M, Hatakeyama N, Kamioka M, Nemoto Y, Hiroi M, Miyoshi I, Taguchi H. Am J Hematol 2001; 67: 200–205.
Dewhurst S, McIntyre K, Schnabel K, Hall CB. J Clin Microbiol 1993; 31: 416–418.
Donati D, Akhyani N, Fogdell-Hahn A, Cermelli C, Cassiani-Ingoni R, Vortmeycr A, Heiss JD, Cogen P, Gaillard WD, Sato S, Theodore WH, Jacobson S. Neurology 2003; 61: 1405–1411.
Dorig RE, Marcil A, Chopra A, Richardson CD. Cell 1993; 75: 295–305.
Eeg-Olofsson O. Brain Dev 2003; 25: 9–13.
Epstein LG, Nordli DR, Hamidullah A, Pellock JM, Frank LM, Lewis DV, Hesdorffer DC, Marmarou A, O'Dell C, Shinnar S, Team TFS. 59th Annual Meeting American Epilepsy Society, Chicago, IL, 2005.
Fogdell-Hahn A, Soldan SS, Shue S, Akhyani N, Refai H, Ahlqvist J, Jacobson S. Virus Res 2005; 110: 57–63.
Fotheringham J, Jacobson S. Herpes 2005; 12: 4–9.
Go T, Nakamura K. Eur J Paediatr Neurol 2002; 6: 221–223.
Goodman AD, Mock DJ, Powers JM, Baker JV, Blumberg BM. J Infect Dis 2003; 187: 1365–1376.
Gordon J, Khalili K. Int J Mol Med 1998; 1: 647–655.
Hall CB, Caserta MT, Schnabel KC, Long C, Epstein LG, Insel RA, Dewhurst S. Clin Infect Dis 1998; 26: 132–137.
Hall CB, Long CE, Schnabel KC, Caserta MT, McIntyre KM, Costanzo MA, Knott A, Dewhurst S, Insel RA, Epstein LG. N Engl J Med 1994; 331: 432–438.
Hill AE, Hicks EM, Coyle PV. Dev Med Child Neurol 1994; 36: 651–652.
Hou J, Major EO. J Neurovirol 2000; 6(Suppl 2): S98–S100.
Ito M, Baker JV, Mock DJ, Goodman AD, Blumberg BM, Shrier DA, Powers JM. Acta Neuropathol (Berl) 2000; 100: 337–341.
Johnson RT. Ann Neurol 1994; 36(Suppl): S54–S60.
Kawano M, Seya T, Koni I, Mabuchi H. Clin Exp Immunol 1999; 116: 542–546.

Kim JS, Lee KS, Park JH, Kim MY, Shin WS. Eur Neurol 2000; 43: 170–173.
Kleinschmidt-DeMasters BK, Tyler KL. N Engl J Med 2005; 353: 369–374.
Knox KK, Carrigan DR. J Acquir Immune Defic Syndr Hum Retrovirol 1995; 9: 69–73.
Kondo K, Nagafuji H, Hata A, Tomomori C, Yamanishi K. J Infect Dis 1993; 167: 1197–1200.
Lanari M, Papa I, Venturi V, Lazzarotto T, Faldella G, Gabrielli L, Guerra B, Landini MP, Salvioli GP. J Med Virol 2003; 70: 628–632.
Langer-Gould A, Atlas SW, Green AJ, Bollen AW, Pelletier D. N Engl J Med 2005; 353: 375–381.
Lusso P. Antiviral Res 1996; 31: 1–21.
Major EO, Amemiya K, Tornatore CS, Houff SA, Berger JR. Clin Microbiol Rev 1992; 5: 49–73.
Marie P. New Sydenham Society 1895; CL11: 102–153.
McCarthy M, Auger D, He J, Wood C. J Neurovirol 1998; 4: 495–511.
Mock DJ, Powers JM, Goodman AD, Blumenthal SR, Ergin N, Baker JV, Mattson DH, Assouline JG, Bergey EJ, Chen B, Epstein LG, Blumberg BM. J Neurovirol 1999; 5: 363–373.
Opsahl ML, Kennedy PG. Brain 2005; 128: 516–527.
Padgett BL, Walker DL, ZuRhein GM, Eckroade RJ, Dessel BH. Lancet 1971; 1: 1257–1260.
Post MJ, Yiannoutsos C, Simpson D, Booss J, Clifford DB, Cohen B, McArthur JC, Hall CD. AJNR Am J Neuroradiol 1999; 20: 1896–1906.
Pugliatti M, Sotgiu S, Rosati G. Clin Neurol Neurosurg 2002; 104: 182–191.
Rantala H, Mannonen L, Ahtiluoto S, Linnavuori K, Herva R, Vaheri A, Koskiniemi M. Dev Med Child Neurol 2000; 42: 418–421.
Represa A, Niquet J, Pollard H, Ben-Ari Y. J Neurobiol 1995; 26: 413–425.
Rotola A, Merlotti I, Caniatti L, Caselli E, Granieri E, Tola MR, Di Luca D, Cassai E. Mult Scler 2004; 10: 348–354.
Safak M, Khalili K. J Neurovirol 2003; 9(Suppl 1): 3–9.
Santoro F, Kennedy PE, Locatelli G, Malnati MS, Berger EA, Lusso P. Cell 1999; 99: 817–827.
Seth P, Diaz F, Tao-Cheng JH, Major EO. J Virol 2004; 78: 4884–4891.
Soldan SS, Berti R, Salem N, Secchiero P, Flamand L, Calabresi PA, Brennan MB, Maloni HW, McFarland HF, Lin HC, Patnaik M, Jacobson S. Nat Med 1997; 3: 1394–1397.
Soldan SS, Fogdell-Hahn A, Brennan MB, Mittleman BB, Ballerini C, Massacesi L, Seya T, McFarland HF, Jacobson S. Ann Neurol 2001; 50: 486–493.
Soldan SS, Jacobson S. Adv Virus Res 2001; 56: 517–555.
Soldan SS, Leist TP, Juhng KN, McFarland HF, Jacobson S. Ann Neurol 2000; 47: 306–313.
Suga S, Suzuki K, Ihira M, Yoshikawa T, Kajita Y, Ozaki T, Iida K, Saito Y, Asano Y. Arch Dis Child 2000; 82: 62–66.
Tejada-Simon MV, Zang YC, Hong J, Rivera VM, Killian JM, Zhang JZ. J Virol 2002; 76: 6147–6154.
Tejada-Simon MV, Zang YC, Hong J, Rivera VM, Zhang JZ. Ann Neurol 2003; 53: 189–197.
Uesugi H, Shimizu H, Maehara T, Arai N, Nakayama H. Psychiatry Clin Neurosci 2000; 54: 589–593.

Virtanen JO, Zabriskie JB, Siren V, Friedman JE, Lyons MJ, Edgar M, Vaheri A, Koskiniemi M. Med Sci Monit 2005; 11: BR84–BR87.
Wainwright MS, Martin PL, Morse RP, Lacaze M, Provenzale JM, Coleman RE, Morgan MA, Hulette C, Kurtzberg J, Bushnell C, Epstein L, Lewis DV. Ann Neurol 2001; 50: 612–619.
Ward KN, Andrews NJ, Verity CM, Miller E, Ross EM. Arch Dis Child 2005; 90: 619–623.
White MK, Gordon J, Reiss K, Del Valle L, Croul S, Giordano A, Darbinyan A, Khalili K. Brain Res Brain Res Rev 2005; 50: 69–85.
Yamanishi K. Microbiol Immunol 1992; 36: 551–561.
Yamanishi K, Okuno T, Shiraki K, Takahashi M, Kondo T, Asano Y, Kurata T. Lancet 1988; 1: 1065–1067.
Yoshikawa T, Asano Y. Brain Dev 2000; 22: 307–314.
Zerr DM, Yeung LC, Obrigewitch RM, Huang ML, Frenkel LM, Corey L. Med Virol 2002; 66: 384–387.

Human Herpesvirus-6: General Virology, Epidemiology and Clinical Pathology
Perspectives in Medical Virology
Gerhard Krueger and Dharam Ablashi (Editors)

DOI 10.1016/S0168-7069(06)12017-0

HHV-6 and the Cardiovascular System

Satoki Fukae, Naoto Ashizawa
Division of Cardiovascular Medicine, Graduate School of Biomedical Sciences, Nagasaki University, Sakamoto 1-7-1, Nagasaki 852-8501, Japan

Introduction

Myocarditis is usually viral in origin and the course in most patients is subclinical. However, in susceptible patients, viral myocarditis may lead to sudden cardiac death or heart failure with a clinical picture similar to that of idiopathic-dilated cardiomyopathy. Enteroviruses, especially coxsackie viruses of group B, are the most common infectious agents. Other viruses, such as adenovirus, cytomegalovirus (CMV), Epstein–Barr virus, influenza virus, mumps virus, rubella virus, human immunodeficiency virus-1, parvovirus B19, hepatitis C virus, herpes simplex virus, and human herpesvirus-6 (HHV-6) have also been linked with myocarditis.

HHV-6 has been proven to be a cause of exanthem subitum, a common disease during infancy, with symptoms of a high-grade fever and skin rash (Yamanishi et al., 1988). A primary HHV-6 infection in cardiovascular system in infancy is rare and only two such cases have been reported from Japan. A 5-month-old girl who suffered from fatal acute myocarditis caused by a primary HHV-6 infection was reported by Yoshikawa et al. (2001). A positive IgM antibody to HHV-6 was detected in the patient's serum and HHV-6 variant B DNA, but not variant A DNA, was detected in several tissue specimens including the myocardium by polymerase chain reaction (PCR). However, no HHV-6 antigen was identified in the myocardium. Large vessel arteritis associated with an HHV-6 infection in a 9-month-old boy was also described by Toyabe, et al. (2002). Neither of these cases had a past history suggesting the presence of a preexisting immunodeficiency. In healthy individuals, HHV-6 persists in a latent state in monocytes/macrophages after a primary infection (Kondo et al., 1991). Rohayen and collaborators reported of an 11-year-old boy who developed fulminant myocarditis due to an acute co-infection with parvovirus B19 and HHV-6 (Rohayem et al., 2001). They could not detect any serological markers using the IgM responses to either virus because the IgM responses to parvovirus B19 and HHV-6 can only be detected from 3 to 4 days and from 5 to 7 days following the onset of symptoms, respectively (Braun et al., 1997). A postmortem examination of the heart showed typical viral myocarditis and *in situ* hybridization confirmed myocardial invasion with parvovirus B19. HHV-6 variant B DNA was detected in the pharynx, spleen, and lung by PCR. HHV-6 was thus speculated to induce a severe immunosuppression that enhanced the dissemination of parvovirus B19, thus leading to fulminant myocarditis. They also proposed the possibility that a primary coinfection with two viruses, each of which by itself usually causes a benign infection, can result in fatal myocarditis. Interestingly, primary HHV-6 infections are caused exclusively by HHV-6 variant B (Dewhurst et al., 1993).

A reactivation from latency or an increased viral replication may occur and thereafter develop into a life-threatening disease in immunocompromised individuals. During immunosuppressive periods, such as after bone marrow transplantation, HHV-6 is a pathogen of CMV-like illness (Dockrell and Paya, 2001). On the other hand, probably because HHV-6 is susceptible to reactivation by cytokines and stress-related mechanisms, the reactivation of predominant HHV-6 variant A, but not HHV-6 variant B and CMV has been documented among critically ill patients (Razonable et al., 2002). The exact mechanisms of reactivation of HHV-6 is unclear, endothelial cells might play an important role in viral dissemination (Caruso et al., 2002; Takatsuka et al., 2003). HHV-6 is capable of infecting human aortic and heart endothelial cells and it also induces endothelial cells to secrete chemoattractants, such as monocyte chemoattractant protein-1 (MCP-1) and interleukin-8 (IL-8), thus leading to the development of inflammatory processes (Caruso et al., 2002). HHV-6 infection causes endothelial cell damage and the resultant elevation of thrombomodulin and plasminogen activator inhibitor-1 (PAI-1) levels thus leads to the development of thrombotic microangiopathy (Takatsuka et al., 2003).

We experienced a patient with fulminant fatal congestive heart failure due to diffuse myocarditis probably caused by an HHV-6 infection (Fukae et al., 2000).

Case report

A 47-year-old man was referred to our hospital on December 29, 1997 because of an exacerbation of fulminant myocarditis after undergoing steroid pulse therapy with methylprednisolone and plasma apheresis for acute hepatitis and acute myocarditis at another hospital. About two weeks earlier, he presented with a low-grade fever, general fatigue, cough, anorexia, and nausea. On December 20, 1997, he was admitted to a hospital because the symptoms worsened. A physical examination revealed a heart rate of 90 beats/min and a blood pressure of 92/70 mmHg. The results of hematological and other blood chemical tests (Fig. 1) showed both liver and cardiac dysfunction (WBC, 10,800/mL; asparate aminotransferase (AST), 1015 IU/L; alanine aminotransferase (ALT), 829 IU/L; lactate dehydrogenase (LDH), 2,691 IU/L (LDH5 > LDH1); and creatine kinase (CK), 284 IU/L), and mild pericardial effusion was also detected by echocardiography. On December 22, the patient was diagnosed to have acute hepatitis, severe type (PT (prothrombin time), 14%; HGF (hepatocyte growth factor), 20.5 ng/ml). He was treated with steroid pulse therapy (methylprednisolone 500 mg/day for three days). In addition, plasma apheresis was also performed. On December 25, the symptoms disappeared and a chest X-ray revealed only mild cardiomegaly with neither pleural effusion nor congestion. On December 29, the CK level became elevated again (348 IU/L) and an electrocardiogram (ECG) showed marked ST-T changes, a wide QRS complex with left axis deviation. The patient was then transferred to our hospital for further evaluation. On admission, he had a body temperature of 37.6°C, blood pressure 90/50 mmHg, heart rate 96 beats/min, and respiratory rate 18/min. Heart auscultation revealed the presence of a gallop rhythm, but no

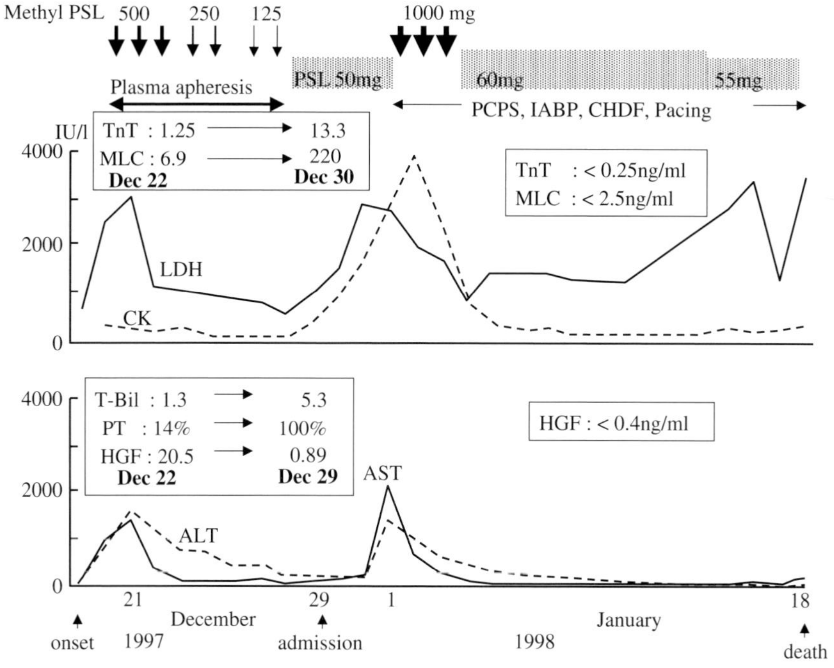

Fig. 1 Clinical course of the patient. ALT: alanine aminotransferase; AST: asparate aminotransferase; CHDF: continuous hemodiafiltration; CK: creatine kinase; HGF: hepatocyte growth factor; IABP: intraaortic balloon pumping; LDH: lactate dehydrogenase; MLC: myosin light chain; PCPS: percutaneous cardiopulmonary support; PSL: prednisolone; PT: prothrombin time; T. Bil: total bilirubin; TnT: troponin T.

pericardial knock sounds were audible. In addition, no peripheral edema was observed. Echocardiography revealed diffuse mild hypokinesis (LVDd, 43 mm; LVDs, 31 mm; EF, 55%) and hypertrophy (IVS, 16 mm; LVPW, 12 mm) of the left ventricle (LV) (Fig. 2A). Mild pericardial effusion and mild mitral regurgitation were also documented. To evaluate the hemodynamic state, a Swan–Ganz catheter was inserted. The hemodynamic study showed Forrester subset I: cardiac output of 4.2 L/min, cardiac index of 2.4 L/min/m^2, and pulmonary capillary wedge pressure of 9 mmHg, which we observed without administering any drugs. On December 31, the patient's condition suddenly deteriorated, with a low blood pressure (70/ mmHg), low cardiac output (2.2 L/min, cardiac index of 1.3 L/min/m^2), and complete AV block on ECG. Full supportive treatment with temporary pacing (VVI mode, 70/min), an intraaortic balloon pump (IABP), and percutaneous cardiopulmonary support (3.0–4.0 L/min) was urgently administered since the patient had fallen into a state of shock. Echocardiography performed on January 2, 1998 revealed an extremely edematous LV (IVS, 20 mm; LVPW, 16 mm) and severe hypokinesis of LV (Fig. 2B). After the readministration of 1000 mg/day of methylprednisolone for three days, the LV thickness (edema) decreased (IVS, 13 mm); however, the LV wall remained almost akinetic. On January 4, the loss of

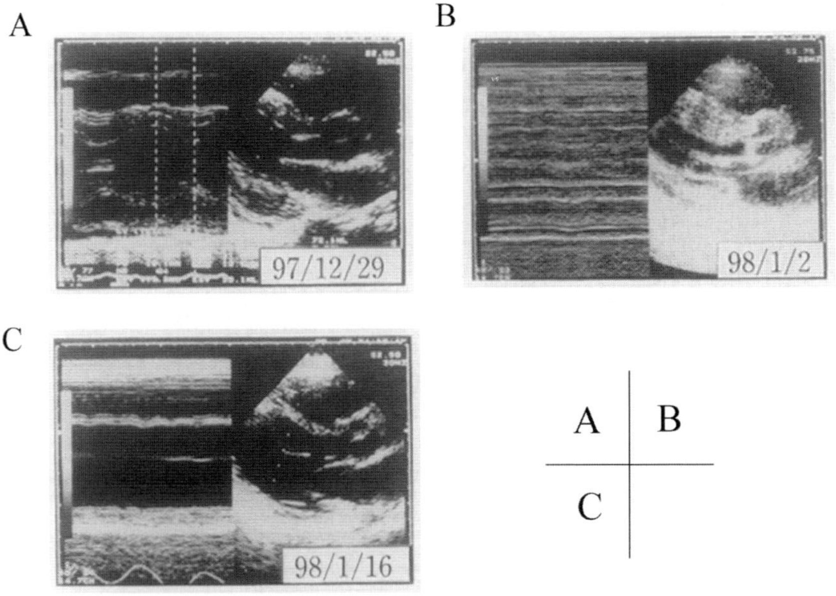

Fig. 2 Clinical course of the echocardiogram findings. (A) On admission (12/29/97): diffuse mild hypokinesis (LVDd, 43 mm; LVDs, 31 mm; EF, 55%) and hypertrophy (IVS, 16 mm; LVPW, 12 mm) of the LV; (B) Four days after admission (1/2/98): extremely edematous LV (IVS, 20 mm; LVPW, 16 mm) and LV wall motion was reduced (EF, 41%); (C) Two days before death (1/16/98): dilatation of the LV cavity (52 mm) and thinning of the LV (IVS, 7 mm; LVPW, 8 mm) with akinetic LV. EF: ejection fraction, IVS: intraventricular septum, LV: left ventricle.

ventricular capture on VVI pacing was observed despite programing for the maximum output, which thus prevented IABP from functioning. Subsequently a gradual suppression of the cardiac electrical activity was noted: including an intraventricular conduction disturbance, a persistent complete AV block, and finally a ventricular standstill. The cardiogenic shock state of the patient did not improve even with the use of ventricular assist devices and the patient died on January 18, 1998.

According to the autopsy findings, although dilatation of sinusoidal capillaries and mild fibrosis of the portal area were observed, no evidence of hepatitis was suggested in the hepatic cells. A dilatation of the LV cavity and thinning of the ventricle (LV, 5 mm; IVS, 9 mm) were observed, which were compatible with the echocardiographic observations performed two days before death (Fig. 2C), and severe fibrosis and granulation changes were observed in the ventricles and the left atrium (Fig. 3A). Only about 10–15% of the normal ventricular myocardium remained at the lateral and posterior wall of the LV. A loss of myofibers with interstitial fibrosis (Fig. 3B) and inflammatory cellular infiltration (mainly rounded cells) were observed in the LV (Fig. 3C). The HHV-6 IgG antibody titers were found to have risen from × 40 (12/22/97) to × 160 (12/30/97) based on an analysis using fluorescent antibody (FA). The antibody titers to other viruses, which can cause either myocarditis or hepatitis, were not found to have increased (Tables 1

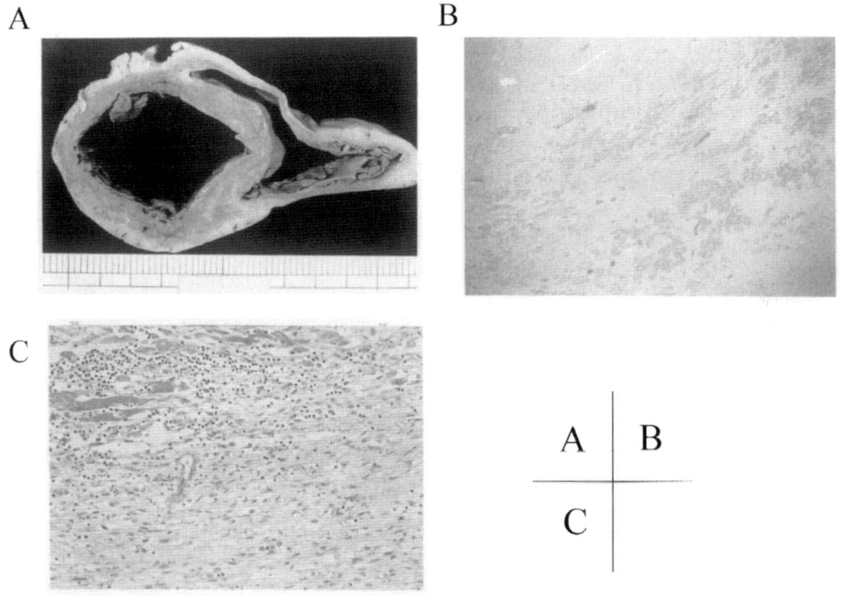

Fig. 3(A) Postmortem specimen showing a dilatation of the left ventricular cavity and thinning of the left ventricle and intraventricular septum; (B) A loss of myofibers with interstitial fibrosis (HE stain, × 40); and (C) Inflammatory cellular infiltration (mainly rounded cell) (HE stain, × 100) was observed in the left ventricle. (for colour version: see colour section on page 356).

Table 1

Serum antibody titer (CF)

	12/22/97	12/30/97
Coxsackie virus B 1–6	<4	<4
Influenza A	<4	×4
Influenza B	×4	×4
Adenovirus	×4	<4
Cytomegalovirus	×16	×8
Herpes simplex virus	×16	×16
Varicella	<4	<4

CF, Complement fixation.

Table 2

Serum anit-hepatits viral antibody titer

	1/1/98
IgM-HA Ab (RIA)	0.2 (<0.9)
HBe Ag (RIA)	0.2 (<0.9)
HBe Ab (RIA)	24% (<29%)
IgM-HBc Ab (RIA)	0.2 (<0.9)
HBs Ab (PHA)	×4 (<4)
HCV Ab (RIA)	0 (<0.9)

Ab, antibody; Ag, antigen; HA, hepatitis A; HBc, hepatitis B core; HBe, hepatitis Be; HBs, hepatitis B surface; HCV, hepatitis C virus; PHA, passive hemagglutination; RIA, radio-immunoassay.

and 2). We also detected HHV-6 DNA in the patient's serum before plasma apheresis by PCR. The DNA extracted from the liver and the heart were found to be positive for HHV-6 by PCR (Fig. 4).

Discussion

We herein describe a fatal case of an adult patient who developed fulminant myocarditis after steroid pulse therapy for severe acute hepatitis. Although the cause of the myocarditis in the patient was unclear, we suspected that the steroid therapy may have played a role on his outcome. As shown in Fig. 1, there are two peaks of CK, ALT, AST, and LDH. The first peak demonstrated the main focus to be the liver because of the superiority of the enzymes derived from the liver. The reduced (14%) PT and elevated HGF, which were taken before plasma apheresis, showed the liver condition to be severe enough to result in a diagnosis of acute hepatitis. In the second peak, however, the main focus was the heart because the elevation of CK, TnT (troponin T), and MLC (myosin light chain) was prominent. It is therefore conceivable that the patient first suffered from severe acute hepatitis, with mild

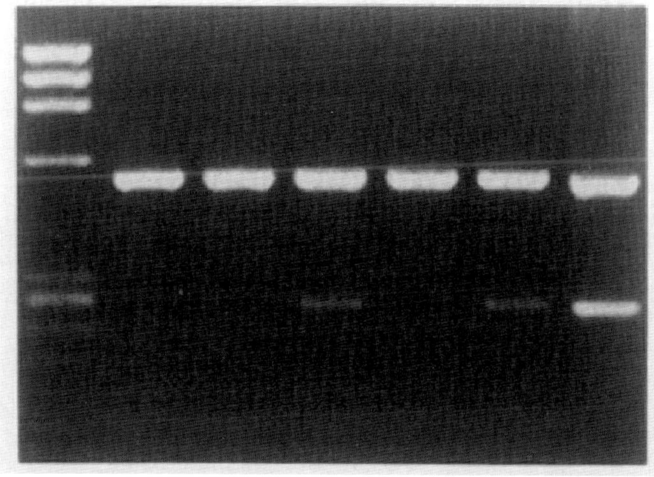

Fig. 4 DNA of HHV-6 isolated from the liver and the heart of the patient. The presence of HHV-6 DNA was documented in the Liv, RA, and RV by PCR. HHV-6: human herpesvirus-6, Liv: liver, PCR: polymerase chain reaction, RA: right atrium, RV: right ventricle.

myocarditis, which was initially almost cured by plasma apheresis and steroid therapy. However, the myocarditis worsened after steroid therapy, according to the clinical course. The evidence from echocardiography (Fig. 2) and histopathology (Fig. 3) also support this speculation. The HHV-6 gene sequences were detected by the PCR in the liver, right atrium, and the right ventricle during necropsy (Fig. 4). Furthermore, both the patient and his daughter (14 years of age) had demonstrated skin eruptions on the forearms with characteristics different from those of typical herpes zoster or herpes simplex. Interestingly, the HHV-6 IgG of his daughter was × 160 on February 6, 1998, about 5 weeks after she visited her father at the hospital. The detection of HHV-6 serum DNA thus suggests an association with HHV-6.

Taken together, we believe that the agent responsible for the initial acute hepatitis and myocarditis has not yet been identified. The later exacerbation of fulminant myocarditis was probably associated with HHV-6. Recently, Sekiguchi et al. (2005) reported a 66-year-old male, who showed fever, a generalized pruritic rash, liver dysfunction, and leukocytosis about one month after mexiletine hydrochloride administration. The patient was thus diagnosed to have drug-induced hypersensitivity syndrome (DIHS) and he died of myocarditis associated with HHV-6 and CMV reactivation. Their case was also administered by systemic corticosteroid which may have modified the clinical course. Because our patient was also treated with steroid pulse therapy, the immunosuppressive state of the host may have caused a reactivation of the HHV-6. The controversy regarding the role of immunosuppression for myocarditis has not yet been resolved (Mason et al., 1995; Garg et al., 1998). However, we sometimes experience patients showing a dramatic improvement after treatment with steroids. This case will hopefully provide further insight into the mechanisms of virus–myocardium–immune interactions.

References

Braun DK, Dominguez G, Pellet PE. Clin Microbiol Rev 1997; 10: 521–567.

Caruso A, Rotola A, Comar M, Favilli F, Galvan M, Tosetti M, Campello C, Caselli E, Alessandri G, Grassi M, Garrafa E, Cassai E, Luca DD. J Med Virol 2002; 67: 528–533.

Dewhurst S, McIntyre K, Schnabel K, Hall CB. J Clin Microbiol 1993; 31: 416–418.

Dockrell DH, Paya CV. Rev Med Virol 2001; 11: 23–36.

Fukae S, Ashizawa N, Morikawa S, Yano K. Intern Med 2000; 39: 632–636.

Garg A, Shiau J, Guyatt G. Ann Intern Med 1998; 129: 317–322.

Kondo K, Kondo T, Okuno T, Takahashi M, Yamanishi K. J Gen Virol 1991; 72: 1401–1408.

Mason JW, O'Connell JB, Herskowitz A, Rose NR, McManus BM, Billingham ME, Moon TE. N Engl J Med 1995; 333: 269–275.

Razonable RR, Fanning C, Brown RA, Epsy MJ, Rivero A, Wilson J, Kremers W, Smith TA, Paya C. J Infect Dis 2002; 185: 110–113.

Rohayem J, Dinger J, Fischier R, Klingel K, Kandolf R, Rethwilm A. J Clin Microbiol 2001; 39: 4585–4587.

Sekiguchi A, Kashiwagi T, Ishida-Yamamoto A, Takahashi H, Hashimoto Y, Kimura H, Tohyama M, Hashimoto K, Iizuka H. J Dermatol 2005; 32: 278–281.

Takatsuka H, Wakae T, Mori A, Okada M, Fujimori Y, Takemoto Y, Okamoto T, Kanamaru A, Kakishita E. Viral infections. Bone Marrow Transplant 2003; 31: 475–479.

Toyabe S, Harada W, Suzuki H, Hirokawa T, Uchiyama M. Clin Rheumatol 2002; 21: 528–532.

Yamanishi K, Okuno T, Shiraki K, Takahashi M, Kondo T, Asano Y, Kurita T. Lancet 1988; I: 1065–1067.

Yoshikawa T, Ihira M, Suga S, Kito H, Iwasaki T, Kurata T, Tanaka T, Saito Y, Asano Y. J Clin Pathol 2001; 54: 792–795.

Perspectives in Medical Virology
Gerhard Krueger and Dharam Ablashi (Editors)

DOI 10.1016/S0168-7069(06)12018-2

HHV-6 in Cardiovascular Pathology

L. Maximilian Buja[a,b]
[a]*Department of Pathology and Laboratory Medicine and Office of the Executive Vice President for Academic Affairs, The University of Texas, Health Science Center at Houston, 7000 Fannin Street, Suite 1715, Houston, TX 77030, USA*
[b]*Department of Cardiovascular Pathology, Texas Heart Institute, St. Luke's Episcopal Hospital, Texas Medical Center, Houston, TX 77030, USA*

Introduction

Human herpesvirus-6 (HHV-6) infects cells, which express the CD46 complement receptor on their cellular membranes (Santoro et al., 1999). CD46 is strongly expressed in salivary gland epithelial cells and in renal tubular cells, moderately well expressed in lymphoid cells, monocytes, and vascular endothelial cells, and weakly expressed in muscle cells (Ablashi and Krueger, 2003; Loveland, 2003). Although this distribution of HHV-6 cellular receptor molecules may indicate the possible targets for HHV-6 infection, the frequency and distribution of diseases reported so far does not closely follow this pattern (Krueger and Ablashi, 2003).

Among major clinical symptoms described in 1994 for active HHV-6 infection were palpitations, tachycardia, and arrhythmia as well as not-further specified "myocarditis" (Krueger et al., 1994). It appears reasonable, therefore, to assume that the cardiovascular system (CVS) may be a possible target for HHV-6 infections. Unfortunately, sufficient systematic studies are not available to date to document the prevalence of HHV-6 infection in cardiovascular diseases, and only observational reports and relatively small case series are currently available to support this assumption.

Diseases of blood vessels

Replicative and latent infection of vascular endothelium by HHV-6 was repeatedly shown *in vivo* and *in vitro* using immunohistochemical and molecular techniques (Wu and Shanley, 1998; Rotola et al., 2000; Caruso et al., 2003). *In vitro* infection of human umbilical vein endothelium (HUVEC) with HHV-6 was followed by the expression of early and late viral antigens in 37.6 and 6.5% of HUVEC, respectively, with persistence of the antigens for up to 27 days. Although virus was not obviously released from these cells, it could be recovered by cocultivation (Wu and Shanley, 1998). Endothelial cells obtained from the aorta, vasa vasorum, or from cardiac microvessels of immunocompetent patients with aortic insufficiency or aneurysm revealed HHV-6 by nested polymerase chain reaction (PCR) even in cases where viral DNA was not recovered from peripheral blood mononuclear cells (Rotola et al., 2000). Viral transcripts from immediate-early (U91, U42) and late (U22) genes were detected in aortic endothelial cells by nested reverse transcription PCR. Since no p41 early antigen was demonstrable by immunohistochemistry, the authors think that there exists only a low-level viral replication at these sites thus confirming the earlier data.

Both *in vitro* and *in vivo* infections by HHV-6 can occur without causing visible cytopathic effects, yet the increase in factors such as thrombomodulin, plasminogen activator–inhibitor-1, and cyclic GMP in infected persons may signal vascular injury (Takatsuka et al., 2003). In addition, HHV-6 upregulates monocyte chemoattractant protein-1 (MCP-1) and interleukin-8 (IL-8) and induces the *de novo* synthesis of RANTES CC chemokine (Caruso et al., 2002, 2003). HHV-6-infected endothelial cells, thus, may well support the attraction of immunocompetent cells and the initiation of an inflammatory reaction.

In contrast to the frequent demonstration of HHV-6 DNA and antigen in vascular endothelial cells stands its rather infrequent association with vascular disease. We have observed increased levels of HHV-6A antigens and DNA in various endothelial cells of patients with HHV-6 reactivation (Fig. 1; unpublished data). In no case, there was any evidence of "endothelitis" or of other forms of vascular inflammation. Okano and coworkers (1989) found 81% of patients with Kawasaki's disease to have elevated IgM and IgG antibodies to HHV-6 and theorized that the virus may add to the immunologic alterations of this disease.

In another study, however, the etiologic role of HHV-6 for Kawasaki's syndrome was not confirmed (Marchette et al., 1990). Both of these studies use only serologic investigations for determining any possible association of the virus with the vasculitis and thus remain inconclusive. Toyabe and collaborators (2002) reported on a 9-month-old boy with large vessel arteritis and active HHV-6 infection. IgM and IgG antibodies to HHV-6 were elevated in the patient's serum and viral DNA was isolated from the peripheral blood and from blood mononuclear cells. No attempts were made, though, to show viral components in the inflamed vessel itself. Three other studies using PCR and real-time PCR on vascular biopsies were unable to prove an association of HHV-6 and giant cell arteritis (Helweg-Larsen et al., 2002; Rodriguez-Pla et al., 2004; Alvarez-Lafuente et al., 2005).

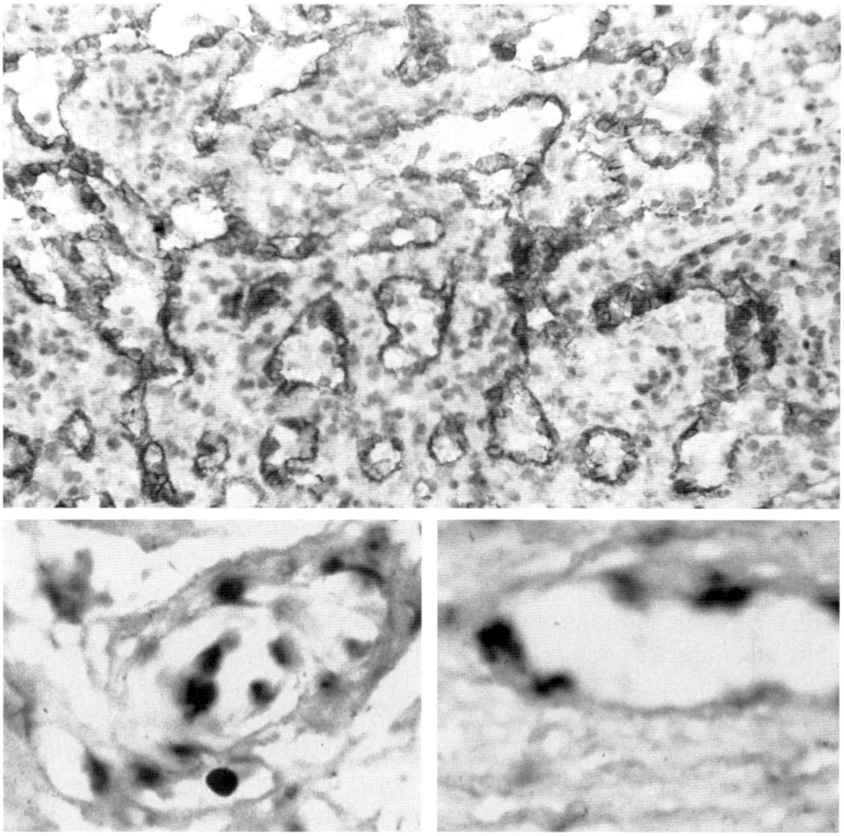

Fig. 1 HHV-6 infection of vascular endothelial cells. *Top*: Splenic sinusoidal endothelial cells containing HHV-6 late antigens (red-stained cells; APAAP reaction using HAR 1-3 antibody). *Bottom*: HHV-6 DNA in endothelial cells of cardiac arteriole in an AIDS patient (left) and of a brain venule (right) in a case of necrotizing encephalitis in a child with active HHV-6 infection (black cells; *in situ* hybridization with pZVH14 probe). (for colour version: see colour section on page 357).

Matsuda and coworkers (1999) reported a case of thrombotic microangiopathy in a patient with high-dose chemotherapy, bone marrow transplantation, and HHV-6 reactivation. Although HHV-6 may theoretically have contributed to the pathogenesis of this disorder, there are many other "pathogens" in the patient's history that must be considered, such as cyclosporine A or other chemotherapy, graft-versus-host reaction, and irradiation. All are known to damage vascular endothelium, and thus be able to initiate some kind of microangiopathy.

Infectious agents, including herpesviruses and *Chlamydia pneumoniae*, have been identified with some frequency in atherosclerotic lesions. Although the primary role of these agents in the pathogenesis of atherosclerosis has not been determined, infectious agents have the potential to participate as promoters of inflammation, which is important in the progression of atherosclerosis (Kol and

Libby, 1998; Anderson, 2005). As discussed above, HHV-6 infection of endothelial cells has the potential of triggering an inflammatory reaction. However, the prevalence of HHV-6 in atherosclerotic lesions is not known, and there is no evidence to support the role of HHV-6 in atherogenesis.

Diseases of the heart

As described above, vascular endothelial cells contain the HHV-6 receptor CD46 and may carry HHV-6 antigens and DNA. This includes the blood vessels in the heart (Fig. 1). CD46 is weakly expressed only in striated muscle cells (Loveland, 2003). The distribution of viral receptors in the myocardium for well-documented cardiotropic viruses (coxsackie virus, adenovirus) is highly variable with a frequent increase in patients with dilated cardiomyopathy (Poller et al., 2002). No systematic studies to this effect are available yet for the HHV-6. However, it is possible that increased expression of the HHV-6 receptor CD46 in cardiomyocytes may occur in myopathic conditions (Poller et al., 2002).

Active myocarditis has been diagnosed in endomyocardial biopsies from a substantial percentage of patients with various kinds of heart failure, with a variable prevalence of less than 20% to over 50% (Nippoldt et al., 1982; Feldman and McNamara, 2000). Viral infection has been implicated as a leading cause of myocarditis. In a large series, viral genome was amplified by PCR from 239 (38%) of 624 patients with clinical myocarditis, and 30 (20%) of 149 patients with dilated cardiomyopathy (Bowles et al., 2003). Most frequently detected viruses were adenovirus and enterovirus. However, reports of HHV-6-associated myocarditis are rather scarce. Fukae and coworkers (2000) describe a case of fulminant myocarditis in which HHV-6 DNA was demonstrated by PCR in the myocardium and serum IgG antibodies increased to fourfold levels. One year later, Yoshikawa and collaborators (2001) reported on a 5-month-old girl with acute myocarditis, HHV-6B isolation from the blood and positive HHV-6 DNA by PCR in various tissues including the heart. Other viral infections were excluded in both cases. We have observed a similar, although focal, myocarditis at autopsy of sporadic cases of AIDS with HHV-6 reactivation (Fig. 2), yet HHV-6 DNA was demonstrable by *in situ* hybridization only in vascular endothelial cells (as shown in Fig. 1). Except for HIV-1, no other cardiotropic virus was identified in our cases. Rohayem et al. (2001) reported a fatal case of fulminant myocarditis due to acute coinfection with parvovirus B19 and HHV-6 in the absence of an antiviral immune response. By nested PCR, parvovirus B19 and HHV-6 DNA were detected in multiple tissues, but only parvovirus B19 DNA was found in the myocardium. The authors suggested that immunosuppression induced by HHV-6 enhanced the dissemination of parvovirus B19, which led to fatal myocarditis.

In a case series of patients, Mahrholdt and colleagues (2004) reported that myocarditis as diagnosed by magnetic resonance imaging and endomyocardial biopsy was linked to HHV-6 in six, and parvovirus B19 in 12 of the 32 patients. All but one diagnostic biopsies was taken from regions of contrast enhancement.

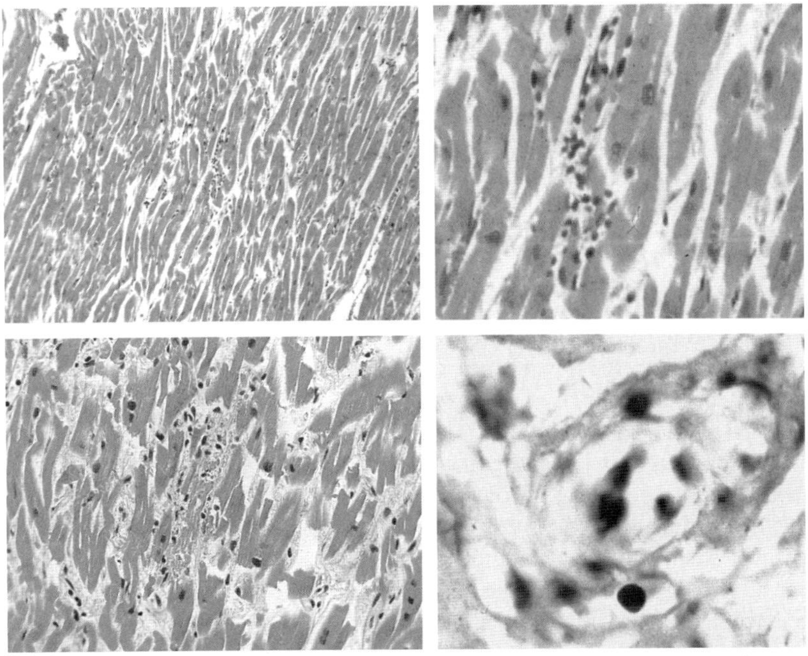

Fig. 2 Various forms of myocarditis accompanying HHV-6 reactivation in AIDS patients (hematoxylin and eosin stain of autopsy specimens). *Lower right* shows an interstitial cardiac arteriole from such cases containing HHV-6 DNA (*in situ* hybridization with pZVH14 probe). (for colour version: see colour section on page 358).

Myocarditis occurred predominantly in the lateral free wall. The left ventricular ejection fraction was reduced in these patients to $47 \pm 19\%$.

Although HHV-6 may reside latently in cardiac vessels, cardiac transplantation followed by immunosuppression appears not associated with a higher risk of HHV-6 myocarditis. Adenovirus and other viruses have been implicated in coronary vasculopathy, chronic graft failure, and acute rejection (Bowles et al., 2001; Shirali et al., 2001). To date, we know of only two reports of reactivated HHV-6 causing disease after cardiac transplantation (Randhawa et al., 1997; Nash et al., 2004). In both cases, the heart alone was not affected, but rather gastroduodenitis, pancreatitis, and encephalitis were described by these authors.

Established criteria for the pathological diagnosis of myocarditis requires the presence of inflammatory cellular infiltrates in association with degeneration and necrosis of cardiomyocytes, the Dallas criteria (Aretz, 1987). The pathogenesis of viral myocarditis involves a variable interaction of direct viral damage to cells, virus-induced immune-mediated cellular damage, and viral persistence (Feldman and McNamara, 2000; Esfandiarei et al., 2004). Viral myocarditis is considered to be an important precursor in the pathogenesis of acquired cardiomyopathy, which is recognized as established or chronic heart muscle disease, often manifested as hypertrophy and dilatation of the heart (Feldman and McNamara, 2000; Poller et al., 2005). However, there is mounting evidence that viral infection can produce myocardial disease in the absence of the classic features of myocarditis (Feldman and McNamara, 2000; Poller et al., 2005). This may involve more subtle forms of cellular injury and inflammation. Adenovirus infection of the myocardium has been found to be associated with significantly less inflammation than enterovirus infection of the myocardium (Bowles et al., 2003). Recent studies discussed below indicate that HHV-6 infection may produce myocardial disease without classical features of myocarditis.

In 245 patients with dilated cardiomyopathy characterized by unexplained left ventricular dysfunction, viral genomes were amplified by PCR in endomyocardial biopsy samples from 165 (67.4%) of the 245 patients (Kühl et al., 2005). HHV-6 was detected in 53 (21.6%), and this was second in frequency to parvovirus B19, which occurred in 126 patients (51.4%). Other viruses, including adenovirus and enterovirus, occurred at lower frequency. Multiple infections occurred in 45 samples (27.3%) with 26 (15.8%) showing coinfection with parvovirus B19 and HHV-6. Active or borderline myocarditis according to the Dallas criteria was not present in any case. Also, lymphocyte and macrophage infiltrates were not significantly different in virus-positive and virus-negative patients. Similar recovery of virus has not been reported in normal hearts of multiorgan donors or patients with valvular heart disease. The findings in the patients with idiopathic left ventricular dysfunction suggest that viral persistence, often presenting as multiple infection, may play a role in the pathogenesis of dilated cardiomyopathy (Kühl et al., 2005).

In another study, 37 of 70 patients presenting with exertional dyspnea and/or reduced exercise tolerance were confirmed to have isolated diastolic left ventricular dysfunction by echocardiography and cardiac catheterization, and 35 (95%) of these patients had cardiotropic viral genomes, which were detected in endomyocardial biopsies (Tschöpe et al., 2005). Of these 37 patients, 24 (65%) had a parvovirus B19 monoinfection, and 6 of 37 (16%) had a coinfection with HHV-6. Of the 37 patients, 24 (65%) had atypical angina pectoris, and 13 of these 24 patients (54%) had evidence of coronary endothelial dysfunction on the basis of abnormal response to provocative testing with acetylcholine. Ten of these 13 patients (77%) also had diastolic dysfunction, and were all positive for parvovirus B19 genomes with 73% having a monoinfection and 27% coinfection with HHV-6.

A group of 71 patients with nonischemic cardiomyopathy were evaluated for coronary microcirculatory dysfunction in relationship to viral persistence and myocardial inflammation with endothelial activation (Vallbracht et al., 2005). Coronary endothelial microcirculatory function was measured by vasoreactive response to acetylcholine challenge, with a positive finding characterized by abnormal vasoconstriction or impaired vasodilation in response to acetylcholine. Myocardial inflammation was measured by increased numbers of lymphocytes and expression of endothelial adhesion molecules (HLA-1, HLA-DR, ICAM-1). PCR showed evidence of viral persistence in 43 patients, with parvovirus B19 in 33 and HHV-6 in 12. Endothelial dysfunction occurred in patients with viral persistence independently of myocardial inflammation and endothelial activation but was more pronounced in patients with concurrent inflammation. These findings indicate that viral-induced microcirculatory dysfunction with impaired coronary blood flow can be important in the pathogenesis of cardiomyopathy. This and other viral-induced abnormalities can occur in the absence of classical myocarditis.

HHV-6 may directly damage cells, particularly endothelial cells, or induce immune or autoimmune reactions. HHV-6 is able to activate other viral infections, including Epstein–Barr virus and parvovirus B19, and the coinfection may enhance the pathogenicity of other viruses (Krueger and Ablashi, 2003). Transient viral infection may be sufficient to trigger enough insult to lead to chronic disease. Conversely, viral persistence may lead to progressive disease. Potential mechanisms include direct cytopathic effects of the viruses and indirect effects, including low-grade inflammation, release of cytokines, alterations of cell signaling pathways, alterations in cytoskeletal proteins of myocytes, and alterations of the extracellular matrix (Kühl et al., 2005; Vallbracht et al., 2005). These mechanisms may be particularly important in HHV-6-induced myocardial disease.

Conclusions

The CVS may be the target of HHV-6 infection with the primary target being endothelial cells, which express the CD46 complement receptor on their cell membranes. Isolated cases of vascular disease and myocarditis related to HHV-6 infection have been reported. More recent evidence has indicated a high rate of recovery of HHV-6 viral genome in myocardium of patients with cardiomyopathy and unexplained left ventricular dysfunction. A consistent pattern of coinfection with HHV-6 and parvovirus B19 has been reported. Endothelial dysfunction and microcirculatory ischemia induced by HHV-6 and other viruses in the absence of classical myocarditis may be important in the pathogenesis of the acquired type of cardiomyopathy. Other direct and indirect effects of persistent viral infection also may contribute to the pathogenesis of cardiomyopathy. Further systematic study is needed of the role of HHV-6 infection in cardiovascular disease.

References

Ablashi DV, Krueger GRF. Human herpesviruses 6, 7 and 8. In: Viral Infections and Treatment (Ruebsamen-Weigman H, Deres K, Hewlett G, Welker R, editors). New York: Marcel Dekker; 2003; pp. 659–705.

Alvarez-Lafuente R, Fernandez-Gutierrez B, Jover JA, Judez E, Loza E, Clemente D, Garcia-Asenjo JA, Lamas JR. Ann Rheuma Dis 2005; 64: 780–782.

Anderson JL. N Engl J Med 2005; 352: 1706–1709.

Aretz HT. Hum Pathol 1987; 18: 619–624.

Bowles N, Radovancevic B, Thomas CD, Thuy V, Frazier OH, Towbin JA, J Heart Lung Transplant 2001; 20: 198. (Abstract)

Bowles NE, Ni J, Kearney DL, Pauschinger M, Schultheiss H- P, McCarthy R, Hare J, Bricker JT, Bowles KR, Towbin JA. J Am Coll Cardiol 2003; 42: 466–472.
Caruso A, Favilli F, Rotola A, Comar M, Horejsh D, Alessandri G, Grassi M, Di Luca D, Fiorentini S. J Med Virol 2003; 70: 451–458.
Caruso A, Rotola A, Comar M, Favilli F, Galvan M, Tosetti M, Campello C, Caselli E, Alessandri G, Grassi M, Garaffa E, Cassai E, Di Luca D. J Med Virol 2002; 67: 528–533.
Esfandiarei M, Luo H, Yanagawa B, Suarez A, Dabiri D, Zhang J, McManus BM. J Virol 2004; 78: 4289–4298.
Feldman AM, McNamara D. N Engl J Med 2000; 343: 1388–1398.
Fukae S, Ashizawa N, Morikawa S, Yano K. Intern Med 2000; 39: 632–636.
Helweg-Larsen J, Tharp B, Obel N, Baslund B. Rheumatology 2002; 41: 445–449.
Kol A, Libby P. Trends Cardiovasc Med 1998; 8: 191–199.
Krueger GRF, Ablashi DV. Intervirology 2003; 46: 257–269.
Krueger GRF, Klueppelberg U, Hoffmann A, Ablashi DV. In vivo 1994; 8: 457–485.
Kühl U, Pauschinger M, Noutsias M, Seeberg B, Bock T, Lassner D, Poller W, Kandolf R, Schultheiss H- P. Circulation 2005; 111: 887–893.
Loveland B. Protein Reviews on the Web; 2003. Available at http://www.ncbi.nlm.nih.gov/PROW/guide/2027814670g.htm.
Mahrholdt H, Goedecke C, Wagner A, Meinhardt G, Athanasiades A, Vogelsberg H, Fritz P, Klingel K, Kandolf R, Sechtem U. Circulation 2004; 109: 1250–1258.
Marchette NJ, Melish ME, Hicks R, Kihara S, Sam E, Ching D. J Infect Dis 1990; 161: 680–684.
Matsuda Y, Hara J, Miyoshi H, Osugi Y, Fujisaki H, Takai K, Ohta H, Tanaka-Taya K, Yamanishi K, Okada S. Bone Marrow Transplant 1999; 24: 919–923.
Nash PJ, Avery RK, Tang WHW, Starling RC, Taege AJ, Yamani MH. Am J Transplant 2004; 4: 1200–1203.
Nippoldt TB, Edwards WD, Holmes DR Jr GS, Hartzler GO, Smith HC. Mayo Clin Proc 1982; 57: 407–418.
Okano M, Luka J, Thiele GM, Sakiyama Y, Matsumoto S, Purtilo DT. J Clin Microbiol 1989; 27: 2379–2380.
Poller W, Fechner H, Noutsias M, Tschoepe C, Schultheiss H- P. Z Kardiol 2002; 91: 978–991.
Poller W, Kühl U, Tschöepe C, Pauschinger M, Fechner H, Schultheiss H-P. J Mol Med 2005. Published online 2 June 2005.
Randhawa PS, Jenkins FJ, Nalesnik MA, Martens J, Williams PA, Ries A, Pham S, Demetris AJ. Am J Surg Pathol 1997; 21: 847–853.
Rodriguez-Pla A, Bosch-Gil JA, Echevarria-Mayo JE, Rosello-Urgell J, Solans-Laque R, Huguet-Redecilla P, Stone JH, Villardell-Tarres M. J Clin Virol 2004; 31: 11–15.
Rohayem J, Dinger J, Fischer R, Klingel K, Kandolf R, Rethwilm A. J Clin Microbiol 2001; 39: 4585–4587.
Rotola A, Di Luca D, Cassai E, Ricotta D, Alessandri G, Turano A, Caruso A, Muneretto C. J Clin Microbiol 2000; 38: 3135–3136.
Santoro F, Kennedy PE, Locatelli G, Malnati MS, Berger EA, Lusso P. Cell 1999; 99: 817–827.
Shirali GS, Ni J, Chinnock RE, Johnston JK, Rosenthal GL, Bowles NE, Towbin JA. N Engl J Med 2001; 344: 1498–1503.
Takatsuka H, Wakae T, Mori A, Okada M, Fujimori Y, Takemoto Y, Okamoto T, Kanamaru A, Kakishita E. Bone Marrow Transplant 2003; 31: 475–479.

Toyabe S, Harada W, Suzuki H, Hirokawa T, Uchiyama M. Clin Rheumatol 2002; 21: 528–532.

Tschöpe C, Bock C- T, Kasner M, Noutsias M, Westermann D, Schwimmbeck P- L, Pauschinger M, Poller W- C, Kühl U, Kandolf R, Schultheiss H- P. Circulation 2005; 111: 879–886.

Vallbracht KB, Schwimmbeck PL, Kühl U, Rauch V, Seeberg B, Schultheiss H- P. Circulation 2005; 111: 1784–1791.

Wu CA, Shanley JD. J Gen Virol 1998; 79: 1247–1256.

Yoshikawa T, Ihira M, Suzuki K, Suga S, Kito H, Iwasaki T, Kurata T, Tanaka T, Saito Y, Asano Y. J Clin Pathol 2001; 54: 792–795.

Human Herpesvirus-6: General Virology, Epidemiology and Clinical Pathology
Perspectives in Medical Virology
Gerhard Krueger and Dharam Ablashi (Editors)

DOI 10.1016/S0168-7069(06)12019-4

HHV-6, the Liver and the Gastrointestinal Tract

Tetsushi Yoshikawa
Department of Pediatrics, Fujita Health University School of Medicine, Toyoake, Aichi, 4701192, Japan

Introduction

Human herpesvirus-6 (HHV-6), a member of the β-herpesvirinae subfamily, shares many biological properties with the other members of this family, human cytomegalovirus (CMV) and HHV-7. Primary CMV infection in healthy children is associated with hepatomegaly and mild elevations of hepatic transaminase. Moreover, CMV infection causes hepatitis and gastrointestinal ulceration in immunocompromised patients. After the discovery of HHV-6, direct virological examination demonstrated that it causes similar clinical manifestations in the liver and gastrointestinal tract as CMV.

HHV-6 and hepatitis

Primary HHV-6 B infection causes exanthem subitum, a common febrile infant disease (Yamanishi et al., 1988). Although only 17% of American children with primary HHV-6 infection develop exanthem subitum (Pruksananonda et al., 1992), in Japan, 70–80% of infants with primary viral infection exhibit a typical clinical course of exanthem subitum (Asano et al., 1991a). Most affected patients visit a pediatric outpatient clinic, and some receive hematological examination to rule out severe infectious diseases such as bacterial meningitis. Some patients experience a mild elevation in hepatic transaminase levels during infection. Indeed, approximately 4% (four cases) of 89 infants with primary HHV-6 infection had signs of hepatic injury during viral infection (Asano et al., 1991a). In these cases, patients had a transient elevation in serum hepatic transaminase levels that completely resolved without any specific treatments. However, one case report described chronic hepatitis caused by HHV-6 infection in one 20 month old, otherwise healthy boy (Tajiri et al., 1997). Chronic hepatitis was confirmed by pathological analysis, and viral DNA was detected in liver tissue by polymerase chain reaction (PCR) and *in situ* hybridization. In these samples, HHV-6 DNA is localized to the nuclei of hepatocytes, the sinusoidal cells, and the nuclei of the epithelial cells of the intra-hepatic bile ducts. Chronic hepatitis in young children can be caused by a variety of viruses, such as hepatitis B virus, hepatitis C virus, and CMV. In addition to these viruses, HHV-6 appears to be an etiologic agent for chronic hepatitis in young children.

In addition to chronic hepatitis, HHV-6 also appears to cause some cases of fulminant hepatitis, a much more severe disease entity. We reported fatal fulminant hepatitis resulting from primary HHV-6 infection in a previously healthy 3-month-old boy in 1990. Despite full supportive care and exchange transfusion, the patient died 7 days after admission to our hospital. We confirmed primary HHV-6 infection by viral isolation and serological analysis (Asano et al., 1990). We also recently reported a non-fatal case of HHV-6-associated fulminant hepatitis treated with living donor liver transplantation (Ohashi et al., 2004). Individuals with fulminant hepatitis ultimately die from liver failure, and living donor liver transplantation could improve the prognosis of HHV-6-associated fulminant hepatitis. Three additional cases of HHV-6-associated fulminant hepatitis from other institutions have been reported. Aita et al. (2001) observed a clinical course similar to those we have reported, and this patient recovered following supportive treatments. The onset of hepatic injury in these three patients corresponded to the exanthematous period of the disease. Histopathological analysis was carried out in the case reported by Aita et al., and showed microvesicular steatosis resembling that of Reye's syndrome. Moreover, HHV-6 DNA was detected in the endothelium of the portal vein but not in hepatocytes by *in situ* hybridization. Two additional cases of neonatal HHV-6 infection-associated fulminant hepatitis were not as clinically severe (Mendel et al., 1995).

HHV-6 infection has also been associated with fulminant hepatitis in adults (Sobue et al., 1991; Ishikawa et al., 2002). Six Japanese adults (aged between 26 and 56 years old) and five children (aged between 10 month and 13 years old) with non-A, non-B, and non-C fulminant hepatitis were analyzed using PCR, *in situ* hybridization, and immunohistochemistry to identify the etiologic agent (Ishikawa et al., 2002). HHV-6 DNA was detected in all five child patients and two out of the six adult patients by PCR. Although histological analysis was performed on samples from only one child (13-year-old boy), HHV-6 DNA, RNA, and antigens were detected in the nuclei of hepatocytes suggesting active viral replication in the hepatocytes. Conversely, no HHV-6 DNA was detected in the hepatic tissue obtained from 10 North American and 40 European non-A, non-B fulminant hepatitis patients (Mason et al., 1996). There are no clear reasons for these differences, but the sensitivity of PCR as well as ethnic background may lead to differences in the detection of HHV-6 in cases of non-A and non-B fulminant hepatitis. A larger number of cases should be prospectively analyzed to more clearly identify the association of HHV-6 and fulminant hepatitis and clarify the clinical features of HHV-6-associated fulminant hepatitis.

HHV-6 association with other liver disease

As shown in the Table 1, HHV-6 is associated with a variety of diseases causing liver damage. Because of the different pathophysiology of these conditions, the role and mechanism of HHV-6 in hepatic damage in these diseases is unclear.

HHV-6 may be associated with Gianotii–Crosti syndrome, a popular acrodermatitis of childhood causing mild elevations of hepatic transaminase, and hepatitis B virus plays an important role in the pathogenesis of this disease. A case report described an 8-month-old infant with Gianotii–Crosti syndrome, in whom HHV-6 infection was diagnosed by rising antibody titers (Yasumoto et al., 1996). A positive association between HHV-6 and this disease was also observed in a case-control analysis (Chuh et al., 2002). In this study, patients diagnosed with Gianotii–Crosti syndrome and age- and sex-matched control patients with unrelated symptoms were examined for virologic evidence of HHV-6 infection in peripheral blood leukocytes and plasma using PCR for viral DNA, reverse transcriptase PCR for viral transcription, and serology. In contrast to the 10 controls, two patients (both infants) with clinically diagnosed Gianotii–Crosti syndrome had evidence of active HHV-6 infection.

Hemophagocytic syndrome (HSP) is a clinicopathologic syndrome characterized by the systemic activation of macrophages leading to the phagocytosis of hematopoietic cells. HSP is characterized by the association of fever, hepatosplenomegaly, lymphadenopathy, skin rash, lung infiltration, and jaundice. Laboratory findings frequently demonstrate pancytopenia, elevated levels of hepatic transaminase, coagulopathy with low fibrinogen, hypertriglyceridemia, and elevated ferritin levels. Several viral infections, such as Epstein–Barr virus, have

Table 1

Spectrum of HHV-6-associated hepatic diseases

Acute hepatitis
Chronic hepatitis
Fulminant hepatitis
Diseases with hepatic damage
Gianotii–Crosti syndrome
Virus-associated hemophagocytic syndrome
Drug-induced hypersensitivity syndrome

been associated with the development of HSP, which is called virus-associated hemophagocytic syndrome (VAHS). Several reports have linked VAHS to HHV-6 infection. HHV-6-associated VAHS can affect immunocompetent children (Huang et al., 1990; Syruckova et al., 1996; Takagi et al., 1996) and adults (Tanaka et al., 2002) as well as immunocompromised patients, such as transplant recipients (Karras et al., 2004). The clinical course of VAHS is severe and often fatal. Therefore, better treatment regimens are necessary to improve disease prognosis. Further research is needed to identify the pathogenesis of HHV-6-associated VAHS and the efficacy of antiviral drugs in its treatment.

Hepatic injury associated with HHV-6 infection is also seen in drug-induced hypersensitivity syndrome (DHS), defined as the clinical triad of fever, rash, and internal organ involvement including hepatic dysfunction due to drug exposure. The precise pathogenesis of this syndrome remains unknown, although several hypotheses have been proposed. Clearly, drug exposure is a key factor in the etiology of these reactions. HHV-6 reactivation has also been observed around the time of onset of DHS (Komura et al., 1993; Suzuki et al., 1998; Tohyama et al., 1998; Aihara et al., 2003; Zeller et al., 2003; Kano et al., 2004; Sekiguchi et al., 2005), and HHV-6 reactivation may lead to a prolonged course, slow resolution, and recurrence of signs and symptoms of DHS. However, the causal role of HHV-6 reactivation in the pathogenesis of DHS including the mechanism of hepatic damage remains unclear.

Pathogenesis of HHV-6-associated hepatitis

As described above, HHV-6 can directly cause hepatitis, but its mechanism of liver injury is unknown. HHV-6 may directly damage hepatocytes, or the immune and inflammatory response to the virus may cause liver damage. HHV-6 DNA, RNA, and antigens were detected in the hepatocytes of affected livers by *in situ* hybridization analysis (Mason et al., 1996; Ozaki et al., 2001; Ishikawa et al., 2002). These observations strongly suggest HHV-6 has the potential to directly damage hepatocytes during active infection. HepG2 cells, a well-differentiated liver cell line, are permissive for HHV-6 infection (Cermelli et al., 1996; Inagi et al., 1996). Infected HepG2 cells produce HHV-6 antigens and infectious viral progeny. Moreover, HHV-6 infection leads to the release of transaminase, a marker of liver cell damage (Cermelli et al., 1996). Inflammatory cytokines likely play an important role in the observed liver dysfunction, and the development of viral hepatitis is closely associated with the production of inflammatory cytokines. In particular, IL-8, a potent chemoattractant for neutrophils and leukocytes, is thought to be an important factor contributing to hepatic injury. Inagi et al. (1996) showed that infection of HepG2 cells with HHV-6 directly induced IL-8 synthesis. These observations strongly suggest that HHV-6 infection induces inflammatory cytokines such as IL-8, leading to liver dysfunction due to cytokine-mediated inflammation.

The association of HHV-6 infection with hepatitis is becoming better established, but the precise mechanism of liver injury by HHV-6 is unclear. It is likely that HHV-6 both directly injures liver tissue as well as induces a liver damaging inflammatory response. More clearly, identifying a role for HHV-6 in hepatitis may provide new treatment strategies for fulminant viral hepatitis, an often fatal disease. Pathological analyses of patient tissue samples and *in vitro* studies are needed to address this issue.

HHV-6 and gastrointestinal tract

Patients with exanthem subitum frequently also develop mild diarrhea. The frequency of diarrhea at the time of primary HHV-6 infection is between 29 and 68% (Asano et al., 1994). Among 176 infants with primary HHV-6 infection, mild diarrhea developed on day 1.5 ± 1.6 (ranged between day 2 and day 6) and persisted for 5.2 ± 2.5 days (ranged between 1 day and more than 10 days). However, a recently published population-based study of primary HHV-6 infection demonstrated that the incidence of diarrhea was not statistically different between the HHV-6 positive and negative groups (Zerr et al., 2005). Clinical features of exanthem subitum (primary HHV-6 infection) including the incidence of diarrhea seem to be slightly different between Japanese and North American patients. While ethnic background may give rise to these differences, international collaborative studies are needed to clarify this issue.

HHV-6 and CMV are highly related pathogens, and like CMV, HHV-6 is associated with a variety of clinical conditions in immunocompromised patients, such as transplant recipients (Yoshikawa, 2004). One report examining hematopoietic stem cell transplant (HSCT) recipients suggests an association between HHV-6 infection and bloody diarrhea (Amo et al., 2003). HHV-6 DNA was detected in both large intestine tissue specimens and peripheral blood mononuclear cells from four HSCT recipients with bloody diarrhea. Viral DNA was detected in the nuclei of goblet cells and, in some cases, in the histiocytes in submucosal lesions of the large intestine. The authors proposed that either HHV-6 infection directly caused tissue damage and the resultant colitis, or the patients were suffering from graft versus host disease that caused a reactivation of latent HHV-6 infection in the intestinal tissue.

The precise mechanism of diarrhea caused by HHV-6 infection remains unclear. The degree of viremia was not associated with the presence of diarrhea at the time of primary HHV-6 infection (Asano et al., 1991a), but viral DNA can be found in the stool of affected patients. Moreover, four of 20 patients excreted viral DNA in stool for more than 30 days after the onset of disease (Suga et al., 1998). As described previously, HHV-6 DNA was detected in the nuclei of goblet cells in transplant recipients with bloody diarrhea (Amo et al., 2003). These *in vivo* data strongly suggest that HHV-6 can directly and persistently infect intestinal tissue. Although an epidermal cell line is susceptible to HHV-6 infection (Yoshikawa et al., 2003), no studies have reported the infection of either intestinal epithelial or endothelial cells with HHV-6. Further, *in vitro* analyses will allow us to understand the pathogenesis of diarrhea caused by HHV-6 infection.

Viral enteritis, such as adenovirus and rotavirus infections, is associated with intussusception, a well-known infantile enteric disease requiring emergency treatment. We first reported three cases of intussusception caused by primary HHV-6 infection in 1991 (Asano et al., 1991b). As it was thought that hyperplasia of intestinal lymphoid tissue caused by adenovirus and rotavirus infections is responsible for the pathogenesis of this disease, we proposed the same mechanism for the pathogenesis of intussusception due to HHV-6 infection. Komura et al. (1993) confirmed this hypothesis. During the surgical treatment of intussusception, the authors found enlarged mesenteric lymph nodes that were positive for HHV-6 DNA by PCR. Thus, primary HHV-6 infection is one of several causes of intussusception in infancy.

References

Aihara Y, Ito SI, Kobayashi Y, Yamakawa Y, Aihara M, Yokota S. Br J Dermatol 2003; 149: 165–169.

Aita K, Jin Y, Irie H, Takahashi I, Kobori K, Nakasato Y, Kodama H, Yanagawa Y, Yoshikawa T, Shiga J. Hum Pathol 2001; 32: 887–889.

Amo K, Tanaka-Taya K, Inagi R, Miyagawa H, Miyoshi H, Okusu I, Sashihara J, Hara J, Nakayama M, Yamanishi K, Okada S. Clin Infect Dis 2003; 36: 120–123.

Asano Y, Nakashima T, Yoshikawa T, Suga S, Yazaki T. J Pediatr 1991a; 118: 891–895.

Asano Y, Yoshikawa T, Suga S, Hata T, Yamazaki T, Yazaki T. Pediatr Infect Dis J 1991b; 10: 335–337.

Asano Y, Yoshikawa T, Suga S, Kobayashi I, Nakashima T, Yazaki T, Kajita Y, Ozaki T. Pediatrics 1994; 93: 104–108.

Asano Y, Yoshikawa T, Suga S, Yazaki T, Kondo K, Yamanishi K. Lancet 1990; 335: 862–863.

Cermelli C, Concari M, Carubbi F, Fabio G, Sabbatini AM, Pecorari M, Pietrosemoli P, Meacci M, Guicciardi E, Carulli N, Portolani M. Virus Res 1996; 45: 75–85.

Chuh AA, Chan HH, Chiu SS, Ng HY, Peiris JS. Pediatr Dermatol 2002; 19: 492–497.

Huang LM, Lee CY, Lin KH, Chuu WM, Lee PI, Chen RL, Chen JM, Lin DT. Lancet 1990; 336: 60–61.

Inagi R, Guntapong R, Nakao M, Ishino Y, Kawanishi K, Isegawa Y, Yamanishi K. J Med Virol 1996; 49: 34–40.

Ishikawa K, Hasegawa K, Naritomi T, Kanai N, Ogawa M, Kato Y, Kobayashi M, Torii N, Hayashi N. J Gastroenterol 2002; 37: 523–530.

Kano Y, Inaoka M, Shiohara T. Arch Dermatol 2004; 140: 183–188.

Karras A, Thervet E, Legendre C. Groupe cooperatif de transplantation d'Ile de France. Transplantation 2004; 77: 238–243.

Komura E, Hashida T, Otsuka T, Kondo K, Yamanishi K, Hibi S, Imashuku S. Pediatr Infect Dis J 1993; 12: 788–789.

Mason A, Sallie R, Perrillo R, Rayner A, Xu L, Dohner DE, Dehner M, Naoumov N, Gelb L, Saha B, O'Grady J, Williams R. Hepatology 1996; 24: 1361–1365.

Mendel I, de Matteis M, Bertin C, Delaporte B, Maguer D, Collandre H, Buffet-Janvresse C. Pediatr Infect Dis J 1995; 14: 993–997.

Ohashi M, Yoshikawa T, Asonuma K, Iwasaki T, Nishiyama Y, Asano Y, Kimoto Y, Yagi T, Urushihara N, Tanaka N, Baba K. Pediatr Int 2004; 46: 730–732.

Ozaki Y, Tajiri H, Tanaka-Taya K, Mushiake S, Kimoto A, Yamanishi K, Okada S. J Clin Microbiol 2001; 39: 2173–2177.
Pruksananonda P, Hall CB, Insel RA, McIntyre K, Pellett PE, Long CE, Schnabel KC, Pincus PH, Stamey FR, Dambaugh TR, Stewart JA. N Engl J Med 1992; 326: 1445–1450.
Sekiguchi A, Kashiwagi T, Ishida-Yamamoto A, Takahashi H, Hashimoto Y, Kimura H, Tohyama M, Hashimoto K, Iizuka H. J Dermatol 2005; 32: 278–281.
Sobue R, Miyazaki H, Okamoto M, Hirano M, Yoshikawa T, Suga S, Asano Y. N Engl J Med 1991; 324: 1290.
Suga S, Yoshikawa T, Kajita Y, Ozaki T, Asano Y. Pediatrics 1998; 102: 900–904.
Suzuki Y, Inagi R, Aono T, Yamanishi K, Shiohara T. Arch Dermatol 1998; 134: 1108–1112.
Syrucková Z, Stary J, Sedlacek P, Smisek P, Vavrinec J, Komrska V, Roubalova K, Vandasova J, Sintakova B, Houskova J, Hassan M. Pediatr Hematol Oncol 1996; 13: 143–150.
Tajiri H, Tanaka-Taya K, Ozaki Y, Okada S, Mushiake S, Yamanishi K. J Pediatr 1997; 131: 473–475.
Takagi M, Unno A, Maruyama T, Kaneko K, Obinata K. Pediatr Hematol Oncol 1996; 13: 451–456.
Tanaka H, Nishimura T, Hakui M, Sugimoto H, Tanaka-Taya K, Yamanishi K. Emerg Infect Dis 2002; 8: 87–88.
Tohyama M, Yahata Y, Yasukawa M, Inagi R, Urano Y, Yamanishi K, Hashimoto K. Arch Dermatol 1998; 134: 1113–1117.
Yamanishi K, Okuno T, Shiraki K, Takahashi M, Kondo T, Asano Y, Kurata T. Lancet 1988; 8594: 1065–1067.
Yasumoto S, Tsujita J, Imayama S, Hori Y. J Dermatol 1996; 23: 499–501.
Yoshikawa T. Br J Haematol 2004; 124: 421–432.
Yoshikawa T, Goshima F, Akimoto S, Ozaki T, Iwasaki T, Kurata T, Asano Y, Nishiyama Y. J Med Virol 2003; 71: 62–68.
Zeller A, Schaub N, Steffen I, Battegay E, Hirsch HH, Bircher AJ. Infection 2003; 31: 254–256.
Zerr DM, Meier AS, Selke SS, Frenkel LM, Huang ML, Wald A, Rhoads MP, Nguy L, Bornemann R, Morrow RA, Corey L. N Engl J Med 2005; 352: 768–776.

Human Herpesvirus-6: General Virology, Epidemiology and Clinical Pathology
Perspectives in Medical Virology
Gerhard Krueger and Dharam Ablashi (Editors)

DOI 10.1016/S0168-7069(06)12020-0

HHV-6 in Chronic Fatigue Syndrome

Daniel Peterson, Lisa Atwell
Sierra Internal Medicine, 865 Tahoe Blvd., #306, Incline Village, NV 89451, USA

Chronic fatigue syndrome

Chronic fatigue syndrome (CFS) is often referred to by other names or used interchangeably with similar disorders, such as chronic fatigue and immune dysfunction syndrome (CFIDS), fibromyalgia (FM), myalgic encephalomyelitis (ME), Gulf War Syndrome, and chronic Epstein–Barr disease. Recently, the Centers for Disease Control (CDC) has published epidemiologic figures estimating that approximately 800,000 Americans are affected by CFS. The associated economic loss is estimated in the billions of dollars, due to disability, medical expenses, and loss of wages. CFS is an incompletely understood, yet severely disabling disease of unknown etiology. It is characterized by profound, debilitating fatigue, of greater than 6 months duration that cannot be resolved with rest. Associated symptoms include fever, sore throat, myalgias, lymphadenopathy, sleep disturbance, headaches, neurocognitive difficulties (such as memory and concentration impairment and "mental fog"), and symptoms associated with autonomic dysfunction.

Frequently, patients report a sudden onset of CFS symptoms following an acute flu-like illness. CFS was initially thought to be caused by a prolonged viral infection, possibly by Epstein–Barr virus (EBV). However, after extensive study, it became clear that CFS is not caused exclusively by any one viral agent. Instead, CFS might represent a common pathogenic end point, and multiple viruses or other infectious agents might have a contributory role in the onset or persistence of symptoms in CFS (CDC, 2005).

Since the initial reports and the definition of CFS was established by the CDC in 1994, many viruses have been implicated and studied, including enteroviruses, retroviruses, and the human herpesviruses (HHVs). A substantial body of literature documents the association of HHV-6 with CFS. Some studies examining the relationship between CFS and HHV-6 have produced ambiguous findings. This might be due in part to overly broad selection criteria for the patient sample, small sample sizes, failure to match control subjects with patients, and the use of inappropriate techniques for detecting active HHV-6 infection. Furthermore, because latent HHV-6 infection is nearly universal in adults, only tests that can differentiate between active and latent infection are likely to produce meaningful results. Additionally, early published studies in particular not only failed to differentiate active and latent infection, but they also did not differentiate between HHV-6 variant A or variant B.

Immune dysfunction

CFS patients often display abnormal or dysfunctional immune responses. For example, natural killer (NK) cells are either reduced in number or cytotoxicity, suppressor T lymphocytes (CD8, CD11b) are reduced (Landay et al., 1991), and IgG subclasses 1 and 3 are decreased. Recent studies also demonstrate that the 2,5-A/RNase L antiviral pathway is over-activated (Suhadolnik et al., 1994a). Common measurements of immune stimulation in these patients include increases in immune-activation markers (CD38 and HLA-DR, T lymphocytes) and abnormal circulating cytokines (Landay et al., 1991; Patarca, 2001; Natelson et al., 2005). Evidence for a viral role in the pathogenesis of CFS was reported initially by Suhadolnik et al. (1994b), who demonstrated that the 2,5-A synthetase/RNase L antiviral pathway was both upregulated and abnormal. This innate antiviral pathway uses an activated form of the molecule 2,5-A synthetase to activate the enzyme RNase L, which in turn degrades viral RNA and prevents protein synthesis. RNase L does not distinguish between cellular and viral RNA, and disrupts both viral and cellular protein synthesis. In a subset of CFS patients, this important immunologic pathway is not only upregulated, but dysfunctional: 2,5-A and RNase L are activated, suggesting a viral stimulus, but RNase L is highly bioactive, perhaps due to an inoperative inhibitor site (De Meirleir et al., 2000), and occurs as an abnormally small molecule (37 kDa as opposed to the normal 80 kDa). Subsequently, other investigators have found CFS patients who demonstrate elevated levels of 2,5-A synthetase and the presence of the abnormally low molecular weight (37 kDa) RNase L, particularly the most severely disabled CFS patients.

The highly bioactive RNase L and decreased suppressor T-cell function noted in CFS can lead to a "hyperimmune response" and production of CFS symptoms. Supporting the observation of a hyperimmune response, several reports have associated HHV-6 encephalitis with drug hypersensitivity syndromes. CFS patients frequently report and demonstrate increased sensitivity to many medications and experience known side effects in increased intensity and at lower than standard doses of medications (Kunisaki et al., 2003). Recently, gene-expression profiling studies have also supported these observations of the disturbance of immune regulation and antiviral defenses (Gow et al., 2005).

Viral implication

From the clinical perspective, viral infection is often implicated as the cause of CFS due to the onset of symptoms with a flu-like episode, waxing and waning clinical course, history of geographic outbreaks, and effectiveness of some antiviral therapies. Approximately 50% of patients with CFS report the onset of their illness after a viral-like infection. Worldwide, CFS has been reported following acute infectious mononucleosis, Lyme disease, Q-fever, and enteroviral infections. Additionally, patients more frequently report the onset of CFS in winter months when viral infections are prevalent (Jason et al., 2005). However, no study has established one virus or agent as a specific cause of CFS (Ablashi et al., 2000).

Proposed pathogenic model of CFS

One etiologic model of CFS integrates genetic predisposition, immune perturbation, viral reactivation, and autonomic and immunoregulatory dysfunction to produce the symptom complex of CFS (Fig. 1). A persistent viral infection is one potential

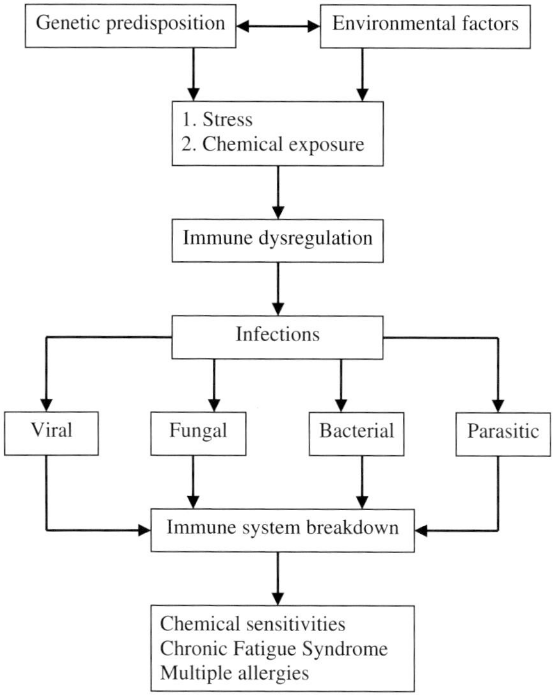

Fig. 1 Proposed model of chronic fatigue syndrome etiology, including the potential role of viral infection.

pathogenetic cofactor in this model. Immune activation (due to viral infection) can result in the release of cytokines and precipitate the observed changes in central nervous system (CNS) symptoms (required by the CDC criteria of CFS). Production of pro-inflammatory cytokines such as IL-1b and IL-6 has been correlated with acute sickness behavior, and many signs and symptoms of CFS, including fever, malaise, pain, and impaired concentration (Vollmer-Conna et al., 2004).

Alternatively, viruses and bacteria that have been maintained in a latent state and controlled by an intact immune system might be allowed to replicate and exacerbate symptoms of illness under circumstances of stress or co-morbid infections. It is currently not known whether the presence of HHV-6 (variant A or B) indicates a primary causative role in CFS, or simply the opportunistic exploitation of a suppressed immune system.

As early as 1994, Ablashi postulated that CFS might essentially represent an immunological disturbance, which allows reactivation of latent herpesviruses, such as HHV-6 (Ablashi, 1994). Ablashi concluded that the evidence at that time showed a much stronger association of CFS with HHV-6 than with other herpesviruses, including EBV, cytomegalovirus (CMV, or HHV-5), or herpes simplex viruses (HSV-1 and HSV-2).

Once reactivated, these viruses could directly contribute to existing morbidity and produce abnormal immune responses. Recently, Smith et al. (2005) reported evidence from an *in vitro* model that the immunosuppressive action of HHV-6 on CD4 lymphocytes is related to the suppression of IL-12 by dendritic cells. It has also been postulated that an infectious organism could have a "hit and run" effect by entering a host and triggering an abnormal immune response. The pathogen then resolves, but leaves behind self-propagating immune abnormalities that result in the symptoms of CFS.

Identifying subsets among the CFS population

A recent review article by Jason et al. (2005) underscored the need to clarify the patient definition utilized for entrance into CFS research and treatment protocols. Jason et al. compared differences among the 1988 CFS "Holmes" criteria (Holmes et al., 1988), the 1994 CDC criteria (Fukuda et al., 1994), and the 2003 Canadian consensus document criteria (Carruthers et al., 2003). They concluded that the Canadian consensus criteria selected cases with less psychiatric co-morbidity, more functional impairment, and greater frequency of fatigue/weakness and neurological symptoms. No single variable has yet been identified that effectively differentiates subgroups of CFS causes or symptoms, or identifies subgroups that might respond to specific therapies. One of these identifiable subgroups could be patients suffering from persistent reactivation of HHV-6, particularly in the CNS.

HHV-6 association with CFS

One of the first reports of isolated post-infectious fatigue associated with encephalitis, lymph proliferation, and the presence of HHV-6 infection, was made by Buchwald et al. (1990). Daugherty et al. (1991) also reported a group of patients with profound fatigue lymphadenopathy and cognitive dysfunction associated with evidence of HHV-6 reactivation, although variant analysis was not available at that time.

Early studies of CFS patients demonstrated an increase in serum IgG and IgM for HHV-6 in a large number of patients compared with control subjects. However, increases in antibodies to other viruses, particularly other herpesviruses, were also detected. Serological measurements of IgG and IgM titers have limited ability to suggest active infection, because most adults have been infected with HHV-6. These immunoglobulins might only indicate exposure, but not active or persistent infection. Studies using molecular analysis (in subsequent years) generally showed higher prevalence of HHV-6 variant A in CFS patients than in control subjects.

Research performed by Knox et al. (1998) demonstrated active HHV-6 infection in 37% of CFS patients; this is significantly higher than in control subjects. Longitudinal observations from this study indicated that active HHV-6 infection was intermittent and the viral load variable. When patients with prominent CNS complaints were considered separately, 56% had evidence of active HHV-6 infection, suggesting that selection for CFS "with CNS involvement" co-selected for active HHV-6 infection. The cerebrospinal fluid examination of these patients demonstrated that 20% (7/35) were positive for HHV-6 DNA, and because the spinal fluid specimens were acellular, the presence of HHV-6 DNA suggested active CNS infection.

While Ablashi et al. (2000) found that 54% of CFS patients were positive for a specific HHV-6 IgM early antigen (p41/38) compared to just 8% of healthy controls, Patnaik et al. (1995) found that 77% of CFS patients were positive for HHV-6 IgG early antigen versus 12% of controls. Conversely, Wallace et al. (1999) and Reeves et al. (2000) found no greater incidence of HHV-6 infection in CFS patients compared with control patients using an enzyme immunoassay for seroreactivity and lymphocyte co-cultivation to detect HHV-6.

Ablashi et al. (2000) also reported that HHV-6 isolates from CFS patients were predominately variant A (70%), whereas the isolates from healthy controls were predominately variant B. Subsequently, Yalcin et al. (1994) and Di Luca et al. (1995) also reported that the HHV-6 isolates from CFS patients were predominately variant A. These findings support the suggestion that the A variant of HHV-6 plays a more dominant role in CFS than in other disease states.

In a review of HHV-6, Krueger and Ablashi (2003) underscored that the natural history of HHV-6 variant A infection is not known, because until recently, serological testing, most often employed to detect HHV-6, could not differentiate variant A from variant B. A variant-specific serological testing for HHV-6 is not currently commercially available but is under development.

In another study of patients, who met the CDC criteria for CFS, 25% (20/81) were found to have active HHV-6 infection in peripheral blood leukocytes compared to 2% (1/55) of healthy control subjects (Knox et al., 1998). When cerebrospinal fluids from CFS patients were analyzed for HHV-6, 9% were found to be positive. These studies again illustrate that while active HHV-6 is not found in the blood or spinal fluid of all CFS patients, it is found in a significantly higher proportion than in healthy immunocompetent adults. Several virologists (D. Ablashi, C. Knox, and D. Carrigan, personal communication) have also noted persistent positivity for HHV-6, both in peripheral samples and cerebrospinal fluid, for prolonged periods of up to 6 years. However, very few patients have been followed rigorously with variant-specific tests.

In a summary of 27 published studies on the association of HHV-6 with CFS, 15 studies employed detection methods that did not differentiate between active and latent HHV-6 infection. These methods of detecting HHV-6 can be particularly problematic given the ubiquitous nature of latent HHV-6 in adults. Of the 12 studies employing methods that did differentiate active from latent HHV-6 infection, 10 (83%) showed a positive and statistically significant association between active HHV-6 and CFS.

HHV-6 in CNS of CFS patients

Various studies have documented abnormalities in the brains of CFS patients, as evidenced by increased T2-weighted images in the high white matter tracts (Fig. 2), functional abnormalities, as demonstrated by an alteration of the hypothalamic pituitary-adrenal (HPA) axis, cognitive impairment as measured by psychometric testing, regional hypoperfusion as demonstrated by SPECT scanning (Fig. 3), and hypometabolism as demonstrated by PET scanning. Clinically, cognitive impairment, non-restorative sleep, and autonomic dysfunction are characteristic symptoms included in the CDC and Canadian consensus case definitions for CFS.

Owing to the prevalence of neurological complaints in a subset of patients, spinal fluid analysis was performed on 145 patients, of which 20% proved positive for viruses in the cerebrospinal fluid, predominantly HHV-6 variant A (DL Peterson, unpublished). In order to study and ultimately treat these patients, a clinical algorithm for CNS infection with HHV-6 variant A was designed (Fig. 4).

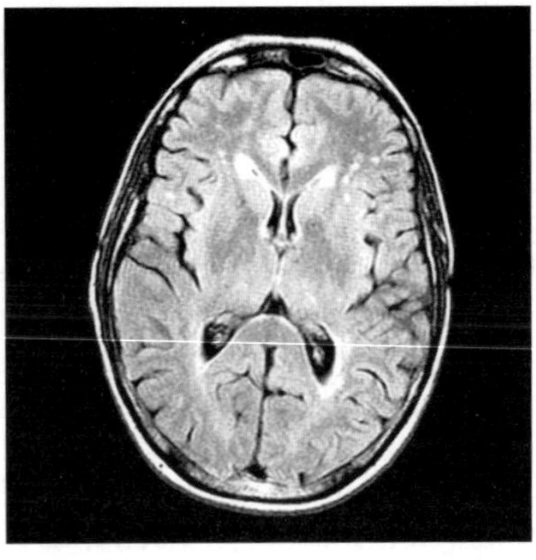

Fig. 2 Magnetic resonance image (MRI) showing increased T2-weighted images in the high white matter tracts.

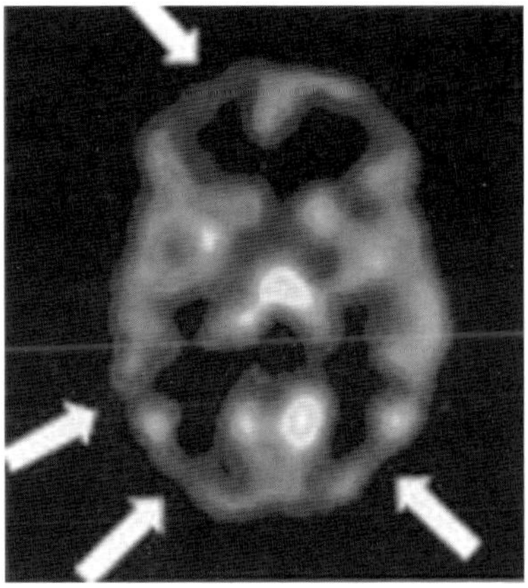

Fig. 3 Axial SPECT image showing multiple foci of decreased perfusion (arrows) in the brain. (for colour version: see colour section on page 358).

Patient meets Canadian Consensus case definition

⇩

Prominent CNS symptoms

⇩

Abnormal MRI/SPECT scan

⇩

Abnormal cerebral spinal fluid

- Increased opening pressure during lumbar puncture
- Elevated total protein
- Lymphocytosis

⇩

HHV-6 positive spinal fluid

Fig. 4 Clinical algorithm to identify a subgroup of chronic fatigue syndrome patients suffering from HHV-6 infection in the central nervous system.

These selected patients displayed CNS symptoms, abnormal magnetic resonance image (MRI) and SPECT scans, and abnormal cerebrospinal fluid (increased opening pressure, increased total protein, and lymphocytosis). Challenges faced in the diagnosis of this subset of patients have included a great deal of laboratory variability, a costly algorithm, and small patient numbers.

Hall et al. (1998) reported on children who were examined after primary infection and noted persistence of HHV-6 variant A in the spinal fluid more frequently than in peripheral blood mononuclear cells (PBMCs) or saliva.

Buchwald et al. (1992) reported a cohort study of 259 patients with neurological symptoms, abnormal MRI findings, and lymphocyte phenotyping, suggesting chronic immunologically mediated inflammatory changes of the CNS. Additionally, active replication of HHV-6 (most likely representing reactivation of latent infection) was demonstrated in approximately 70% of the patients and only 20% of controls using monoclonal antibodies specific for HHV-6 and polymerase chain reaction (PCR).

There is persuasive evidence that the HHV-6 variant A infection in the CNS is a particularly serious condition. While HHV-6 presence in the blood is ubiquitous, and thus the relevance of a reactivated infection is of questionable clinical significance, HHV-6 in the CNS is relatively rare, and its presence there is often concurrent with MRI and SPECT changes in the brain, and neurocognitive dysfunction. As pointed out by Shor (2003), if a chronic HHV-6 variant A CNS infection is clearly identified, this subset of patients would no longer technically be diagnosed with CFS as currently defined by the CDC, because they would fall into the exclusionary category of "other chronically fatiguing illness." Again, defining subsets of CFS patients is of utmost importance because this HHV-6 variant A positive group could be a more homogenous subset for study and treatment protocols.

HHV-6 therapy in CFS patients

CFS patients with active infection by HHV-6 (variants A or B) can be treated with antivirals or immune modulatory agents in order to relieve the symptoms of fatigue and minimize CNS complaints. Acyclovir has remained the gold standard of treatment for herpes viral infections in general. However, pilot studies using acyclovir and ganciclovir showed persistence of HHV-6 variant A in spinal fluid even after treatment (Peterson, unpublished studies). HHV-6 does not encode thymidine kinase, and thus is not highly sensitive to acyclovir and its analogs (Gomples et al., 1995). Non-guanosine derivatives, however, have been shown *in vitro* to be of greater efficacy against HHV-6 specifically. De Clercq et al. (2001) demonstrated increased efficacy of the non-guanosine compounds S2242, cidofovir, and foscarnet, both in T lymphoblast cells, and in fresh blood lymphocytes, though it should be noted that only foscarnet and cidofovir are commercially available. Cidofovir, the first nucleotide analog available for clinical use, is highly active against CMV and demonstrates activity against HHV-6 as well *in vitro*. Cidofovir also has a substantially more favorable administration and side-effect profile. Current pilot studies are being conducted with cidofovir in a standard dosage of 5 mg/kg intravenously for patients positive for HHV-6 in their spinal fluid, as detected by two different tests (PCR and tissue culture). Preliminary results suggest that HHV-6 infection is suppressed by cidofovir, but recurs when treatment is stopped. In a recent case study, acute infection by HHV-6 variant A in the CNS of an immunocompetent adult was successfully treated with cidofovir and ganciclovir (Denes et al., 2004).

The only herpesvirus vaccine licensed for human use in the United States is for VZV, thus this preventive strategy is not available for the clinician treating HHV-6 patients. HSV-1, HSV-2, and EBV vaccines are in development and a specific HHV-6 vaccine could be forthcoming as well.

Antiviral therapies that have been employed for CFS patients who demonstrate evidence of persistent peripheral and CNS infection include Poly I: Poly $C_{12}U$, isoprinosine, ganciclovir, foscarnet, and cidofovir. Valacyclovir (a prodrug of acyclovir) and ganciclovir were found not to result in prolonged viral remission, and oral antiviral therapy (such as acylcovir and famvir) has yielded minimal results. Parenteral antiviral therapies have concentrated on patients who demonstrate severe disabling fatigue and prominent CNS clinical complaints and immune dysfunction. Neither oral nor parenteral therapy has been studied in rigorous double-blind placebo-controlled studies.

Immunomodulatory therapy for HHV-6 in CFS patients

The current clinical rationale for studies of immunomodulatory therapy is based on findings that CFS patients can exhibit low NK cell activity, activated RNase L, abnormal 37 kDa RNase L, and evidence of viral reactivation. Poly I: Poly $C_{12}U$ (Ampligen) has been used in phases 1, 2, and 3 clinical trials in patients with severely debilitating CFS. One of its effects is to modulate the RNase L/2,5-A synthetase antiviral pathway. Initial results from these clinical trials have shown improvement in exercise performance, normalization of the RNase L/2,5-A synthetase antiviral pathway, and decrease in viral load (Strayer et al., 1995). Therapy with interferon and IV gammaglobulin, though, has produced mixed results.

HHV-6 diagnostic testing in CFS patients

Historically, clinicians have used serology, tissue culture, and PCR to detect HHV-6 in CFS patients. Serological testing for HHV-6 is of extremely limited clinical utility—for example, conventional serological testing, including IgG and IgM, does not differentiate A from B variants. Because the vast majority of adult Americans are seropositive due to early exposure to HHV-6 variant B, such screening is of little use. On the other hand, tissue culture techniques have shown the greatest sensitivity for the detection of HHV-6 variant A in small pilot studies, as indicated by JH Brewer (personal communication). However, because this technique is labor-intensive, time-consuming, expensive, and requires skilled technicians, it is not amenable to mass commercial production.

PCR amplification has been widely used recently and is capable of differentiating variant A from variant B, but does not inherently differentiate between active and latent infection, so specimens undergoing PCR must be acellular in order to provide evidence of active infection. Acellular DNA suggests release from cells lysed by the active replication of HHV-6. PCR also varies greatly in specificity and sensitivity, especially with regard to specificity among commercial laboratories providing this technology. Standard PCR testing can also fail to detect low-grade infections, such as those found in the CNS. Currently, the most sensitive assay for non-acute active infection is thought to be IgM antibody to HHV-6 early antigen.

Techniques are evolving to detect an immediate early antigen of HHV-6, a protein produced early in the replication phase of HHV-6. This might offer mass reproducibility at a reasonable cost with high specificity and sensitivity for HHV-6 variant A or B in peripheral blood, tissue, and spinal fluid specimens. At the present time, however, these tests are not commercially available.

Future research

Future research concerning the role of chronic HHV-6 variant A or B infections and CFS should focus upon:

- The application of diagnostic technologies with sufficient sensitivity to detect low levels of active infection.
- Establishment of banks of biopsy- and autopsy-derived tissues from CFS patients to facilitate studies that will help define organ-specific pathology.
- Systematic evaluation of antiviral therapy in well-defined subsets of CFS patients who demonstrate evidence of viral reactivation or persistence.

Summary

CFS patients often show evidence of immune dysfunction and/or dysregulation, such as low suppressor cell numbers, low numbers and efficacy of NK cells, and abnormal RNase L. These dysfunctions are manifest when the immune system is challenged by a virus or other infection, suggesting that there might be a viral trigger that initiates and/or perpetuates CFS. Active infection by HHV-6 is found in association with patients suffering from CFS in significantly greater proportion than in the healthy immunocompetent population. In this "subgroup" of CFS patients, HHV-6 might be the viral trigger for CFS. The fact that HHV6 is not found universally in CFS patients might be because (a) other viruses/agents or events can also precipitate CFS by triggering faulty immune response; or (b) the nature of HHV-6, as with all herpesviruses, is cyclical and thus a "snapshot" of a group of CFS patients would also include those patients in the latent phase of chronic, recurring HHV-6 infection. Anecdotal evidence from some CFS patients we have observed, in some cases over many years, does reveal chronic, recurring HHV-6 reactivation. It must be noted that it is not clear whether HHV-6 actually plays a causative role in the development of CFS or is simply released from suppression when CFS occurs. There are at least three hypotheses that could explain the presence of active HHV-6 infection in a subset of CFS patients:

Hypothesis 1: HHV-6 is a primary co-factor (along with genetic, environmental, and/or co-infection conditions) in the development of CFS in a subset of patients.

Hypothesis 2: HHV-6 has no direct causative role in CFS, but exacerbates existing symptoms and further affects immune function; i.e. an indirect or supporting role in the condition of CFS.

Hypothesis 3: HHV-6 has no role in CFS, but is often found co-existing with CFS due to suppressed immune function in these patients; i.e. HHV-6 exploits a CFS-depressed immune system, but is otherwise unrelated to the symptoms of CFS.

Another HHV-6-related subgroup of CFS is that in which patients display CNS symptoms, including neurocognitive difficulties, such as memory impairment, "mental fog," and diminished concentration. These patients can display evidence of active HHV-6 infection in the spinal fluid. A higher proportion of HHV-6 in the spinal fluid appears to be of the A variant (compared with blood), suggesting that variant A is more neurotropic than variant B. It should also be noted that peripheral results for HHV-6 might not correlate with the HHV-6 status of the CNS (Hall et al., 1998). In CFS patients with HHV-6 in the spinal fluid, it is postulated that HHV-6 has a direct role in the illness, and the therapy to reduce the HHV-6 infection could be an effective treatment.

Treatment of CFS with antivirals (e.g. acyclovir, ganciclovir, foscarnet, cidofovir) or immunomodulating therapy (e.g. Poly I: Poly $C_{12}U$, and IV gammaglobulin) has met with mixed success, further suggesting the need to identify CFS subgroups for whom specific effective treatment can be targeted.

Currently, all CFS patients are combined under this one diagnostic umbrella. Not surprisingly, universal causes and treatments have not been found, often leading to the conclusion that any one theory of cause or treatment is invalid. Here, two possible subgroups of HHV-6-related CFS are proposed: chronic, recurring HHV-6 reactivation in the periphery that precipitates or exacerbates CFS symptoms; and acute or chronic HHV-6 infection in the CNS.

References

Ablashi DV. Clin Infect Dis 1994; 18(Suppl 1): S–130.

Ablashi DV, Eastman HB, Owen CV, Roman MM, Friedman J, Zabriskie JB, Peterson DL, Pearson GR, Whitman JE. J Clin Virol 2000; 16: 179.

Buchwald D, Cheney PR, Peterson DL, Henry B, Wormsley SB, Geiger A, Ablashi DV, Salahuddin SZ, Saxinger C, Biddle R, Kikinis R, Jolesz FA, Folks T, Balachandran N, Peters JB, Gallo RC, Komaroff AL. Ann Intern Med 1992; 116: 103.

Buchwald D, Freedman AS, Ablashi DV, Sullivan JL, Caligiuri M, Weinberg DS, Hall CG, Ashley RI, Saxinger C, Balachandran N, Ritz J, Nadler LM, Komaroff AL. J Clin Immunol 1990; 10: 335.

Carruthers BM, Jain AK, De Meirleir KL, Peterson DL, Klimas NG, Lerner M, Bested AC, Flor-Henry P, Joshi P, Powels P, Sherkey JA, van de Sande MI. J Chronic Fatigue Syndr 2003; 11: 7.

CDC, 2005. Available from: http://www.cdc.gov/ncidod/diseases/cfs/index.htm. Accessed August 18, 2005.

Daugherty SA, Henry BE, Peterson DL, Swarts RL, Bastein S, Thomas RS. Rev Infect Dis 1991; 13: S39.

De Clercq E, Naesens L, De Bolle L, Schols D, Zhang Y, Neyts J. Rev Med Virol 2001; 11: 381–395.

De Meirleir K, Bisbal C, Campine I, De Becker P, Salehzada T, Demettre E, LeBleu B. Am J Med 2000; 108: 99.

Denes E, Magy L, Pradeau K, Alain S, Weinbreck P, Ranger-Rogez S. Emerg Infect Dis 2004; 10: 729.

Di Luca D, Zorzenon M, Mirandola P, Colle R, Botta GA, Cassai E. J Clin Microbiol 1995; 33: 1660–1661.
Fukuda K, Straus SE, Hickie I, Sharpe MC, Dobbins JG, Komaroff A. Ann Intern Med 1994; 121: 953.
Gompels UA, Nicholas J, Lawrence G, Jones M, Thompson BJ, Martin ME, Efstathiou S, Craxton M, Macauly HA. Virol 1995; 209: 29.
Gow JW, Cannon C, Behan WMH, Herzyk P, Keir S, Riboldi-Tunnicliffe G, Behan PO, Chaudhuri A. Internat Conf Fatigue Sci, Karuizawa, Japan; February 9–11, 2005.
Hall CB, Caserta MT, Schnabel KC, Long C, Epstein LG, Insel RA. Clin Infect Dis 1998; 26: 132.
Holmes GP, Kaplan JE, Gantz NM, Komaroff AL, Schonberger LB, Straus SE, Jones JF, Dubois RE, Cunningham-Rundles C, Pahwa S. Ann Intern Med 1988; 108: 387.
Jason LA, Corradi K, Torres-Harding S, Taylor RR, King C. Neuropsychology Rev 2005; 15: 29.
Knox KK, Brewer JH, Carrigan DR. Fourth Internat Am Assoc Chronic Fatigue Syndr Conf. Cambridge: MA; 1998.
Krueger GR, Ablashi DV. Intervirology 2003; 46: 257.
Kunisaki Y, Goto H, Kitagawa K, Nagamo M. Internal Med 2003; 42(2): 203–207.
Landay AL, Jessop C, Lennette ET, Levy JA. Lancet 1991; 338: 707.
Natelson BH, Weaver SA, Tseng CL, Ottenweiler JE. Clin Diagn Lab Immunol 2005; 12: 52.
Patarca R. Ann New York Acad Sci 2001; 933: 185.
Patnaik M, Komaroff AL, Conley E, Ojo-Amaize EA, Peter JB. J Infect Dis 1995; 172: 1364.
Reeves WC, Stamey FR, Black JB, Mawle AC, Stewart JA, Pellett PE. Clin Infect Dis 2000; 31: 48.
Shor S. J Chronic Fatigue Syndr 2003; 11: 51.
Smith AP, Paolucci C, Di Lullo G, Burastero SE, Santoro F, Lusso P. J Virol 2005; 79: 2807.
Strayer DR, Carter W, Straus KI, Brodsky I, Suhadolnik RJ, Ablashi D, Henry B, Mitchell WM, Bastein S, Peterson D. J Chronic Fatigue Syndr 1995; 1: 35.
Suhadolnik RJ, Reichenbach NL, Hitzges P, Adelson ME, Peterson DL, Cheney P, Salvato P, Thompson C, Loveless M, Muller WE, Schroder H, Strayer D, Carter WA. In Vivo 1994a; 8: 599.
Suhadolnik RJ, Reichenbach NL, Hitzges P, Sobol RW, Peterson DL, Henry B, Ablashi DV, Muller WE, Schroder HC, Carter WA, Strayer D. Clin Infect Dis 1994b; 18: S96.
Vollmer-Conna U, Fazou C, Cameron B, Li H, Brennan C, Luck L, Davenport T, Wakefield D, Hickie I, Lloyd A. Psychol Med 2004; 34: 1289.
Wallace HL, Natelson D, Gause W, Hay J. Clin Diagn Lab Immunol 1999; 6: 216.
Yalcin S, Kuratsune H, Yamaguchi K, Kitani T, Yamanishi K. Microbiol Immunol 1994; 38: 587.

Human Herpesvirus-6: General Virology, Epidemiology and Clinical Pathology
Perspectives in Medical Virology
Gerhard Krueger and Dharam Ablashi (Editors)

DOI 10.1016/S0168-7069(06)12021-2

HHV-6 and HIV-1 Infection

Paolo Lusso
Unit of Human Virology, Department of Biological and Technological Research (DIBIT), San Raffaele Scientific Institute, 20132 Milan, Italy

Introduction

HHV-6 was discovered at the outset of the AIDS era, and was indeed initially isolated from several immunosuppressed patients, including two with human immunodeficiency virus type 1 (HIV-1) disease (Salahuddin et al., 1986). The privileged relationship between HHV-6 and HIV-1 was immediately evident following the demonstration that these two viruses share a primary tropism for CD4$^+$ T cells (Lusso et al., 1988). The hypothesis that HHV-6 may be an important player in the pathogenesis of the immunological derangement that accompanies the progression of HIV-1 disease to full-blown AIDS dates back to 1989, when a synergic cytopathic effect of the two viruses was documented in coinfected primary CD4$^+$ T cells (Fig. 1) (Lusso et al., 1989a). Despite the subsequent identification of multiple mechanisms whereby HHV-6 can directly damage the immune system and/or enhance the virulence and pathogenicity of HIV-1, conclusive evidence of the role played by HHV-6 in AIDS is still wanting today. This uncertainty stems in part from technical limitations to the study of HHV-6 infection *in vivo*, as well as from the inherent difficulties in dissecting the impact of a single infectious agent in a disease characterized by a wide multiplicity of opportunistic infections. Nevertheless and in spite of some conflicting evidence, several *in vivo* observations, both in naturally infected patients and in experimental animal models, corroborate the concept that HHV-6, particularly subgroup A, may play a significant role in the pathogenesis of the immunological damage that leads to AIDS (reviewed in Lusso

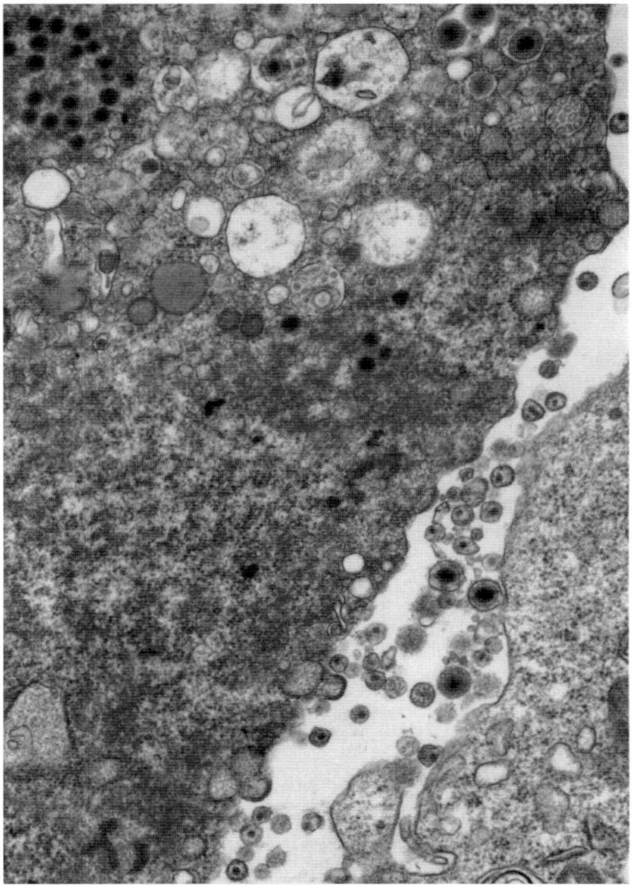

Fig. 1 Electron micrograph showing primary $CD4^+$ T lymphocytes productively coinfected with HHV-6 and HIV-1.

and Gallo, 1995). A precise and definitive understanding of this role may have critical implications not only for our understanding of AIDS pathogenesis, but also for the implementation of effective preventive and therapeutic measures for the control of HIV infection.

In vitro interactions between HHV-6 and HIV-1

The first evidence suggesting that HHV-6 may be implicated in AIDS came from studies on its *in vitro* interactions with HIV-1, which led to the identification of several mechanisms whereby HHV-6 can modulate the growth, cellular tropism and pathogenicity of HIV-1 (Fig. 2).

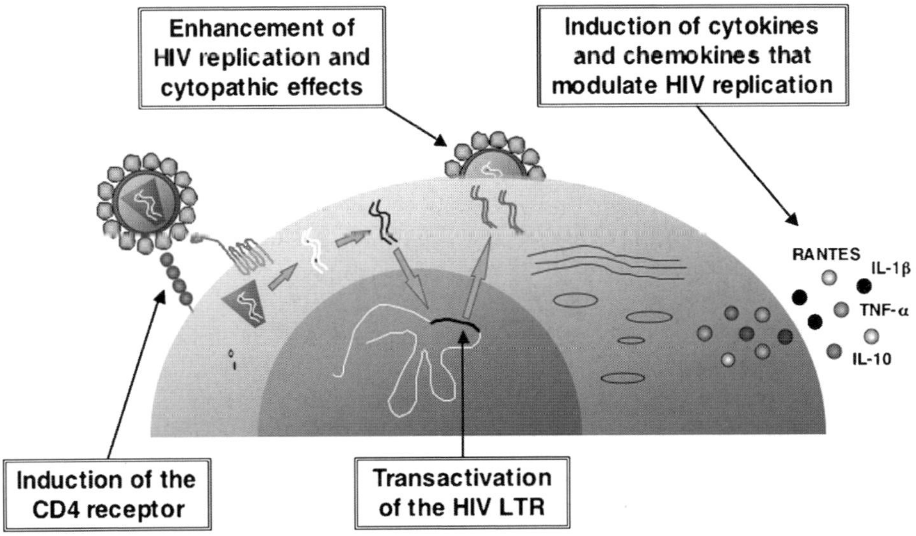

Fig. 2 Mechanisms of interaction between HHV-6 and HIV-1.

Productive coinfection of CD4⁺ T cells by HHV-6 and HIV-1

Consistent with their shared cellular tropism, HHV-6 and HIV-1 can coinfect and simultaneously replicate within the same $CD4^+$ T cell (Lusso et al., 1989a). No viral interference occurs at the entry level, since the primary cellular receptor for HIV-1, CD4, is not involved in the HHV-6 receptor mechanism (Lusso et al., 1989b). Studies on the biological consequences of HHV-6–HIV-1 coinfection have yielded conflicting results, which may reflect differences in the experimental conditions used. Some authors have documented a dramatic acceleration in the kinetics of HIV-1 expression in primary T-lymphocyte cultures coinfected with HHV-6, as well as a synergistic destruction of $CD4^+$ T cells; both phenomena are consistent with the ability of HHV-6 to activate transcription driven by the long terminal repeat (LTR) of HIV-1 (see below). Increased HIV-1 replication upon HHV-6 infection was also observed in continuous $CD4^+$ T-cell lines (Ongradi et al., 1999; Lusso et al., unpublished). Moreover, positive interactions between HHV-6 and simian immunodeficiency virus (SIV), resulting in increased levels of SIV replication, were documented in macaque T-lymphocyte cultures (Lusso et al., 1994). In contrast, other authors have reported a suppressive, rather than enhancing, effect of HHV-6 on the *in vitro* replication of HIV-1 both in T lymphocytes (Carrigan et al., 1990; Levy et al., 1990) and in dendritic cells (Asada et al., 1999). Although Levy and colleagues observed an early (day 4) enhancement of HIV-1 replication, reverse transcriptase activity was later suppressed, in parallel with an increasing loss of cell viability. Even in these studies, however, the cytopathic effect was accelerated in the presence of HHV-6, with respect to cultures infected with HIV-1 alone, suggesting that the observed suppression of HIV-1 derived from a rapid loss of target $CD4^+$ T cells capable of supporting its replication. Of note, the effects of HIV-1 on HHV-6 replication were also stunningly divergent in the various reports, ranging from HHV-6 enhancement (Carrigan et al., 1990) to no significant effects (Lusso et al., 1989a) to suppression (Levy et al., 1990). These discrepancies emphasize the difficulties in interpreting the results obtained in different study models, as even subtle differences in the experimental conditions (viral strains, multiplicities of infection for each virus, cell activation status, timing of infection, etc.) may have a major impact on the outcome of the experiments. Consistent with this concept, in one study, inhibition of HIV-1 replication in primary lymphocyte cultures was only seen when HHV-6 was used at a high multiplicity and HIV-1 at a low multiplicity (Bonura et al., 1999). As discussed in detail below, one of the critical factors is the biological variant of HIV-1 used in coinfection studies, as illustrated by the divergent effects observed in lymphoid tissue *ex vivo*, with a selective suppression of CCR5-dependent (R5) strains and an overall enhancement of CXCR4-using strains (Grivel et al., 2001). Another important variable to be considered is the potential effect of bacterial endotoxin, a common cell culture contaminant, which was reported to abrogate the enhancing effects of HHV-6 on HIV-1 (Ongradi et al., 1999).

HIV-1 LTR transactivation by HHV-6

Consistent with the increased HIV-1 replication observed in coinfected cultures, HHV-6 is a potent transactivator of the LTR of HIV-l, HIV-2 and SIV (Ensoli et al., 1989; Horvat et al., 1989; Lusso et al., 1989a). Using LTR deletion mutants, the HHV-6-responsive elements were mapped to the NFκB- and Sp1-binding sites (Ensoli et al., 1989; Horvat et al., 1989) and shown to function independently of the TAR element (the sequence responsive to the HIV transactivator Tat), as also proven by the additive effect observed in the simultaneous presence of HHV-6 and Tat (Ensoli et al., 1989). Likewise, Di Luca et al. (1991) demonstrated a synergistic activation of the HIV-LTR in a human $CD4^+$ T cell line constitutively expressing Tat (Jurkat-Tat), upon infection with HHV-6. Interestingly, Csoma et al. (2002) demonstrated that nonproductive HHV-6 infection of primary human syncytiotrophoblasts is sufficient to induce activation of HIV-1 replication from latency, suggesting a possible role of HHV-6 in the vertical transmission of HIV-1. Similar results were reported using a latently HIV-infected promonocytic cell line, U1 (Knox and Carrigan, 1996). At least seven genes of HHV-6 code for proteins that act as transactivators; five of them, DR7, U16, U27, U89 and U94, were shown to activate the HIV-1 LTR *in vitro* (Gompels et al., 1995).

Effects of HIV-1 on HHV-6 replication

The reciprocal effects of HIV-1 on HHV-6 replication were investigated in a few studies. An enhanced HHV-6 replication was reported in coinfected primary T lymphocytes and T-cell lines (Carrigan et al., 1990; Sieczkowski et al., 1995). A similar trend, albeit not statistically significant, was observed in coinfected human lymphoid tissue *ex vivo* (Grivel et al., 2001). Sieczkowski and colleagues (1995) demonstrated that the Tat protein of HIV-1 is sufficient to increase HHV-6 protein expression and virus release. In contrast, another study demonstrated a 4- to 15-fold reduction in HHV-6 replication in Jurkat-Tat, compared to wild-type Jurkat (Di Luca et al., 1991). However, this experiment was performed using two different cell lines, the parental Jurkat cell line and the stably transfected Jurkat-Tat; the latter was cloned under selection and grown independently for several generations. Thus, these results should be interpreted with caution.

Direct virion–virion interactions

The discovery that the ubiquitous complement regulatory protein CD46 is the cellular receptor for HHV-6 (Santoro et al., 1999) and the recognition that CD46 is passively taken up by HIV-1, along with other cell surface proteins, during the process of budding from the cellular membrane (Saifuddin et al., 1997) raised the possibility of a direct interaction between mature HHV-6 and HIV-1 virions, mediated by HHV-6 binding to CD46 mounted on the HIV-1 virion surface. Albeit intriguing, however, this hypothesis remains unproven.

De nova *induction of CD4 expression*

The most striking interaction documented between HHV-6 and HIV is the ability of HHV-6 to regulate the expression of the primary HIV receptor, CD4. Since the early phenotypic studies on HHV-6 A-infected cells (Lusso et al., 1988), it was noticed that a variable proportion of them coexpress CD4 and CD8. Subsequently, using purified T-cell populations, it was possible to demonstrate that infection with HHV-6 A positively regulates the expression of CD4: not only is CD4 upregulated in cells that already express it at low levels, but it is even induced in cells that physiologically do not express it. *De novo* CD4 expression was documented in cytotoxic effector cells such as $CD8^+$ T cells (Lusso et al., 1991a), NK cells (Lusso et al., 1993) and $\gamma\delta$ T cells (Lusso et al., 1995), as well as in hematopoietic precursor cells (Furlini et al., 1996). Since CD3 is transcriptionally downregulated by HHV-6 (Lusso et al., 1991b), infected $CD8^+$ T cells reacquire a cortical thymic phenotype ($CD3^-CD4^+CD8^+$). The induction of CD4 occurs at the transcriptional level, as indicated by studies with the CD4 promoter (Flamand et al., 1998), and is mediated by early gene products of HHV-6 A, as indicated by studies with the viral DNA polymerase inhibitor phosphonoformic acid (PFA) (Lusso et al., 1991a). Of importance, the newly induced CD4 receptor is functional, as it can mediate productive HIV infection in otherwise refractory cells. Through this unique mechanism, HHV-6 may significantly expand the range of cells susceptible to HIV-1 *in vivo* and, thereby, favor its spread in coinfected hosts.

The ability of HHV-6 B to induce *de novo* CD4 expression is still uncertain owing to the inefficient growth of these strains in CD4-negative cells. However, the increased levels of CD4 expression observed in HHV-6 B-infected Jurkat cells

(Lusso, unpublished), which constitutively express low CD4 levels, seems to confirm that both HHV-6 variants upregulate CD4.

HHV-6 infection has also been suggested to modulate the expression of another cellular receptor for HIV-1, the coreceptor CXCR4, but the evidence is not univocal. Yasukawa et al. (1999) have reported downmodulation of CXCR4 expression in HHV-6-infected T cells, but other studies with either $CD4^+$ T cells (Lusso et al., unpublished) or dendritic cells (Asada et al., 1999) failed to confirm this finding. The enhanced replication of CXCR4-using HIV-1 strains observed in human lymphoid tissue coinfected with HHV-6 (see below) provides additional evidence that this putative effect has a limited, if any, biological impact on the interactions between the two viruses.

Induction of inflammatory cytokines that modulate HIV replication

Several studies have shown that HHV-6 infection increases the production of inflammatory cytokines, such as TNF-α, IL-1β and IL-8 (Flamand et al., 1991; Inagi et al., 1996; Arena et al., 1997), that enhance the *in vitro* expression of HIV-1; only IL-2 among the factors that promote HIV-1 growth is reduced (Flamand et al., 1995). On the other side, the secretion of cytokines that exert direct or indirect inhibitory effects on HIV-1 infection, including IFN-α, IL-10, IL-15 and RANTES, is also enhanced (Kikuta et al., 1990; Flamand et al., 1996; Arena et al., 1999; Grivel et al., 2001, 2003), with the exception of IFN-γ, whose levels are reduced in HHV-6-infected cultures (Arena et al., 1999). Upregulation of IL-10 coupled with a reduction of IFN-γ, IL-2 and IL-12 (see below) may unbalance the immune response toward a Th2 predominance, which is believed to be favorable for the spread of HIV-1 *in vivo* (Clerici and Shearer, 1993).

Another intriguing mechanism whereby HHV-6 may modulate the cytokine system in HIV-infected subjects is related to the presence of two chemokine (U22 and U83) and two chemokine-receptor (U12 and U51) gene homologs in its genome (Isegawa et al., 1998; French et al., 1999; Milne et al., 2000; Kondo et al., 2002). U83 encodes an effective CCR2 agonist, which can attract CCR2-expressing monocytes, increasing the local pool of HIV-susceptible cells. U12 encodes a functional CCR5-related chemokine receptor, which is sensitive to activation by the natural HIV inhibitors, RANTES, MIP-1α, MIP-1β, while the U51 product specifically binds RANTES, without transducing intracellular signals. Interestingly, U51 expression induces downregulation of RANTES expression, which may favor the local growth of CCR5-dependent HIV-1 strains.

Mechanisms of direct immunological damage by HHV-6

In addition to boosting the virulence and pathogenicity of HIV-1, HHV-6 is *per se* capable of causing immunosuppression. In particular, owing to its broad "immunotropism," HHV-6 A has the potential ability to cause significant immunological damage and dysregulation. The diverse mechanisms whereby HHV-6 can induce immunosuppressive effects have been extensively discussed in Chapter 5. In summary, they include: (a) productive infection and killing of CD4$^+$ T cells (Lusso et al., 1988); (b) productive infection and killing of cytotoxic effector cells (especially by HHV-6 A), including CD8$^+$ T lymphocytes, natural killer cells and $\gamma\delta$ T lymphocytes (Lusso et al., 1991a, 1993, 1995); (c) induction of phenotypic and functional defects in mononuclear phagocytic cells and dendritic cells, in particular IL-12 suppression (Smith et al., 2003, 2005); (d) modulating effects on inflammatory and immunosuppressive cytokines (see above); (e) transcriptional downregulation of the CD3–T cell receptor complex (Lusso et al., 1991b); (f) downmodulation of CD46, affecting both infected and bystander uninfected cells (Santoro et al., 1999; Grivel et al., 2003); (g) inhibition of IL-2 secretion (Flamand et al., 1995); and (h) putative induction of T-regulatory type-1 (Tr1) cells mediated by engagement of CD46 (Kemper et al., 2003). Altogether, these mechanisms may concur in determining significant immunological damage during HHV-6 reactivation episodes, which would be independent of and, potentially, synergic with the immunosuppression induced by HIV. In this respect, *in vivo* evidence linking HHV-6 with immunosuppression in the absence of concomitant HIV-1 infection has been reported (Knox et al., 1995b; Wang et al., 2002; Yoshikawa et al., 2002).

In vivo studies of HHV-6 infection in HIV-1 disease

Several studies have investigated the *in vivo* prevalence and replication of HHV-6 in HIV-infected patients with two major aims: on one side, to establish a link with the etiology of organ-specific opportunistic infections in order to shed light on the role of HHV-6 as an opportunistic agent; on the other, to verify the correlation between active HHV-6 infection and the clinical stage of HIV disease in order to add new ground to the concept that HHV-6 acts as a cofactor in the progression of HIV disease. However, the interpretation of *in vivo* studies of HHV-6 infection is complicated by the ubiquitous distribution of this virus (at least subgroup B) in the human population. As a consequence, the significance of certain diagnostic methods, like the detection of specific IgG antibodies in serum or of viral DNA sequences in tissues or blood cells, is significantly diminished. In fact, previous exposure to the virus, as revealed by a seropositive status or the presence of latent viral DNA in blood cells, does not imply active infection, which is the only clinically relevant form of infection. This makes it difficult to interpret the results, often conflicting, of serological studies based on IgG antibody testing (Brown et al., 1988; Fox et al., 1988; Essers et al., 1991; Chen et al., 1992; Spira et al., 1990; Dorrucci et al., 1999). Similarly, the merely qualitative detection of HHV-6 DNA in blood cells or tissues does not provide useful information for elucidating the clinical impact of HHV-6 infection in HIV-infected subjects. In light of the above considerations, there is no doubt that unraveling the *in vivo* role of HHV-6 requires the use of reliable markers of active virus replication. Examples of such tests include the detection of plasma HHV-6 viremia by PCR (Secchiero et al., 1995; Locatelli et al., 2000), late viral gene mRNA by RT–PCR (Van Den Bosch et al., 2001) and serum IgM antibodies (Fox et al., 1988; Suga et al., 1992; Secchiero et al., 1995). The ability to distinguish active infections caused by HHV-6 A and B would be an additional feature of great value since the impact of the two variants on HIV-1 disease may be entirely different.

HHV-6 as an opportunistic agent in AIDS

HHV-6 may cause various opportunistic diseases in immunocompromised people, including HIV-1-infected patients. In AIDS, HHV-6 has been implicated in pneumonitis (Knox and Carrigan, 1994), neurological disorders (Knox and Carrigan, 1995; Knox et al., 1995a; Saito et al., 1995) and retinitis (Qavi et al., 1992; Reux et al., 1992; Fillet et al., 1996). Instead, no correlation was documented between HHV-6 and AIDS-related lymphoma (Dolcetti et al., 1996). Of note, most of the isolates associated with these opportunistic diseases were HHV-6 B. Additional details about the specific organ pathology linked to HHV-6 infection are given in other sections of this book.

Correlation between HHV-6 replication and HIV-1 disease progression

A link between HHV-6 and the progression of HIV-1 infection was documented by several clinical studies, some of which were conducted using specific markers of active infection: (a) HHV-6 was repeatedly isolated from patients with symptomatic HIV infection (Salahuddin et al., 1986; Downing et al., 1987; Tedder et al., 1987; Lopez et al., 1988); (b) HHV-6 DNA was detected at high frequency in circulating leukocytes of HIV-infected patients (Buchbinder et al., 1988; Dolcetti et al., 1994, 1996; Blazquez et al., 1995; Ablashi et al., 1997; Iuliano et al., 1997), with a significant correlation between the viral DNA load and the number of circulating $CD4^+$ T cells (Fairfax et al., 1994; Fabio et al., 1997), suggesting an early activation of HHV-6 *in vivo,* which precedes, instead of following, the depletion of $CD4^+$ T cells; (c) HHV-6 A plasma viremia, a specific marker of active infection, was evidenced in a high proportion of patients with early symptomatic HIV infection or full-blown AIDS, but not in asymptomatic HIV-seropositive individuals (Secchiero et al., 1995), and was temporally correlated with a subsequent dramatic loss of circulating $CD4^+$ cells (Iuliano et al., 1997); (d) HHV-6 A antigenemia, IgM positivity and virus isolation were frequently detected in symptomatic HIV-1-infected subjects (Ablashi et al., 1998); (e) HHV-6 A antigen expression, indicative of active viral replication, was detected in all the lymph nodes obtained from HIV-infected patients, the majority of which were collected during the early symptomatic stage of HIV disease and showed histological features of follicular hyperplasia (Knox and Carrigan, 1996); (f) HHV-6 DNA was detected more frequently during the early histological phases of HIV-related lymphadenopathy (Dolcetti et al., 1996); (g) necropsy studies in AIDS patients documented a widespread and active HHV-6 infection in multiple anatomical sites using both PCR (Corbellino et al., 1993) and immunohistochemistry (Knox and Carrigan, 1994), with HHV-6 DNA loads significantly higher than in control tissues (Clark et al., 1996; Emery et al., 1999); and (h) the presence of HHV-6 in necropsy tissues was associated with an increased HIV-1 viral load (Emery et al., 1999).

Only a few studies so far have investigated the role of HHV-6 in pediatric HIV infection. Although the results obtained *in vitro* with primary human syncytiotrophoblasts suggest a possible facilitating role of HHV-6 in vertical HIV-1 transmission (Csoma et al., 2002), no *in vivo* data are available at present to confirm this hypothesis. HHV-6 can be vertically transmitted from HIV-infected mothers and, in this setting, it actively replicates in the neonates (Joshi et al., 2000). The impact of HHV-6 coinfection on the progression of HIV-1 disease has been investigated by Kositanont et al. (1999) who documented a significant association between HHV-6 infection within the first year of life and early progression of HIV-1 disease. HHV-6 may also play a role in AIDS-associated encephalopathy in children, as suggested by the disseminated infection of oligodendrocytes in the white matter and other neural cell types at postmortem examination (Saito et al., 1995).

Altogether, the above observations confirm the widespread diffusion of HHV-6 infection among HIV-1-infected patients. Strikingly, several studies have permitted to establish that HHV-6 reactivation occurs at a relatively early stage during the progression of HIV infection, before the time when other opportunistic infections typically appear, as reflected by a still high number of circulating $CD4^+$ T cells and a partially preserved lymph node histology. Thus, HHV-6 may start to induce immunological damage, either directly or by enhancing the virulence and pathogenicity of HIV-1, before the development of a clinically relevant immunodeficiency. Such damage can be particularly severe in the case of HHV-6 A, given its "immunotropic" nature. If confirmed in large prospective studies, this order of events, with HHV-6 activation preceding the loss of $CD4^+$ T cells and the overt immunodeficiency, argues in favor of a cofactor role of HHV-6 in the immune system breakdown that leads to the development of full-blown AIDS.

Experimental coinfection models *ex vivo* and *in vivo*

Experimental *in vivo* or *ex vivo* models have provided precious information on the effects of coinfection by HHV-6 and HIV-1 or SIV. Currently, two experimental *in vivo* models are available for investigating the interactions between HHV-6 and HIV-1/SIV: SCID-hu *Thy/Liv* mice and pig-tailed macaques. However, a routine use of such models is hampered by their high costs and technical challenge. An alternative model was recently developed using structurally intact human lymphoid tissue *ex vivo* (Grivel et al., 2003). This system is physiologically relevant because it preserves most of the cellular elements present in immunocompetent, HHV-6-primed adult primary human lymphoid tissue.

Coinfection studies in structurally intact human lymphoid tissue **ex vivo***: does HHV-6 favor the phenotypic switch of HIV-1?*

Coinfection studies in human lymphoid tissue *ex vivo* have recently provided a new paradigm to understand the interactions between HHV-6 and HIV-1, clearly documenting a divergent effect of HHV-6 according to the coreceptor specificity of

HIV-1: while the replication of CXCR4-tropic HIV-1 (either X4 or R5X4) was generally enhanced by HHV-6 coinfection, the growth of CCR5-tropic strains was consistently suppressed (Grivel et al., 2001). In turn, HHV-6 replication was generally enhanced by coinfection with HIV-1. In addition, HHV-6 was shown to potently induce the production of RANTES, the most effective HIV-inhibitory chemokine, providing a potential mechanism for the divergent effects seen on the different HIV-1 variants. Strikingly, RANTES induction by HHV-6 specifically occurs in structurally intact lymphoid tissue, because in mononuclear cell cultures grown in suspension, including cells extracted from minced tonsil or lymph node tissue, the effect is only marginal (Lusso et al., unpublished), suggesting that it requires a triggering action of specific cells, such as stromal cells, that are not present in suspension cultures.

Altogether, the observations made in lymphoid tissue *ex vivo* suggest a novel mechanism whereby HHV-6 can influence the course of HIV-1 infection. Thus, when HHV-6 is reactivated *in vivo* during the progression of HIV-1 disease, it may selectively suppress the dominant CCR5-dependent HIV-1 variants, while favoring the replication of CXCR4-tropic variants, which are kept under tight control, most likely by immunological mechanisms, during the asymptomatic phase of the infection. Thus, HHV-6 replication may be one of the factors driving the coreceptor switch of HIV-1 from CCR5 to CXCR4 usage.

In vivo *coinfection with HHV-6 and HIV-1 in chimeric SCID-hu Thy/Liv mice: lack of reciprocal modulation*

The SCID-hu *Thy/Liv* mouse model has been discussed in Chapter 5. Despite its value for pathogenesis studies, this model did not provide critical information to elucidate the role of HHV-6 in AIDS. When the human thymic implants were coinfected with HHV-6 (A or B) and HIV-1, the two viruses were shown to simultaneously replicate *in vivo* but there were no significant differences in their growth rate and cytopathic effects (Gobbi et al., 2000). A slight reduction of HIV-1 replication was only observed in parallel to a marked cellular depletion, particularly of the main target cells for HIV-1 replication, the intrathymic T-progenitor cells (ITTP). Conversely, there was a trend toward enhanced HHV-6 viral load in coinfected implants. However, the overall lack of dramatic effects induced by coinfection with HHV-6 and HIV-1 in SCID-hu *Thy/Liv* mice suggests that the two viruses can replicate simultaneously in this model system without affecting each other in a positive or negative way, possibly due to infection of different target cells.

In vivo *coinfection with HHV-6 and SIV in pig-tailed macaques (Macaca nemestrina): accelerated immunological and clinical progression to AIDS*

Chimpanzees (*Pan troglodytes*) represent the first choice for *in vivo* coinfection studies due to their evolutionary closeness to humans and their marked susceptibility to HIV-1 (Alter et al., 1984) and, at least *in vitro,* to HHV-6 (Lusso et al., 1990).

However, their high cost, limited supply and endangered status have so far impeded the performance of HHV-6 studies in this species. Since *in vitro* studies have shown that another nonhuman primate, the pig-tailed macaque (*Macaca nemestrina*), is highly susceptible to HHV-6 infection (Lusso et al., 1994), this species was chosen for an *in vivo* study of experimental coinfection performed in 1994–1996 at the National Cancer Institute in Bethesda, Maryland (Lusso et al., unpublished). The prototype HHV-6 subgroup A strain (GS) and a pathogenic SIV strain (smE660) were employed in these experiments. Three groups of four young adult animals were infected by intravenous injection with either SIV alone (group 1), HHV-6 alone (group 2) or both viruses (group 3), and then followed longitudinally for several virological, immunological and clinical parameters. The infections in group 3 were performed sequentially, with a 14-day interval between SIV injection and the subsequent HHV-6 superinfection. As expected, since the animals were immunocompetent before the inoculations, infection with HHV-6 alone resulted in only transient viremia without any evident short- and long-term clinical consequences. Consistent with previous studies using the SIV_{smE660} strain, infection with SIV alone induced a progressive immunological deterioration with a rapid loss of circulating $CD4^+$ T cells, but clinical progression to full-blown AIDS occurred in only one out of four animals within the 32 months of the study. By contrast, HHV-6 coinfection induced a dramatic acceleration of the clinical progression, with the appearance of AIDS-defining conditions in all the animals before termination of the study. The mean time to euthanasia for AIDS-related conditions was 20 ± 6.7 months in coinfected animals vs. >27 months in those infected with SIV alone. The depletion of circulating $CD4^+$ and, more strikingly, $CD8^+$ T lymphocytes in peripheral blood was more rapid in coinfected animals. SIV antigenemia and HHV-6 plasma viremia were both positive during the early phase of the coinfection. Morever, simultaneous replication of SIV and HHV-6 was documented by *in situ* hybridization in lymph nodes. These data provided the first conclusive evidence in a relevant primate model *in vivo* that coinfection with HHV-6 A significantly accelerates the progression of SIV disease.

The potential mechanisms underlying the accelerated disease progression in HHV-6–SIV-coinfected macaques were recently investigated through an extensive biological characterization of the SIV isolates obtained from singly and dually infected animals between 10 and 12 months post-inoculation. Strikingly, while the SIV isolates derived from singly-infected macaques were regularly sensitive to inhibition by RANTES, in agreement with their dependence on CCR5 for entry, all the isolates derived from coinfected macaques showed a marked RANTES resistance, and in two cases even RANTES dependence, as they grew more efficiently in the presence of RANTES (Biancotto et al., unpublished). These results are consistent with the model derived from the observations made in human lymphoid tissue *ex vivo*: HHV-6 may exert a selective pressure through the induction of RANTES *in vivo,* driving SIV to become resistant to RANTES-mediated inhibition. Although the use of alternative coreceptors cannot be totally, ruled out these SIV isolates were unable to grow in CCR5-Δ32/Δ32 T cells and thus may have learnt how to engage CCR5 in a different manner, which is insensitive to or even dependent on the presence of the bound inhibitor. Likewise, HIV-1 resistance to CCR5-targeted inhibitors was shown to develop without a change in coreceptor usage (Kuhmann et al., 2004).

In conclusion, coinfection studies in macaques have provided an intriguing model for elucidating the interactions between HHV-6 and primate immunodeficiency viruses *in vivo*. The accelerated induction of SIV disease observed in co-infected animals represents the first conclusive *in vivo* evidence of the role of HHV-6 as a progression cofactor in AIDS. Moreover, the phenotypic changes detected in SIV isolates derived after about 1 year of HHV-6 coinfection *in vivo* reveal the footprints of RANTES induction by HHV-6, providing a proof-of-principle of the model derived from *ex vivo* studies in lymphoid tissue.

Concluding remarks

The continuing debate on the role played by HHV-6 in the course of HIV-1 infection attests to the inherent difficulties in elucidating the relevance of infectious cofactors in a disease characterized by multiple opportunistic infections. The critical standing issue regarding HHV-6 is whether this virus is directly or indirectly involved in the pathogenesis of the immunological damage that leads to AIDS, particularly the loss of $CD4^+$ T cells. Definitive proof of the role of HHV-6 in AIDS has been difficult to achieve, also as a consequence of the lack of uniformed diagnostic criteria and methods for *in vivo* studies in infected patients. If appropriate markers of active infection are employed, the demonstration of a temporal relation between HHV-6 activation and the clinical/immunological signs of disease progression may provide a key to elucidating whether HHV-6 acts as a primary causative factor or, alternatively, replicates *in vivo* simply as a consequence of the lowered immune surveillance of the host. These studies require a careful longitudinal monitoring of HIV-infected patients during the critical phase that precedes the development of overt disease. Besides *in vivo* studies in HIV-infected patients, other experimental approaches can provide important complementary evidence. These include coinfection studies *in vivo* in suitable animal models, particularly nonhuman primates. Finally, important information may come, *ex juvantibus,* from the treatment of HIV-infected patients with selective anti-HHV-6 drugs during the early symptomatic phase of the disease, monitoring in parallel HHV-6 and HIV replication, as well as immunological parameters of disease progression. In this respect, it is remarkable that treatment with the potent HHV-6 inhibitor PFA (foscarnet) significantly prolonged the survival among AIDS patients (Studies of Ocular Complications of AIDS Research Group, 1992). If successful, this approach may also provide a model of cofactor-targeted therapeutic intervention, which could be an effective complement of the current HIV-specific protocols for the treatment of AIDS.

References

Ablashi DV, Chatlynne LG, Whitman Jr. JE. In Vivo 1997; 11: 383.
Ablashi DV, Marsh S, Kaplan M, Whitman Jr. JE, Pearson GR. Intervirology 1998; 41: 1.

Alter HJ, Eichberg JW, Masu H, Saxinger WC, Gallo RC, Macher AM, Lanc HC, Fauci AS. Science 1984; 226: 549.
Arena A, Liberto MC, Capozza AB, Foca A. New Microbiol 1997; 20: 13.
Arena A, Liberto MC, Iannello D, Capozza AB, Foca A. New Microbiol 1999; 22: 293.
Asada H, Klaus-Kovtun V, Golding H, Katz SI, Blauvelt A. J Virol 1999; 73: 4019.
Blazquez MV, Madueno JA, Jurado R, Fernandez-Arcas N, Munoz E. J AIDS 1995; 9: 389.
Bonura F, Perna AM, Vitale F, Villafrate MR, Viviano E, Guttadauro R, Mazzola G, Romano N. New Microbiol 1999; 22: 161.
Brown NA, Kovacs A, Lui CR, Hur C, Zaia JA, Mosley JW. Lancet 1988; 2: 1146.
Buchbinder A, Josephs SF, Ablashi DV, Salahuddin SZ, Gallo RC. J Virol Meth 1988; 21: 133.
Carrigan DR, Knox KK, Tapper MA. J Infect Dis 1990; 162: 844.
Chen H, Pesce AM, Carbonari M, Ensoli F, Cherchi M, Campitelli G, Sbarigia D, Luzi G, Aiuti F, Fiorilli M. Eur J Epidemiol 1992; 8: 217.
Clark DA, Ait-Khaled M, Wheeler AC, Kidd IM, McLaughlin JE, Johnson MA, Griffiths PD, Emery VC. J Gen Virol 1996; 77: 2271.
Clerici M, Shearer GM. Immunol Today 1993; 14: 107.
Corbellino M, Lusso P, Gallo RC, Parravicini C, Galli M, Moroni M. Lancet 1993; 342: 1242.
Csoma E, Bacsi A, Liu X, Szabo J, Ebbesen P, Beck Z, Konya J, Andirko I, Nagy E, Toth FD. J Med Virol 2002; 67: 67.
Dolcetti R, Di Luca D, Carbone A, Mirandola P, De Vita S, Vaccher E, Sighinolfi L, Gloghini A, Tirelli U, Cassai E, Boiocchi M. J Med Virol 1996; 48: 344.
Dolcetti R, Di Luca D, Mirandola P, De Vita S, De Re V, Carbone A, Tirelli U, Cassai E, Boiocchi M. Lancet 1994; 344: 543.
Dorrucci M, Rezza G, Andreoni M, Pezzotti P, Nicastri E, Ventura L, Zignani M, Alliegro MB, Tarantini G, Salassa B, Colangeli V, Mazzarello G, Ursitti MA, Barbanera M, Pristera R, Castelli F, Ortona L. Eur J Epidemiol 1999; 15: 317.
Downing RG, Sewankambo N, Serwadda D, Honess R, Crawford D, Jarrett R, Griffin BE. Lancet 1987; ii: 390.
Emery VC, Atkins MC, Bowen EF, Clark DA, Johnson MA, Kidd IM, McLaughlin JE, Phillips AN, Strappe PM, Griffiths PD. J Med Virol 1999; 57: 278.
Ensoli B, Lusso P, Schachter F, Josephs SF, Rappaport J, Negro F, Gallo RC, Wong-Staal F. EMBO J 1989; 8l: 3019.
Essers S, Schwinn A, ter Meulen J, von Lips H, Dietz K, Mhalu FS, Shao J, ter Meulen V. Eur J Epidemiol 1991; 7: 658.
Fabio G, Knight SN, Kidd IM, Noibi SM, Johnson MA, Emery VC, Griffiths PD, Clark DA. J Clin Microbiol 1997; 35: 2657.
Fairfax MR, Schacker T, Cone RW, Collier AC, Corey L. J Infect Dis 1994; 169: 1342.
Fillet AM, Reux I, Joberty C, Fournier JG, Hauw JJ, Le Hoang P, Bricaire F, Huraux JM, Agut H. J Med Virol 1996; 49: 289.
Flamand L, Gosselin J, D'Addario M, Hiscott J, Ablashi DV, Gallo RC, Menezes J. J Virol 1991; 65: 5105.
Flamand L, Gosselin J, Stefanescu I, Ablashi D, Menezes J. Blood 1995; 85: 1263.
Flamand L, Romerio F, Reitz MS, Gallo RC. J Virol 1998; 72: 8797.
Flamand L, Stefanescu I, Menezes J. J Clin Invest 1996; 97: 1373.
Fox J, Briggs M, Tedder RS. Lancet 1988; 2: 396.
French C, Menegazzi P, Nicholson L, Macaulay H, DiLuca D, Gompels UA. Virology 1999; 262: 139.
Furlini G, Vignoli M, Ramazzotti E, Re MC, Visani G, La Placa M. Blood 1996; 87: 4737.

Gobbi A, Stoddart CA, Locatelli G, Santoro F, Bare C, Linquist-Stepps V, Moreno MB, Abbey NW, Herndier BG, Malnati MS, McCune JM, Lusso P. J Virol 2000; 74: 8726.
Gompels UA, Nicholas J, Lawrence G, Jones M, Thomson BJ, Martin ME, Efstathiou S, Craxton M, Macaulay HA. Virology 1995; 209: 29.
Grivel JC, Ito Y, Faga G, Santoro F, Shaheen F, Malnati MS, Fitzgerald W, Lusso P, Margolis L. Nat Med 2001; 7: 1232.
Grivel JC, Santoro F, Chen S, Faga G, Malnati MS, Ito Y, Margolis L, Lusso P. J Virol 2003; 77: 8280.
Horvat RT, Wood C, Balachandran N. J Virol 1989; 63: 970.
Inagi R, Guntapong R, Nakao M, Ishino Y, Kawanishi K, Isegawa Y, Yamanishi K. J Med Virol 1996; 49: 34.
Isegawa Y, Ping Z, Nakano K, Sugimoto N, Yamanishi K. J Virol 1998; 72: 6104.
Iuliano R, Trovato R, Lico S, Luppi M, Forastieri G, Barsanti LA, Pizzigallo AM, Mecocci L, Barozzi P, Torelli G, Mazzotta F, Ceccherini-Nelli L. J Med Virol 1997; 51: 259.
Joshi PJ, Merchant RH, Pokharankar SL, Damania KS, Gilada IS, Mukhopadhyaya R. J Hum Virol 2000; 3: 317.
Kemper C, Chan AC, Green JM, Brett KA, Murphy KM, Atkinson JP. Nature 2003; 421: 388.
Kikuta H, Nakane A, Lu H, Taguchi Y, Minagawa T, Matsumoto S. J Infect Dis 1990; 162: 35.
Knox KK, Carrigan DR. Lancet 1994; 343: 577.
Knox KK, Carrigan DR. J AIDS 1995; 9: 69.
Knox KK, Carrigan DR. J AIDS 1996; 11: 370.
Knox KK, Harrington DP, Carrigan DR. J Med Virol 1995a; 45: 288.
Knox KK, Pietryga D, Harrington DJ, Franciosi R, Carrigan DR. Clin Infect Dis 1995b; 20: 406.
Kondo K, Shimada K, Sashihara J, Tanaka-Taya K, Yamanishi K. J Virol 2002; 76: 4145.
Kositanont U, Wasi C, Wanprapar N, Bowonkiratikachorn P, Chokephaibulkit K, Chearskul S, Chimabutra K, Sutthent R, Foongladda S, Inagi R, Kurata T, Yamanishi K. J Infect Dis 1999; 180: 50.
Kuhmann SE, Pugach P, Kunstman KJ, Taylor J, Stanfield RL, Snyder A, Strizki JM, Riley J, Baroudy BM, Wilson IA, Korber BT, Wolinsky SM, Moore JP. J Virol 2004; 78: 2790.
Levy JA, Landay A, Lennette ET. J Clin Microbiol 1990; 28: 2362.
Locatelli G, Santoro F, Veglia F, Gobbi A, Lusso P, Malnati MS. J Clin Microbiol 2000; 38: 4042.
Lopez C, Pellett P, Stewart J, Goldsmith C, Sanderlin K, Black J, Warfield D, Feorino P. J Infect Dis 1988; 157: 1271.
Lusso P, De Maria A, Malnati M, Lori F, DeRocco SE, Baseler M, Gallo RC. Nature 1991a; 349: 533.
Lusso P, Ensoli B, Markham PD, Ablashi DV, Salahuddin SZ, Tschachler E, Wong-Staal F, Gallo RC. Nature 1989a; 337: 368–370.
Lusso P, Gallo RC. Immunol Today 1995; 16: 67–71.
Lusso P, Gallo RC, DeRocco SE, Markham PD. Lancet 1989b; 1: 730.
Lusso P, Garzino-Demo A, Crowley RW, Malnati MS. J Exp Med 1995; 181: 1303.
Lusso P, Malnati M, De Maria A, Balotta C, DeRocco SE, Markham PD, Gallo RC. J Immunol 1991b; 147: 685.

Lusso P, Malnati MS, Garzino-Demo A, Crowley RW, Long EO, Gallo RC. Nature 1993; 362: 458.
Lusso P, Markham PD, DeRocco SE, Gallo RC. J Virol 1990; 64: 2751.
Lusso P, Markham PD, Tschachler E, di Marzo Veronese F, Salahuddin SZ, Ablashi DV, Pahwa S, Krohn K, Gallo RC. J Exp Med 1988; 167: 1659.
Lusso P, Secchiero P, Crowley RW. AIDS Res Hum Retroviruses 1994; 10: 181.
Milne RS, Mattick C, Nicholson L, Devaraj P, Alcami A, Gompels UA. J Immunol 2000; 164: 2396.
Ongradi J, Ceccherini Nelli L, Matteucci D, Bertok L, Bendinelli M. Orv Hetil 1999; 140: 2577.
Qavi HB, Green MT, SeGall GK, Lewis DE, Hollinger FB. Curr Eye Res 1992; 11: 315.
Reux I, Fillet AM, Agut H, Katlama C, Hauw JJ, LeHoang P. Am J Ophthalmol 1992; 114: 375.
Saifuddin M, Hedayati T, Atkinson JP, Holguin MH, Parker CJ, Spear GT. J Gen Virol 1997; 78: 1907.
Saito Y, Sharer LR, Dewhurst S, Blumberg BM, Hall CB, Epstein LG. J Neurovirol 1995; 1: 30.
Salahuddin SZ, Ablashi DV, Markham PD, Josephs SF, Sturzenegger S, Kaplan M, Halligan G, Biberfeld P, Wong-Staal F, Kramarsky B, Gallo RC. Science 1986; 234: 596.
Santoro F, Kennedy PE, Locatelli G, Malnati MS, Berger EA, Lusso P. Cell 1999; 99: 817.
Secchiero P, Carrigan DR, Asano Y, Benedetti L, Crowley RW, Komaroff AL, Gallo RC, Lusso P. J Infect Dis 1995; 171: 273.
Sieczkowski L, Chandran B, Wood C. Virology 1995; 211: 544.
Smith AP, Paolucci C, Di Lullo G, Burastero SE, Santoro F, Lusso P. J Virol 2005; 79: 2807.
Smith A, Santoro F, Di Lullo G, Dagna L, Verani A, Lusso P. Blood 2003; 102: 2877.
Spira TJ, Bozeman LH, Sanderlin KC, Warfield DT, Feorino PM, Holman RC, Kaplan JE, Fishbein DB, Lopez C. J Infect Dis 1990; 161: 567.
Studies of Ocular Complications of AIDS Research Group. N Engl J Med 1992; 326: 213.
Suga S, Yoshikawa T, Asano Y, Nakashima T, Yazaki T, Fukuda M, Kojima S, Matsuyama T, Ono Y. Microbiol Immunol 1992; 36: 495.
Tedder RS, Briggs M, Cameron CH, Honess R, Robertson D, Whittle H. Lancet 1987; 2: 390.
Van Den Bosch G, Locatelli G, Geerts L, Fagà G, Ieven M, Böttiger D, Öberg B, Lusso P, Berneman ZN. J Clin Microbiol 2001; 39: 2308.
Wang FZ, Larsson K, Linde A, Ljungman P. Bone Marrow Transplant 2002; 30: 521.
Yasukawa M, Hasegawa A, Sakai I, Ohminami H, Arai J, Kaneko S, Yakushijin Y, Maeyama K, Nakashima H, Arakaki R, Fujita S. J Immunol 1999; 162: 5417.
Yoshikawa T, Asano Y, Akimoto S, Ozaki T, Iwasaki T, Kurata T, Goshima F, Nishiyama Y. J Med Virol 2002; 66: 497.

Human Herpesvirus-6: General Virology, Epidemiology and Clinical Pathology
Perspectives in Medical Virology
Gerhard Krueger and Dharam Ablashi (Editors)

DOI 10.1016/S0168-7069(06)12022-4

Human Herpesvirus-6 Infection in Solid Organ and Stem Cell Transplant Recipients

P. Ljungman[a], N. Singh[b]
[a]*Karolinska University Hospital, Karolinska Institutet, Stockholm, Sweden*
[b]*University of Pittsburgh Medical Center, Pittsburgh, Pennsylvania, USA*

Epidemiology

Seroepidemiologic studies have shown that infection due to human herpesvirus-6 (HHV-6) is usually acquired during the first year of life; the virus subsequently persists in the host. Seroprevalence in healthy adults exceeds 90%. Serology is unable to differentiate between the two subtypes and therefore it is possible that seropositive patients might get a new infection with the second subtype presumably most commonly subtype A. Most transplant patients will therefore be seropositive prior to transplantation. Previous HHV-6 infection has been documented in 87–91% of solid organ transplant (SOT) recipients by serologic assays (Dockrell et al., 1997; Ihira et al., 2001) and by the detection of HHV-6 DNA sequences in the peripheral blood mononuclear cells in 32% of the patients (Chapenko et al., 2001). Similar percentages have been noted in stem cell transplant (SCT) recipients. Yoshikawa documented pretransplant seropositivity in 100% of SCT recipients (Yoshikawa et al., 2002) while Savolainen et al. showed seropositivity in 78% of autologous and 96% of allogeneic pediatric SCT recipients (Savolainen et al., 2005). They also documented HHV-6 antigenemia in 61% and DNAemia in 40% of the patients pretransplant.

Transmission

Given the high HHV-6 rate of seropositivity, most infections in transplant patients are considered to result from reactivation of a latent virus. The transmission of HHV-6 from the donor with the allograft has been documented. Mononuclear cells latently infected with HHV-6 in the donor allograft are believed to be the likely source of transmission (Lau et al., 1998). Following liver transplantation, primary HHV-6 infection has been reported in 61–100% of the patients who were seronegative for HHV-6 prior to transplantation (Dockrell et al., 1997; Yoshikawa et al., 2001). HHV-6 has been shown to develop latency in the kidney *in vivo* (Asano et al., 1989). Indeed, two renal transplant recipients who received the allografts from the same cadaveric donor were documented to have identical genomic patterns of their HHV-6 isolates (Yoshikawa et al., 1992). Transplanted allograft was the likely source of fatal infection due to HHV-6 variant A in a renal transplant recipient who was seronegative for HHV-6 prior to transplantation (Rossi et al., 2001). Similar reports exist in SCT recipients with one small child developing a definite primary infection (Lau et al., 1998) and one teenage SCT recipient developing fatal HHV-6 subtype A encephalitis and whose donor was polymerase chain reaction (PCR) positive for HHV-6 subtype A in the peripheral blood lymphocyte (PBL) before SCT (Bosi et al., 1998).

Documented infections

Solid organ transplant recipients

Overall, 38–55% of renal, 22–54% of liver, 36% of heart and up to 57% of heart–lung lung transplant recipients have been shown to develop HHV-6 infection (Morris et al., 1989; Okuno et al., 1990; Herbein et al., 1996; Dockrell et al., 1997; Lautenschlager et al., 2000; Rogers et al., 2000; de Ona et al., 2002). Following living related liver transplantation, HHV-6 infection has been documented in 48% of the patients; these included 4/4 patients who were seronegative and 42% (15/36) of those who were seropositive for HHV-6 prior to living related transplantation (Ihira et al., 2001).

Most HHV-6 infections occur between 2 and 4 weeks after SOT; this characteristic timing distinguishes HHV-6 from other betaherpesvirus infections that usually occur later post-transplantation (Singh and Carrigan, 1996). In a study, in liver transplant recipients, where HHV-6 and HHV-7 DNA detection was sought in the plasma, HHV-6 infections occurred in 38% (15/40) patients; 67% of the infections occurred at 2 weeks post-transplantation after which the frequency of viral genome detection in the plasma sharply declined (Ihira et al., 2001). In contrast, only 31% of the HHV-7 infections were detected at 2 weeks with no distinctive peak in time to detection. Further, HHV-7 DNA remained detectable in 10–20% of the patients until 8 weeks post-transplant. A study of the temporal sequence of infections with the 3 betaherpesviruses in organ transplant recipients showed that

HHV-6 reactivation occurred at a median of 20 days, followed by HHV-7 (median 26 days) and cytomegalovirus (CMV) (median 36 days) (Griffiths et al., 1999).

Risk factors for HHV-6 infections in SOT recipients have not been fully defined. Receipt of OKT3 monoclonal antibodies or antithymocyte globulin has been associated with HHV-6 reactivation (Rossi et al., 2001; Nash et al., 2004). An HHV-6 seroconversion in one study was noted more frequently in patients who received immunosuppressive regimens containing sirolimus and IL-12 receptor antibodies as induction therapy (Deborska et al., 2003).

Stem cell transplant recipients

By virus isolation, HHV-6 was documented in 38% of SCT recipients (Yoshikawa et al., 2002). Detection of HHV-6 DNA has been documented in 40–70% by studies of PBL or whole blood (Wang et al., 1996; Imbert-Marcille et al., 2000; Sashihara et al., 2002) and 47–70% by studies of plasma or serum (Zerr et al., 2001, 2005). The peak in viral load does similarly in SOT patients, which occurs early after transplant, usually within the first 4 weeks after SCT (Imbert-Marcille et al., 2000; Ljungman et al., 2000; Maeda et al., 2000; Yoshikawa et al., 2002).

Risk factors associated with HHV-6 infection after SCT has been reported to be allogeneic rather than autologous SCT, transplantation for leukemia or lymphoma, advanced hematological disease, younger age, sex mismatch between donor and recipient and treatment with corticosteroids (Imbert-Marcille et al., 2000; Yoshikawa et al., 2002; Zerr et al., 2005) It has been reported that patients receiving peripheral blood SCT are more frequently PCR positive than patients receiving bone marrow grafts (Maeda et al., 2000) but this has not been confirmed by other studies (Imbert-Marcille et al., 2000; Ljungman et al., 2000; Zerr et al., 2005). A small study reported a high incidence of HHV-6 in patients undergoing cord blood transplantation (Sashihara et al., 2002)

Clinical manifestations

Clinical sequelae of HHV-6 may result from symptoms directly attributable to the virus or from its immunomodulatory effects. Table 1 shows a summary over the clinical syndromes that has been suggested being associated with HHV-6. Symptomatic infections seem to be more common in SCT than in SOT patients although published reports vary from very limited clinical effects of HHV-6 to a contributing effect on overall mortality. A fever of unknown origin; with or without a skin rash; bone marrow suppression, and encephalitis are the most frequently observed clinical features of HHV-6 (Carrigan et al., 1991; Drobyski et al., 1993; Carrigan and Knox, 1995; Wang et al., 1999; Ljungman et al., 2000; Zerr et al., 2001). Less commonly, interstitial pneumonitis, gastrointestinal disease, and hepatitis have been reported (Cone et al., 1993; Singh et al., 1997; Rossi et al., 2001; Hentrich et al., 2005).

Table 1

Association of HHV-6 with clinical sequelae and the level of supportive evidence

I.	Supportive evidence from cohort studies
	A. Febrile syndrome
	B. Encephalitis
	C. Bone marrow suppression
	D. Association with fungal infections
	E. Association with cytomegalovirus infection
	F. Aggressive recurrence of hepatitis C virus after liver transplantation
	G. Mortality
II.	Evidence from case reports
	A. Interstitial pneumonitis
	B. Hepatitis
	C. Gastroduodenitis
	D. Leukocytoclastic vasculitis
III.	Proposed association with conflicting supportive evidence
	A. Allograft rejection
	B. Obliterative bronchitis

A non-specific febrile syndrome occurring in the early post-transplant period is the most frequently observed clinical manifestation of HHV-6 infection in organ transplant recipients. Unexplained fever was documented in 87% of the liver transplant recipients with HHV-6 infections compared to 20% of those without it ($p < 0.01$) (Yoshikawa et al., 2000). The febrile mononucleosis syndrome attributable to CMV in transplant recipients may, in fact, be related to a concurrent infection due to HHV-6 and 7 rather than CMV alone. Eighty-nine percent of the liver transplant recipients with CMV infections had concomitant HHV-6 variant B or HHV-7 infection (Razonable et al., 2003). Bone marrow suppression, most often manifesting as thrombocytopenia or leukopenia has been associated with HHV-6 and has been seen both in SCT recipients (Carrigan and Knox, 1994; Wang et al., 1996; Imbert-Marcille et al., 2000; Ljungman, et al., 2000; Zerr et al., 2005) and in SOT recipients (Singh et al., 1997). HHV-6 can infect hematological progenitor cells and reduce colony formation (Burd et al., 1993; Isomura et al., 1997). Increased levels of HHV-6 DNA are associated with a delayed platelet engraftment (Ljungman et al., 2000; Maeda et al., 2000; Zerr et al., 2005) and increased requirement for platelet transfusions in SCT recipients (Ljungman et al., 2000; Zerr et al., 2005).

HHV-6 has a propensity for the CNS (Caserta et al., 1994; Challoner et al., 1995) and although HHV-6 DNA can occasionally be detected in the CSF of asymptomatic SCT recipients (Wang et al., 1999; Zerr et al., 2001), several case reports and small patient series have given strong support for that HHV-6 is an important cause of encephalitis in SCT recipients. Approximately 30 cases have been published (Drobyski et al., 1994; Mookerjee and Vogelsang, 1997; Bosi et al., 1998; Cole et al., 1998; Rieux et al., 1998; Tsujimura et al., 1998; Bethge et al., 1999; De Almeida Rodrigues et al., 1999; Wang et al., 1999; Ljungman et al., 2000; Tiacci et al., 2000; Zerr et al., 2001; MacLean and Douen, 2002; Yoshida et al., 2002; Hentrich et al., 2005). A summary of published information around these cases regarding patient characteristics, diagnostic findings, and outcome of HHV-6 CNS disease in SCT patients is shown in Table 2. It should be noted that not all data are available for all patients. In SOT patients, a number of well documented case reports and case-controlled studies and at least three studies that included concurrent controls have documented an association between HHV-6 and CNS complications of unidentifiable etiology (Paterson et al., 1999; Rogers et al., 2000; Singh and Paterson, 2000; Bollen et al., 2001). A study of liver transplant recipients showed that 15% (12/80) of the patients had mental status changes of unknown etiology after transplantation and patients with HHV-6 viremia had a significantly higher incidence of mental status changes of unidentifiable etiology (29%, 9/31) as compared to those without HHV-6 viremia (6%, 3/49, $p = 0.008$) (Rogers et al., 2000). The symptoms are frequently uncharacteristic with lethargy, confusion, convulsions, and decreased consciousness as the predominant clinical

Table 2

Patient characteristics, diagnostic findings, therapy and outcome of SCT patients with suspected or proven HHV-6 encephalitis

Patient characteristics	
Median age	36 (12–56)
Type of transplant	
Unrelated or mismatched	21
Sibling donor	6
Autologous	1
Acute GVHD grade II-IV	11/20
CSF findings	
Pleocytosis	11/27
Increased protein	17/27
HHV-6 DNA	30/30
Radiographic findings	
MRI changes	19/29
CT changes	4/17
EEG changes	22/22
Survival of patients receiving therapy	17/28
Ganciclovir	5/6
Foscarnet	8/10
Acyclovir	1/2
Foscarnet + ganciclovir[a]	3/10

[a]Given in combination or consecutively.

manifestations of HHV-6 encephalitis. Focal neurological findings have been reported but are less common. The reports regarding CSF pleocytosis are conflicting. In one review CSF pleocytosis ranging from 6 to 53 cells/mm^3 was present in 50% of the patients with HHV-6 encephalitis (Singh and Paterson, 2000). Zerr et al. reported, however, that increased protein levels were common but CSF pleocytosis was rarely found (Zerr et al., 2001). Magnetic resonance imaging can show abnormalities but it can also be normal. MRI abnormalities were present in 2 of 8 patients in one report (Singh and Paterson, 2000). These changes included multiple, non-enhancing, low attenuation lesions in the gray matter. EEG usually shows diffuse changes. The prognosis is poor unless the encephalitis is treated with antiviral drugs.

There have been reports of HHV-6 as the cause of interstitial pneumonia in SCT patients (Cone et al., 1993; Hentrich et al., 2005). It has been difficult to assess the true relevance of HHV-6 since several other causes of pneumonia exist in SCT patients and in many reported patients several of these causes have not been systematically analyzed. Case reports in SCT recipients have documented an association between obliterative bronchitis and HHV-6 (Nishimaki et al., 2003). Although HHV-6 has been detected in 9.4–14.6% of the bronchoalveolar lavage fluid samples in lung transplant recipients (Ross et al., 2001; Jacobs et al., 2003), its relevance as a pathogen in this setting is controversial. HHV-7, on the other hand, was proposed to have a role in the pathogenesis of bronchiolitis obliterans with organizing pneumonia (BOOP) given that HHV-7 DNA was detected in all (7/7) transbronchial biopsies that showed BOOP while co infection with other betaherpesviruses was rare (Ross et al., 2001).

Other clinical syndromes suggested that associated with HHV-6 infections are hepatitis and gastrointestinal disease (Appleton et al., 1995; Lautenschlager et al., 1998; Hentrich et al., 2005). Several other differential diagnoses for these conditions exist in transplant recipients especially graft-versus-host disease GVHD and other infections making interpretation of HHV-6 s role in these conditions difficult.

Indirect sequelae

HHV-6 is considered an immunomodulatory and immunosuppressive virus that may facilitate superinfections with other opportunistic infections and contribute to higher mortality (Flamand et al., 1995; Singh and Carrigan, 1996; Dockrell et al., 1997; Singh et al., 1997). Zerr et al. (2005) showed in SCT patients that detection of HHV-6 in plasma predicted for subsequent all-cause mortality. Also liver transplant recipients with HHV-6 had a significantly higher mortality; the independent association between HHV-6 and late mortality approached statistical significance (Rogers et al., 2000). Similar observations were made in lung transplant recipients where patients with HHV-6 infections had a significantly higher mortality rate (35% versus 0%, $p = 0.04$) than those without HHV-6 (Jacobs et al., 2003). It is difficult to assess in all these studies whether there was a cause and effect association between HHV-6 and mortality or if they are instead parallel phenomena.

One possibility for the potential effect of HHV-6 on all-cause mortality is the effect of HHV-6 on other infections. A primary HHV-6 infection was identified as a significant risk factor for the development of a symptomatic CMV infection, including tissue-invasive CMV disease in liver transplant recipients (Dockrell et al., 1997). A HHV-6 infection has also been associated with repeated CMV infections and absence of CMV-specific lymphocyte proliferation response in SCT recipients (Wang, et al., 2002). HHV-6 infection has also been reported as an independently significant predictor of invasive fungal infections in liver transplant recipients (Dockrell et al., 1997; Rogers et al., 2000). When controlled for the level of immunosuppression, a HHV-6 infection increased the risk of opportunistic infections in liver transplant recipients by 3.68-fold (Humar et al., 2002).

HHV-6 infection has also been shown to have a contributory role in the pathogenesis of hepatitis C disease (Humar et al., 2002; Singh et al., 2002a). Although the HHV-6 viremia did not affect the overall rate of recurrence of hepatitis C, it was associated with a more severe form of recurrence. The development of severe recurrence, that is, a fibrosis score of >2, was significantly more common in patients with HHV-6 infections in one study (Humar et al., 2002). In another report, patients with HHV-6 viremia tended to have an earlier recurrence and significantly higher fibrosis scores on recurrence than did those with HHV-6 viremia (Singh et al., 2002a). Other variables that may influence the outcome of hepatitis C recurrence such as the hepatitis C virus (HCV) genotype, alcohol use, CMV infections, and augmented immunosuppression, did not differ for patients who did or did not have HHV-6 viremia. Of note, patients who received ganciclovir as preemptive therapy for CMV infection had lower total Knodell scores and a trend toward lower fibrosis scores than those of patients who did not receive ganciclovir. Although the pathophysiologic basis of the association between HHV-6 and severity of HCV, particularly the progression of fibrosis remains to be determined, a number of biologic plausibilities exist. HHV-6 is a potent inducer of cytokines, e.g. TNF-α, that play a role in the development of hepatic fibrosis.

Liver transplant recipients with hepatocellular carcinoma in one report were more likely to develop HHV-6 viremia than patients without it (Rogers et al., 2000). The association between HHV-6 and hepatocellular carcinoma was considered to be mediated through hepatotropic viruses, e.g. hepatitis B and C viruses. It was proposed that hepatotropic viruses may facilitate the emergence of HHV-6 from latency by transactivating its immediate-early proteins (Rogers et al., 2000).

There is conflicting data on the association of HHV-6 with rejection in SOT patients. A HHV-6 infection has been shown to increase the expression of adhesion molecules and the number of human leukocyte antigen class II-positive T cells. HHV-6 infection has been associated with acute allograft rejection in some, but not in all reports (Griffiths et al., 1999; Lautenschlager et al., 2000; Rogers et al., 2000). HHV-6 infection and peak HHV-6 viral load in liver transplant recipients were associated with only those rejection episodes that occurred after day 30 post-transplantation (Rossi et al., 2001). In renal transplant recipients, HHV-7, but not CMV or HHV-6, correlated with biopsy-proven cellular rejection (Kidd et al., 2000).

A common cause of allograft deterioration in renal transplant recipients beyond the first year of transplantation is an entity known as chronic allograft nephropathy (Paul, 1999). Various pathophysiologic mechanisms including long-term calcineurin-inhibitor agent use, alloimmune injury, metabolic factor, and viral infections, e.g. CMV, are considered to contribute to this process. An association between chronic allograft nephropathy and HHV-6 infection has been reported (Tong et al., 2002). Histopathologic findings of chronic allograft nephropathy were observed in late biopsies in renal transplant recipients who had received anti-thymocyte globulin/ALG as induction therapy and had HHV-6 reactivation (24).

HHV-6 has been associated with acute GVHD in SCT patients. Two associations have been reported. The first is that HHV-6 can induce a rash that can be confused with early acute skin GVHD. This association was first reported by Yoshikawa (Yoshikawa et al., 1991) and has since been reported in several subsequent studies (Cone et al., 1999; Yoshikawa et al., 2001; Sashihara, 2005. The other possibility is that HHV-6 interacts with acute GVHD either by aggravating the condition or through that patients with more severe GVHD reactivates HHV-6. Zerr et al. (2005) reported that HHV-6 reactivation was associated with the development of subsequent severe acute GVHD. In another study it was found that HHV-6 DNA was detected in skin and/or rectal biopsies more frequently in allogeneic recipients with severe GVHD (92%) than in those with either moderate (55%) or mild GVHD (22%) (Appleton et al., 1995). One difficulty is the timing of the two events with respect to each other. Hentrich et al. (2005) reported that 22% of patients with both GVHD and HHV-6 had HHV-6 documented before the development of GVHD, 21% had both documented simultaneously while 57% had GVHD diagnosed before HHV-6.

Taken together, the indirect effect of HHV-6 appears to be its predominant clinical manifestation after SOT, whereas tissue-invasive disease occurs less frequently. HHV-6-specific T-cell immune responses were shown to be impaired for up to 1 year after transplantation (Singh et al., 2002a,b). It has been proposed that in the setting of preexistent immunity or seropositivity before transplantation, even a less than optimal T-helper cell response may be sufficient to protect against a progressive infection or tissue-invasive disease (Singh et al., 2002a,b). On the other hand more direct clinical sequelae seems to be more prominent after SCT although indirect effects especially on acute GVHD and persistent impairment of T-helper cell response could contribute to the immunosuppressive effects of HHV-6 and possibly to overall mortality.

Prevention and therapy

In vitro studies show that ganciclovir, cidofovir and foscarnet should be effective against HHV-6. Tokimasa et al. (2002) reported a lower rate of HHV-6 reactivations in patients receiving ganciclovir as CMV prophylaxis. Wang et al. (1996) showed in an epidemiological study that patients who received high-dose acyclovir had lower HHV-6 DNA levels and were less likely to suffer from a delayed marrow engraftment. However, in other studies, no effect on the viral load by acyclovir has been seen (Ljungman et al., 2000; Zerr et al., 2005).

Antiviral therapy with ganciclovir or foscarnet has been shown to lead to reduction in HHV-6 viral load in CSF (Zerr et al., 2002) and blood (Mendez et al., 2001; Zerr et al., 2002). Ganciclovir is also able to reduce HHV-6 viral load in saliva (Ljungman et al., 2001).

Both ganciclovir and foscarnet have been reported being effective against HHV-6 meningo-encephalitis after transplantation and the superiority of either ganciclovir or foscarnet over the other has not been established. In a review of HHV-6 encephalitis in transplant recipients, cure was documented in 7/8 patients who received ganciclovir or foscarnet for at least 7 days as compared to 0/4 in those who did not receive these drugs or received them for less than 7 days ($p = 0.01$) (Singh and Paterson, 2000). Table 2 shows outcome with different antiviral therapy in published and unpublished cases of HHV-6 encephalitis in SCT patients (Drobyski et al., 1994; Mookerjee and Vogelsang, 1997; Bosi et al., 1998; Cole et al., 1998; Rieux et al., 1998; Tsujimura et al., 1998; Bethge et al., 1999; De Almeida Rodrigues et al., 1999; Wang et al., 1999; Ljungman et al., 2000; Tiacci et al., 2000; Zerr et al., 2001; MacLean and Douen, 2002; Yoshida et al., 2002; Hentrich et al., 2005). Co-morbid clinical conditions such as renal failure or marrow suppression may also dictate whether ganciclovir or foscarnet is employed as therapy for HHV-6. The choice of the antiviral drug might be made based on the side-effect profile. Foscarnet may be preferable in patients with marrow suppression because it does not possess the myelosuppressive effect of ganciclovir. All 3 available agents are nephrotoxic so the choice in patients with renal dysfunction is not obvious.

Acknowledgment

The authors acknowledge The Swedish Children's Cancer Foundation for funding the research.

References

Appleton AL, Sviland L, Peiris JS, Taylor CE, Wilkes J, Green MA, Pearson AD, Kelly PJ, Malcolm AJ, Proctor SJ, Hamilton PJ, Cant AJ. Bone Marrow Transplant 1995; 16: 777.

Asano Y, Yoshikawa T, Suga S, Yazaki T, Hirabayashi S, Ono Y, Tsuzuki K, Oshima S. Lancet 1989; 2: 1391.

Bethge W, Beck R, Jahn G, Mundinger P, Kanz L, Einsele H. Bone Marrow Transplant 1999; 24: 1245.

Bollen AE, Wartan AN, Krikke AP, Haaxma-Reiche H. J Neurol 2001; 248: 619.

Bosi A, Zazzi M, Amantini A, Cellerini M, Vannucchi AM, De Milito A, Guidi S, Saccardi R, Lombardini L, Laszlo D, Rossi Ferrini P. Bone Marrow Transplant 1998; 22: 285.

Burd EM, Knox KK, Carrigan DR. Blood 1993; 81: 1645.

Carrigan DR, Drobyski WR, Russler SK, Tapper MA, Knox KK, Ash RC. Lancet 1991; 338: 147.

Carrigan DR, Knox KK. Blood 1994; 84: 3307.
Carrigan DR, Knox KK. Blood 1995; 86: 835.
Caserta MT, Hall CB, Schnabel K, McIntyre K, Long C, Costanzo M, Dewhurst S, Insel R, Epstein LG. J Infect Dis 1994; 170: 1586.
Challoner PB, Smith KT, Parker JD, MacLeod DL, Coulter SN, Rose TM, Schultz ER, Bennett JL, Garber RL, Chang M. Proc Natl Acad Sci USA 1995; 92: 7440.
Chapenko S, Folkmane I, Tomsone V, Kozireva S, Bicans J, Amerika D, Rozentals R, Murovska M. Transplant Proc 2001; 33: 2463.
Cole PD, Stiles J, Boulad F, Small TN, O'Reilly RJ, George D, Szabolcs P, Kiehn TE, Kernan NA. Clin Infect Dis 1998; 27: 653.
Cone RW, Hackman RC, Huang ML, Bowden RA, Meyers JD, Metcalf M, Zeh J, Ashley R, Corey L. N Engl J Med 1993; 329: 156.
Cone RW, Huang ML, Corey L, Zeh J, Ashley R, Bowden R. J Infect Dis 1999; 179: 311.
De Almeida Rodrigues G, Nagendra S, Lee CK, De Magalhaes-Silverman M. Scand J Infect Dis 1999; 31: 313.
de Ona M, Melon S, Rodriguez JL, Sanmartin JC, Bernardo MJ. Transplant Proc 2002; 34: 75.
Deborska D, Durlik M, Sadowska A, Nowacka-Cieciura E, Pazik J, Lewandowski Z, Chmura A, Galazka Z, Paczek L, Lao M. Transplant Proc 2003; 35: 2199.
Dockrell DH, Prada J, Jones MF, Patel R, Badley AD, Harmsen WS, Ilstrup DM, Wiesner RH, Krom RA, Smith TF, Paya CV. J Infect Dis 1997; 176: 1135.
Drobyski WR, Dunne WM, Burd EM, Knox KK, Ash RC, Horowitz MM, Flomenberg N, Carrigan DR. J Infect Dis 1993; 167: 735.
Drobyski WR, Knox KK, Majewski D, Carrigan DR. N Engl J Med 1994; 330: 1356.
Flamand L, Gosselin J, Stefanescu I, Ablashi D, Menezes J. Blood 1995; 85: 1263.
Griffiths PD, Ait-Khaled M, Bearcroft CP, Clark DA, Quaglia A, Davies SE, Burroughs AK, Rolles K, Kidd IM, Knight SN, Noibi SM, Cope AV, Phillips AN, Emery VC. J Med Virol 1999; 59: 496.
Hentrich M, Oruzio D, Jager G, Schlemmer M, Schleuning M, Schiel X, Hiddemann W, Kolb HJ. Br J Haematol 2005; 128: 66.
Herbein G, Strasswimmer J, Altieri M, Woehl-Jaegle ML, Wolf P, Obert G. Clin Infect Dis 1996; 22: 171.
Humar A, Kumar D, Caliendo AM, Moussa G, Ashi-Sulaiman A, Levy G, Mazzulli G. Clinical impact of human herpesvirus-6 infection after liver transplantation. Transplant 2002; 73: 599–604.
Ihira M, Yoshikawa T, Suzuki K, Ohashi M, Suga S, Asonuma K, Tanaka K, Asano Y. Microbiol Immunol 2001; 45: 225.
Imbert-Marcille BM, Tang XW, Lepelletier D, Besse B, Moreau P, Billaudel S, Milpied N. Clin Infect Dis 2000; 31: 881.
Isomura H, Yamada M, Yoshida M, Tanaka H, Kitamura T, Oda M, Nii S, Seino Y. J Med Virol 1997; 52: 406.
Jacobs F, Knoop C, Brancart F, Gilot P, Melot C, Byl B, Delforge ML, Estenne M, Liesnard C. Transplantation 2003; 75: 1996.
Kidd IM, Clark DA, Sabin CA, Andrew D, Hassan-Walker AF, Sweny P, Griffiths PD, Emery VC. Transplantation 2000; 69: 2400.
Lau YL, Peiris M, Chan GC, Chan AC, Chiu D, Ha SY. Bone Marrow Transplant 1998; 21: 1063.

Lautenschlager I, Hockerstedt K, Linnavuori K, Taskinen E. Clin Infect Dis 1998; 26: 702.
Lautenschlager I, Linnavuori K, Hockerstedt K. Transplantation 2000; 69: 2561.
Ljungman P, Wang FZ, Clark DA, Emery VC, Remberger M, Ringden O, Linde A. Br J Haematol 2000; 111: 774.
MacLean HJ, Douen AG. Transplantation 2002; 73: 1086.
Maeda Y, Teshima T, Yamada M, Harada M. Leuk Lymphoma 2000; 39: 229.
Mendez JC, Dockrell DH, Espy MJ, Smith TF, Wilson JA, Harmsen WS, Ilstrup D, Paya CV. J Infect Dis 2001; 183: 179.
Mookerjee BP, Vogelsang G. Bone Marrow Transplant 1997; 20: 905.
Morris DJ, Littler E, Arrand JR, Jordan D, Mallick NP, Johnson RW. N Engl J Med 1989; 320: 1560.
Nash PJ, Avery RK, Tang WH, Starling RC, Taege AJ, Yamani MH. Am J Transplant 2004; 4: 1200.
Nishimaki K, Okada S, Miyamura K, Ohno I, Ashino Y, Sugawara T, Kondo T, Hattori T. Bone Marrow Transplant 2003; 32: 1103.
Okuno T, Higashi K, Shiraki K, Yamanishi K, Takahashi M, Kokado Y, Ishibashi M, Takahara S, Sonoda T, Tanaka K, Baba K, Yabuuchi H, Kurata T. Transplantation 1990; 49: 519.
Paterson DL, Singh N, Gayowski T, Carrigan DR, Marino IR. Liver Transpl Surg 1999; 5: 454.
Paul LC. Kidney Int 1999; 56: 783.
Razonable RR, Rivero A, Brown RA, Hart GD, Espy MJ, van Cruijsen H, Wilson J, Groettum C, Kremers W, Smith TF, Paya CV. Clin Transplant 2003; 17: 114.
Rieux C, Gautheret-Dejean A, Challine-Lehmann D, Kirch C. Agut HVernant JP. Transplantation 1998; 65: 1408.
Rogers J, Rohal S, Carrigan DR, Kusne S, Knox KK, Gayowski T, Wagener MM, Fung JJ, Singh N. Transplantation 2000; 69: 2566.
Ross DJ, Chan RC, Kubak B, Laks H, Nichols WS. Transplant Proc 2001; 33: 2603.
Rossi C, Delforge ML, Jacobs F, Wissing M, Pradier O, Remmelink M, Byl B, Thys JP, Liesnard C. Transplantation 2001; 71: 288.
Sashihara J, Tanaka-Taya K, Tanaka S, Amo K, Miyagawa H, Hosoi G, Taniguchi T, Fukui T, Kasuga N, Aono T, Sako M, Hara J, Yamanishi K, Okada S. Blood 2002; 100: 2005.
Savolainen H, Lautenschlager I, Piiparinen H, Saarinen-Pihkala U, Hovi L, Vettenranta K. Pediatr Blood Cancer 2005; 45: 820–825.
Singh N, Carrigan DR. Annals of Internal Medicine 1996; 124: 1065.
Singh N, Carrigan DR, Gayowski T, Marino IR. Transplantation 1997; 64: 674.
Singh N, Husain S, Carrigan DR, Knox KK, Weck KE, Wagener MM, Gayowski T. Clin Transplant 2002a; 16: 92.
Singh N, Bentlejewski C, Carrigan D, Gayowski T, Knox K, Zeevi A. Transpl Infect Dis 2002b; 4: 59–63.
Singh N, Paterson DL. Transplantation 2000; 69: 2474.
Tiacci E, Luppi M, Barozzi P, Gurdo G, Tabilio A, Ballanti S, Torelli G, Aversa F. Haematologica 2000; 85: 94.
Tokimasa S, Hara J, Osugi Y, Ohta H, Matsuda Y, Fujisaki H, Sawada A, Kim JY, Sashihara J, Amou K, Miyagawa H, Tanaka-Taya K, Yamanishi K, Okada S. Bone Marrow Transplant 2002; 29: 595.

Tong CY, Bakran A, Peiris JS, Muir P, Herrington CS. Transplantation 2002; 74: 576.
Tsujimura H, Iseki T, Date Y, Watanabe J, Kumagai K, Kikuno K, Yonemitsu H, Saisho H. Eur J Haematol 1998; 61: 284.
Wang FZ, Dahl H, Linde A, Brytting M, Ehrnst A, Ljungman P. Blood 1996; 88: 3615.
Wang FZ, Larsson K, Linde A, Ljungman P. Bone Marrow Transplant 2002; 30: 521.
Wang FZ, Linde A, Hagglund H, Testa M, Locasciulli A, Ljungman P. Clin Infect Dis 1999; 28: 562.
Yoshida H, Matsunaga K, Ueda T, Yasumi M, Ishikawa J, Tomiyama Y, Matsuzawa Y. Int J Hematol 2002; 75: 421.
Yoshikawa T, Asano Y, Ihira M, Suzuki K, Ohashi M, Suga S, Kudo K, Horibe K, Kojima S, Kato K, Matsuyama T, Nishiyama Y. J Infect Dis 2002; 185: 847.
Yoshikawa T, Ihira M, Ohashi M, Suga S, Asano Y, Miyazaki H, Hirano M, Suzuki K, Matsunaga K, Horibe K, Kojima S, Kudo K, Kato K. Matsuyama TNishiyama Y. Bone Marrow Transplant 2001; 28: 77.
Yoshikawa T, Ihira M, Suzuki K, Suga S, Asano Y, Asonuma K, Tanaka K, Nishiyama Y. J Pediatr 2001; 138: 921.
Yoshikawa T, Ihira M, Suzuki K, Suga S, Iida K, Saito Y, Asonuma K, Tanaka K, Asano Y. J Med Virol 2000; 62: 52.
Yoshikawa T, Suga S, Asano Y, Nakashima T, Yazaki T, Ono Y, Fujita T, Tsuzuki K, Sugiyama S, Oshima S. Transplantation 1992; 54: 879.
Yoshikawa T, Suga S, Asano Y, Nakashima T, Yazaki T, Sobue R, Hirano M, Fukuda M, Kojima S, Matsuyama T. Blood 1991; 78: 1381.
Zerr DM, Corey L, Kim HW, Huang ML, Nguy L, Boeckh M. Clin Infect Dis 2005; 40: 932.
Zerr DM, Gooley TA, Yeung L, Huang ML, Carpenter P, Wade JC, Corey L, Anasetti C. Clin Infect Dis 2001; 33: 763.
Zerr DM, Gupta D, Huang ML, Carter R, Corey L. Clin Infect Dis 2002; 34: 309.

Human Herpesvirus-6: General Virology, Epidemiology and Clinical Pathology
Perspectives in Medical Virology
Gerhard Krueger and Dharam Ablashi (Editors)

DOI 10.1016/S0168-7069(06)12023-6

Therapeutic Approaches to HHV-6 Infection

Lieve Naesens, Leen De Bolle, Erik De Clercq
Faculty of Pharmaceutical Sciences, Ghent University, B-9000 Gent, Belgium

Introduction

In contrast to other human herpesviruses such as herpes simplex virus (HSV) or cytomegalovirus (CMV), HHV-6 has not been the subject of extensive antiviral screening, the main reason being the uncertainty about the large need for specific anti-HHV-6 therapies. Transplant recipients commonly show laboratory signs of HHV-6 reactivation, but the frequency by which this is associated with serious disease is still ill defined (Yoshikawa, 2004). Even more controversial is the role of HHV-6 in chronic neurological disorders, such as multiple sclerosis (MS) or chronic fatigue syndrome (CFS). In the absence of HHV-6-specific therapies, treatment of HHV-6 infections currently relies on the relatively broad-spectrum antiherpetic agents (val)ganciclovir and foscarnet. Although these drugs offer an indisputable benefit in the therapy or prophylaxis of CMV, their clinical efficacy against HHV-6 can only be estimated from a number of heterogeneous case reports. Long-term administration of these antivirals, as would be required in chronic diseases, is expected to cause dose-limiting side effects, and hence anti-HHV-6 drugs with a better safety and a new mode of action are urgently needed.

Antiviral assays for HHV-6

Development of new antiviral drugs requires a relevant cell culture system for virus replication. In case of HHV-6, some investigators have used the continuous T-lymphoblasts HSB-2 and Molt-3, which are the preferred host cell lines for propagation of HHV-6A or HHV-6B strains, respectively. Other T-lymphoblast lines such as SupT-1 and MT4 are less efficient in supporting HHV-6 replication (De Bolle et al., 2005b). The continuous nature of these T-cell lines ensures that the antiviral data are consistent. On the other hand, the high metabolic rate of these rapidly dividing cell lines increases their sensitivity to the cytotoxic effects exerted by the antiviral test compounds. In addition, these tumor cells may carry mutations that affect basic pathways such as nucleoside metabolism, signal transduction or cell cycle regulation. For antiviral compounds that depend on cellular factors for their activation or antiviral target, these mutations may result in cell-type-dependent antiviral effects. Therefore, primary human lymphocytes isolated from peripheral blood or cord blood provide a more relevant test system to confirm the anti-HHV-6 activity and selectivity of new compounds. Antiviral evaluation in primary or continuous cells from neuroglial origin (i.e. oligodendrocytes, microglia or astrocytes) is not commonly performed, since these cell types show a relatively inefficient replication of laboratory strains of HHV-6 (De Bolle et al., 2005b). Clinical HHV-6 isolates may show a higher neurotropism and be better suited to study the effect of antiviral compounds in these clinically relevant cell types.

Different methods can be used to estimate the inhibitory effect of antiviral compounds on HHV-6 replication. Though relatively easy and inexpensive, microscopic examination of the characteristic cytopathic effect (CPE) of HHV-6 (consisting of large ballooning cells or syncytia) may be subjective to certain extent. A definite proof of anti-HHV-6 activity requires an assay to directly measure the compound's inhibitory effect on HHV-6 replication. In antigen-based assays, HHV-6 proteins are detected with HHV-6-specific monoclonal or polyclonal antibodies. After indirect immunofluorescence staining, the fluorescence intensity in the HHV-6-infected cells is quantified by fluorescence microscopy or fluorescence-activated cell sorter (FACS) analysis (Agut et al., 1989, 1991; Reymen et al., 1995; Manichanh et al., 2000). Alternatively, HHV-6 antigen detection may be performed in an enzyme-linked immunoassay (EIA) (Takahashi et al., 1997) or by immunoblotting ('dot-blot') assay (Yoshida et al., 1998). Unfortunately, the choice of commercial antibodies for HHV-6 is limited. DNA-based assays with custom-made primers or probes may therefore be preferred. We have good experience with a non-radioactive DNA hybridization assay to quantify viral DNA in HHV-6-infected cells (De Clercq et al., 2001). This technique can now be replaced by real-time PCR.

Care should be taken when comparing antiviral data for HHV-6 from different laboratories. The antiviral activity and selectivity may depend on the HHV-6 strain and cell system used. No standardization exists as to what should be the viral load ('multiplicity of infection'). Some compounds loose their anti-HHV-6 activity at higher viral loads. Also, the activity may depend on the assay used to follow HHV-6 replication. For instance, compounds that act at a late stage in the viral replication cycle (such as inhibitors of virus maturation) may show good activity in a CPE assay, but have a minor effect on viral DNA content.

Classical viral DNA polymerase inhibitors

Mechanism of action

Since the discovery of acyclovir, inhibitors of the viral DNA polymerase have been the mainstay in anti-herpesvirus therapy. For the acyclic nucleoside analogs acyclovir and ganciclovir, inhibition of viral DNA synthesis is carried out by their triphosphate metabolite, which is formed in three consecutive phosphorylations (De Bolle et al., 2005a). The first is catalyzed by a virus-encoded kinase such as the HSV thymidine kinase, or the protein kinase encoded by the CMV UL97 gene or the HHV-6 U69 gene. The second and third phosphorylation steps are carried out by cellular enzymes. The resulting triphosphate metabolite inhibits the herpesvirus DNA polymerase by competition with the natural substrate dGTP, by incorporation in the growing DNA (effecting chain termination in case of absence of a hydroxyl group in the acyclic side chain) and/or by dead-end complex (Reardon and Spector, 1989). The selectivity of the acyclic nucleoside analogs relies on a greater affinity of their triphosphates for the viral DNA polymerase compared to cellular polymerases and, more importantly, their selective phosphorylation by a viral kinase. In comparison, the acyclic nucleoside phosphonate analog cidofovir does not require activation by a viral kinase, explaining its broad-spectrum anti-DNA virus activity. Due to the presence of the phosphonate moiety, two subsequent phosphorylations by cellular enzymes are sufficient for cidofovir to reach its active state. Cidofovir diphosphate acts as a competitive inhibitor for dCTP and an alternative substrate for the herpesvirus DNA polymerase. Its selectivity is based on its higher affinity for the viral DNA polymerase compared to cellular DNA polymerases. This is also the case for the pyrophosphate analog foscarnet, for which the inhibition of the herpesvirus DNA polymerase is based on its reversible binding to the pyrophosphate-binding site of the enzyme. Consequently, cleavage of pyrophosphate during incorporation of dNTPs is prevented and viral DNA synthesis is terminated.

In vitro *data*

As summarized in Table 1, several groups have determined the anti-HHV-6 activity of acyclovir, ganciclovir, cidofovir, and foscarnet in cell culture. In general, acyclovir appeared to have poor activity and selectivity, while foscarnet was consistently shown to have a pronounced effect on HHV-6 replication with relatively high selectivity. In contrast, marked differences can be seen in the data reported for ganciclovir and cidofovir, reflecting variations in the experimental conditions (Table 1). Both compounds are considerably more active and less cytotoxic in primary human lymphocytes than in continuous T-cell lines.

In case of ganciclovir, the anti-HHV-6 activity is diminished by two factors: inefficient phosphorylation of ganciclovir by the HHV-6 pU69 protein kinase and relatively low affinity of ganciclovir triphosphate for the HHV-6 DNA polymerase.

Table 1

Activity of viral DNA polymerase inhibitors against HHV-6 in cell culture

Cell system[a]	Assay[b]	Virus strain (variant)	Antiviral EC_{50}[c] (μM)				Selectivity index[d]				Reference
			ACV[e]	GCV[e]	CDV[e]	PFA[e]	ACV[e]	GCV[e]	CDV[e]	PFA[e]	
Studies using primary lymphocytes											
PBL	IFA	SIE (A)	27	1.1	ND	9	ND	ND	ND	ND	Agut et al. (1989)
PBL	IFA	Various A and B strains	12–32	1–2.5	ND	10–22	ND	ND	ND	ND	Agut et al. (1991)
CBL	EIA	U1102 (A)	81	1.7	19	10	2	94	10	89	Takahashi et al. (1997)
		HST (B)	66	1.9	20	13	3	84	9	69	
CBL	Dot-blot	Various (A)	26	2.6	0.9	37	33	308	228	18	Yoshida et al. (1998)
	DNA	Various (B)	100	5.3	3.8	38	9	150	53	8	
CBL	DNA hyb	GS (A)	10	5.8	0.56	9.5	28	17	182	68	De Bolle et al. (2004a)
Studies using continuous T-cell lines											
MT4	FACS	HST (B)	25	6.4	0.95	6	4	14	53	58	Manichanh et al. (2000)
HSB-2	IFA	GS (A)	59	25	ND	49	4	8	ND	31	Akesson-Johansson et al. (1990)
HSB-2	FACS	GS (A)	270	>25	11	5.8	3	ND	13	141	Reymen et al. (1995)
HSB-2	DNA hyb	GS (A)	180	32	9.0	16	4	<2	3	78	De Bolle et al. (2004a)
Molt-3		Z29 (B)	185	69	9.8	25	1	<1	6	40	

Source: Adapted from Manichanh et al. (2000).

[a]PBL, peripheral blood lymphocytes; CBL, cord blood lymphocytes.

[b]IFA, immunofluorescence assay; EIA, enzyme-linked immunoassay; FACS, fluorescence-activated cell sorter analysis; DNA hyb, DNA hybridization.

[c]EC_{50}, compound concentration producing 50% inhibition of virus replication.

[d]Selectivity index, or ratio of cytotoxic concentration over EC_{50}.

[e]ACV, acyclovir; GCV, ganciclovir; CDV, cidofovir; PFA, foscarnet. ND: No data reported.

The role of HHV-6 pU69 in the phosphorylation of ganciclovir was first demonstrated in insect cells infected with a pU69-recombinant baculovirus (Ansari and Emery, 1999). Our studies in human cells infected with recombinant vaccinia viruses showed that the capacity of HHV-6 pU69 to phosphorylate ganciclovir is 10-fold lower as compared to that of HCMV pUL97 (De Bolle et al., 2002). An $M^{318}V$ substitution in HHV-6 pU69 was found to arise after prolonged exposure of HHV-6-infected cells to ganciclovir, and was also detected in a patient (see below), supporting the crucial role of HHV-6 pU69 in the activation of ganciclovir (Manichanh et al., 2001). Likewise, several mutations introduced in the HHV-6 U69 gene were shown to impair ganciclovir phosphorylation in a baculovirus system (Safronetz et al., 2003). Unfortunately, this has not been confirmed in an enzymatic assay using purified HHV-6 pU69 protein. In addition, there are no data on the natural role of this protein kinase in the viral replication cycle. HHV-6 pU69 is expressed in the cell nucleus alike the CMV pUL97 protein (De Bolle et al., 2002), which is known to be indispensable for CMV replication due to a proposed role in CMV DNA synthesis, nuclear egress and capsid assembly, and/or regulation of CMV gene expression (Coen and Schaffer, 2003).

Few biochemical data are available on the effect of these inhibitors at the level of the HHV-6 DNA polymerase. Like all herpesvirus DNA polymerases, the HHV-6 DNA polymerase is a heterodimer consisting of a catalytic subunit (encoded by the HHV-6 U38 gene) and a processivity factor (encoded by HHV-6 U27 gene), which allows synthesis of long DNA stretches without dissociation of pU38 from its DNA template. Using heterodimeric enzymes purified from infected cells, Bapat et al. (1989) determined that the HHV-6 DNA polymerase was 6-fold less sensitive to inhibition by ganciclovir triphosphate than the CMV enzyme, and 800-fold less sensitive to acyclovir triphosphate than HSV DNA polymerase. We have performed enzyme inhibition studies with the pU38 catalytic subunit of HHV-6 DNA polymerase that was prepared by coupled transcription–translation from an expression plasmid containing the HHV-6 U38 gene (De Bolle et al., 2004b). Site-directed mutagenesis on this plasmid provides a rapid tool to determine the inhibitor sensitivity of mutant forms of the HHV-6 DNA polymerase, as new mutations in the HHV-6 U38 gene may be identified during prolonged use of these inhibitors in HHV-6-infected patients (Manichanh et al., 2001).

Clinical data

Primary HHV-6 infections in immunocompetent children are self-limiting and do not require antiviral treatment. In immunocompromised persons, such as transplant recipients receiving immunosuppressive therapy, HHV-6 reactivation can cause life-threatening complications, mainly due to CNS disease. The drugs used for these HHV-6 infections are the same as those used for CMV therapy. No controlled trials of antiviral therapy against HHV-6 have yet been conducted. Thus, conclusions on the efficacy of antiherpetic drugs rely on a compilation of case studies (reviewed in Yoshikawa, 2004; De Bolle et al., 2005a). The interpretation is difficult due to the heterogeneous clinical background of the patients (i.e. bone marrow, stem cell, or solid-organ transplant recipients), and the differences in drug-dosing schedules and criteria used to estimate drug activity (i.e. based on viral load reductions or clinical improvement). Most case reports favor the effectiveness of ganciclovir although some cases of fulminant HHV-6 infection showed no response. The oral prodrug valganciclovir (the valine ester of ganciclovir) offers an obvious advantage over intravenous ganciclovir, yet both the oral and intravenous formulations can cause severe neutropenia as a result of bone marrow suppression. Acyclovir and its oral prodrug valacyclovir have an excellent therapeutic index, but, unfortunately, appear to be ineffective against HHV-6. Foscarnet, which is used as second-line therapy for CMV, appears to be effective against HHV-6, but can cause renal toxicity. A few authors reported suppression of HHV-6 after combined or successive use of ganciclovir and foscarnet. No reports are yet available on the clinical efficacy of cidofovir against HHV-6. In order to prevent nephrotoxicity associated with cidofovir therapy, sufficient hydration and co-administration of the anion transport inhibitor probenecid are mandatory.

While the studies mentioned above refer to therapies initiated at the onset of HHV-6 disease, the beneficial effects of prophylactic antiviral administration are less evident. Long-term use of these potentially toxic antiviral drugs may cause severe side effects and facilitate emergence of drug-resistant HHV-6 mutants. Preemptive antiviral therapy (i.e. initiated after laboratory detection of the virus, but prior to disease onset) may be preferred. Only one clinical report has been published on resistance of HHV-6 to antiviral compounds (Manichanh et al., 2001). This HHV-6 mutant was isolated from an AIDS patient after long-term ganciclovir therapy, and was shown to carry a mutation in the HHV-6 U69 gene. HHV-6 mutants with changes in the HHV-6 DNA polymerase have been detected *in vitro* but not (yet) in patients.

Antiviral studies in MS patients

Clinical trials with antiviral therapeutics may be helpful in solving the controversy about the role of HHV-6 in MS. Three placebo-controlled double-blind studies have been performed in MS patients to investigate the efficacy and safety of long-term (up to two years) therapy with acyclovir or its oral prodrug valacyclovir. Despite its relatively weak activity against HHV-6, acyclovir was preferred to other antiherpetic drugs because of its excellent safety profile. Still, levels of acyclovir obtained in the cerebrospinal fluid are about 5-fold lower than in serum, which is probably too low to adequately suppress HHV-6 replication in the CNS. In one study, acyclovir reduced the exacerbation rate in relapsing–remitting MS patients, yet did not affect on the overall neurological deterioration (Lycke et al., 1996). In another study, valacyclovir effected a significant reduction in new lesion formation in patients with high disease activity (Bech et al., 2002). A similar trend toward disease stabilization in severe MS was seen in a recent valacyclovir trial, although drug effects were not significant in this study (Friedman et al., 2005).

New nucleoside analogs

The clinical successes with antiviral nucleoside analogs have evoked the chemical synthesis and antiviral evaluation of a vast number of new derivatives. Although a high-throughput screening for HHV-6 remains to be established, small-scale antiviral studies have revealed a number of experimental nucleoside analogs that are worth mentioning here (De Clercq et al., 2001).

S2242 is an N7-substituted purine derivative with an acyclic side chain, which is independent of a viral kinase for its conversion to the active triphosphate, explaining its broad-spectrum of anti-DNA virus activity. In cell culture, S2242 is active against HHV-6 at very low concentrations (EC_{50} ~0.004 μg/ml) that are well below the concentrations producing toxicity (selectivity index ~80) (De Clercq et al., 2001).

A-5021 is a nucleoside analog with a cyclopropyl entity in the sugar portion of its structure. It is phosphorylated by HSV thymidine kinase and is in preclinical development for HSV keratitis. A-5021 also has a favorable and selective *in vitro* activity against other human herpesviruses, including HHV-6, suggesting that it may be useful as a broad-spectrum anti-herpesvirus compound (Neyts et al., 2001).

A good activity against HHV-6 has been obtained with the methylenecyclopropane analog cyclopropavir, which is under development for CMV therapy (Kern et al., 2005). Cyclopropavir is phosphorylated by CMV pUL97, and, possibly, also by the homologous HHV-6 pU69 protein kinase. Hence, there is a possibility that ganciclovir-resistant virus isolates with mutations in the CMV UL97 or HHV-6 U69 gene may show cross-resistance to cyclopropavir.

Several lipophilic ester prodrugs of cidofovir [i.e. hexadecyloxypropyl cidofovir (HDP-CDV)] have been synthesized to overcome the low oral bioavailability of cidofovir (which requires intravenous administration) (Beadle et al., 2002). After surviving first-pass degradation by the liver, these cidofovir esters are transported to the target cells, in which they show a higher uptake than cidofovir. Once inside the cells, the esters are cleaved to free cidofovir. In addition, these cidofovir esters appear to have a lower potential for renal toxicity than cidofovir. Although current interest in these cidofovir esters stems from their potential use in the treatment of poxvirus infections, their future development for treatment of herpesvirus (including HHV-6) infections may be warranted (Beadle et al., 2002).

Another class of interesting new antivirals is represented by the acyclic nucleoside phosphonate derivatives of 2,4-diaminopyrimidine (Balzarini et al., 2004). Our ongoing studies on their activation pathway, mode of action and therapeutic safety *in vitro* and *in vivo* should demonstrate their potential as antiviral agents for the treatment of DNA viruses, including HHV-6.

New targets for anti-herpesvirus therapy

The 4-oxo-dihydroquinoline-3-carboxamide PNU-183792 is the prototype of a new class of non-nucleoside inhibitors of herpesvirus DNA polymerases (Brideau et al.,

2002). In enzyme assays with HSV DNA polymerase, these compounds inhibit the enzyme by competition with the dNTPs. The region in the viral polymerase with which they interact is highly conserved in the human herpesviruses, except in HHV-6 (and HHV-7), and this explains their lack of activity against HHV-6. Nevertheless, these compounds are a proof of concept that non-nucleoside inhibitors of herpesvirus DNA polymerases could a priori be developed.

We found several arylsulfone derivatives to have potent and relatively selective *in vitro* activity against HHV-6 and CMV (Naesens et al., 2001). Mechanistic studies suggested that these compounds may suppress HHV-6 DNA synthesis. However, since no direct inhibition was seen in an enzyme assay with the catalytic subunit of the HHV-6 DNA polymerase, these arylsulfones seem to interact with another viral protein involved in viral DNA synthesis. An original approach to inhibition of viral DNA synthesis is seen with small molecules that disrupt the protein–protein complex between the herpesvirus DNA polymerase and its accessory protein (the so-called 'processivity factor'). High-throughput screening has already led to some promising lead structures for HSV or CMV therapy (Loregian and Palu, 2005).

The herpesvirus helicase–primase is another protein complex that is indispensable for herpesvirus DNA synthesis. It is composed of three viral proteins, which act together to unwind the double-stranded viral DNA and generate primers for viral DNA synthesis. Three classes of inhibitors of herpesvirus helicase–primase have been reported (prototypes: BAY 57-1293, BILS 179 BS and T-0902611) (Coen and Schaffer, 2003). Their oral bioavailability, and low toxicity and superior activity in animals may give them an advantage over existing anti-herpesvirus therapies. Compound BAY 57-1293 is in Phase I clinical trials for HSV therapy. Its activity against CMV or HHV-6 has not been reported. Compound T-0902611 exhibits potent activity against CMV in cell culture, but no information is available on its activity against HHV-6. This compound was discontinued after initial clinical trials.

Maribavir (1263W94) is another promising anti-CMV compound that has advanced to Phase I/II trials (Williams et al., 2003). It specifically targets the CMV pUL97 protein kinase, which, as already explained, is indispensable for virus growth but whose precise role in the replication cycle is still disputed. The observation that maribavir is inactive against HHV-6 in cell culture suggests that HHV-6 pU69 is functionally different from CMV pUL97, as also supported by our own data that these two protein kinases have a different capacity to phosphorylate ganciclovir (De Bolle et al., 2002). Likewise, the indolocarbazole Gö 6976 was found to inhibit CMV pUL97, explaining its *in vitro* activity against CMV, yet was proven to be inactive against HHV-6 (De Bolle et al., 2004a).

Although the benzimidazole riboside 5,6-dichloro-2-bromo-1-beta-D-ribofuranosylbenzimidazole (BDCRB) is structurally related to maribavir, it has a completely different mode of action. 5,6-dichloro-2-bromo-1-beta-D-ribofuranorylbenzimidazole (BDCRB) acts at a later stage in the CMV replication cycle by inhibition of the viral 'terminase' complex that is involved in encapsidation and cleavage of viral DNA during virus maturation (Williams et al., 2003). BDCRB was found to be metabolically instable in animals, and clinical trials are now focusing on its derivative GW-273175X. The phenylenediamine sulfonamide BAY 38-4766 is another CMV maturation inhibitor under clinical investigation. Unfortunately, neither BDCRB nor BAY 38-4766 showed *in vitro* activity against HHV-6 (Reefschlaeger et al., 2001; Williams et al., 2003).

All the above-mentioned compounds target a virus-specific protein, explaining their favorable selectivity. In the clinical setting, long-term therapy with these antiviral drugs may favor emergence of drug-resistant virus mutants, resulting in virus rebound and/or inadequate clinical response. This explains the increasing interest in antiviral compounds targeted at a cellular protein that is of crucial importance for virus replication. The central role of diverse cellular protein kinases in the regulation of cell replication and differentiation has boosted an enormous investment in protein kinase inhibitors, mainly directed to anti-cancer or immunosuppressive therapy. However, specific protein kinase inhibitors may also find their way to anti-herpesvirus therapy. For example, the cyclin-dependent kinase inhibitor roscovitine, and the protein tyrosine kinase inhibitors leflunomide and herbimycin have been reported to have some *in vitro* activity against human herpesviruses, although their selectivity appears to be low (Schang, 2002; De Bolle et al., 2004a). Development of more selective compounds will require a better knowledge of the cellular and/or viral protein kinases involved in regulation of herpesvirus gene expression. This should aid to unravel the precise mode of action of several new compounds with ill-defined anti-herpesvirus activities (reviewed by Coen and Schaffer, 2003; De Clercq, 2003). We have reported on the potent and remarkably selective activity of compound CMV423 against CMV and HHV-6 (selectivity index in HHV-6-infected cord blood lymphocytes: >1700) (De Bolle et al., 2004a). We provided evidence that CMV423 indirectly inhibits one or more cellular protein tyrosine kinases, but were thus far unable to identify its precise target. The activity of the HHV-6 pU69 protein kinase was not affected. CMV423 was shown to act on a step prior to HHV-6 DNA synthesis, and to lose its antiviral activity in certain T-cell lines or at high viral loads. However, it seems unlikely that these high viral loads are relevant for the *in vivo* situation. We thus consider CMV423 to be one of the most promising anti-HHV-6 compounds reported to this date.

Conclusions

The relatively broad-spectrum antiherpetic agents (val)ganciclovir and foscarnet appear to have some efficacy against HHV-6 infections in transplant recipients. Unfortunately, their potential for serious side effects excludes long-term use of these drugs in patients with chronic HHV-6-associated disorders. In order to develop safer anti-HHV-6 drugs, approaches other than 'classical' inhibitors of the viral DNA polymerase should also be explored. The feasibility of this strategy is shown by the recent discoveries of new leads for HSV or CMV therapy, some of

PART IV: NEW APPROACHES TO HHV-6 RESEARCH

Human Herpesvirus-6: General Virology, Epidemiology and Clinical Pathology
Perspectives in Medical Virology
Gerhard Krueger and Dharam Ablashi (Editors)

DOI 10.1016/S0168-7069(06)12024-8

Animal Models

Claude P. Genain
Department of Neurology, University of California, 513 Parnassus Avenue, San Francisco, CA 94143-0435, USA

Summary

Human herpesvirus-6 (HHV-6) has been implicated in the etiology of multiple sclerosis (MS), and epidemiological studies indeed suggest that viruses may trigger MS or influence its course. A direct link of causality to MS has so far been lacking, however, and the question of a causal relationship between HHV-6 and MS can only be resolved by studying experimental systems that recapitulates the pathogenesis of disease following exposure to this ubiquitous virus under specific conditions.
A laboratory model of HHV-6-induced central nervous system (CNS) demyelination has been created in the common marmoset *Callithrix jacchus jacchus*, a New World non-human primate that develops spontaneous autoimmunity and is also used in studies of experimental allergic encephalomyelitis. CNS autoimmune demyelination appears associated with repeated exposures of adult marmosets to HHV-6A. Preliminary evidence indicates that this HHV-6 variant is potentially capable of inducing apoptosis in CNS glial cells, and that chronic disease ensues with appearance of T-cell reactivity to myelin antigens. This novel model will now allow the definition of factors of pathogenesis for MS lesions, a preamble to the discovery of strategies to remedy HHV-6-induced pathology.

Introduction

Multiple sclerosis (MS) designates a group of immune-mediated chronic demyelinating disorders of the central nervous system (CNS) affecting 350,000

Americans and over 1 million individuals worldwide. MS affects women twice as often as men, and thus also represents a significant women's health issue. Pathologically, MS is characterized by plaques of perivascular infiltration comprised of mononuclear cells and macrophages, accompanied by concentric destruction of the myelin sheaths (demyelination), death of oligodendrocytes, proliferation of astrocytes, and axonal damage (Raine, 1983; Lassmann, 1998; Trapp et al., 1998). Studies of the immunopathology of MS lesions suggest that heterogeneity of mechanisms is present, as specific patterns of lesion with various proportions of inflammation, demyelination, and oligodendrocyte and axonal pathology have been described (Lassmann, 1998; Storch et al., 1998; Lucchinetti et al., 2000). Effector mechanisms of tissue damage in MS include direct toxicity of infiltrating T cells, secretion of pro-inflammatory cytokines, antibody-mediated toxicity, and complement and macrophage activation (reviewed in Brosnan and Raine, 1996).

Strong circumstantial evidence suggests that MS is an autoimmune disorder arising in a genetically susceptible host under the pressure of environmental triggers (Oksenberg and Hauser, 1996; Hohlfeld, 1997). A viral etiology has been long suspected, based on epidemiologic studies (Kurtzke and Hyllested, 1986; Kurtzke, 1993) and CNS demyelinating diseases that occur in the context of infection with neurotropic viruses (Gilden et al., 1996; Raine, 1997a; Stohlman and Hinton, 2001). The association between certain viral infections or vaccinations (for example, measles, varicella zoster, vaccinia, Epstein–Barr virus (EBV), and HTLV-I) and cases of acute disseminated encephalomyelitis, encephalitis, or myelitis is well recognized. It is also widely perceived that viral infections may trigger MS attacks. Higher antibody titers against neurotropic viruses are reported for MS serum or cerebrospinal fluid (CSF) compared to controls (Johnson et al., 1984; Johnson, 1994). The presence of an antigen-driven CNS restricted immune response in MS owing to underlying viral infections of the CNS is supported by findings of specific oligoclonal bands in patients' CSF (Tourtelotte et al., 1980). Such clonally expanded antibodies are directed against viral antigens in viral encephalitis (Gilden et al., 1996). The specificity of oligoclonal bands in MS has not been firmly established although they may react to some component of EBV (Cepok et al., 2005). The number of viruses that have been incriminated in MS pathogenesis is constantly growing, and in fact interferon (IFN)-γ was first tried as a treatment for MS owing to its antiviral activity.

Existing models of viral pathogenesis

One difficulty in establishing direct relationships between viral exposures and MS is that appropriate *in vivo* experimental systems for validation of such associations are lacking. Examples of virus capable of inducing acute or chronic demyelinating disease include canine distemper virus, the JHM strain of mouse hepatitis virus, murine Semliki Forest virus, sheep Visna, caprine arthritis–encephalitis virus, SV40 in macaque monkeys, Theiler's murine encephalomyelitis virus (TMEV) (Johnson, 1994), and lymphocytic choriomeningitis virus (LCMV) (Evans et al., 1996). Viral proteins can also be expressed in the CNS of transgenic mice, which renders the animals susceptible to infection (Evans et al., 1996). Disease pathogenesis varies between these models, and may include a component linked to viral persistence (monophasic disease), or secondary CNS inflammation and destruction not associated with virus infestation. Infection of mice with TMEV produces a gastroenteritis, which is rapidly cleared. Only inbred susceptible strains subsequently develop an unrelenting, and severe progressive demyelinating disease with what is believed to be the "bystander" damage to myelin (Dal Canto et al., 1995; Stohlman and Hinton, 2001). Infection of mice with LCMV produces a cytotoxic, $CD8^+$-mediated response that directly destroys CNS cell targets. It is important to understand that although these models have provided the first (and only existing) insights into the relationships between autoimmune CNS demyelination and viral infections, they are still insufficient for proving direct association of MS with any of the viruses that ubiquitously infect humans without adverse consequences. CNS complications of TMEV infections are under the restrictive control of genetic influence and it is difficult to extrapolate their mechanisms to outbred human populations. Many of these disease models require intracranial injection of viruses in neonatal animals, an artificial situation that, similar to experimental allergic encephalomyelitis (EAE), does not mimic natural exposure of humans to pathogens.

Epidemiology of HHV-6 infections

Human herpesvirus-6 (HHV-6) is a β-herpesvirus that possesses a 159–170 kbp long genome with seven gene blocks common to all *Herpesviridae*. There are two distinct HHV-6 variants, designated A and B. HHV-6B contains 119 open reading frames (ORFs) in comparison with 110 for HHV-6A (Dockrell, 2003). Despite being very similar, the two HHV-6 variants have very different cell tropism and disease manifestations, which support the concept that they are different herpesviruses.

HHV-6B causes *exanthema subitum* in children, or initial exposure may be asymptomatic. Practically, all individuals get infected prior to age 2 (Caserta et al., 2004; Zerr et al., 2005). HHV-6B is found in a wide variety of tissues including lymphoid organs, brain, serum, and salivary glands (Ablashi et al., 1988; Levy et al., 1990a,b; Lusso and Gallo, 1995a). HHV-6A has a particular tropism for the CNS and skin (Hall et al., 1998). This variant has so far rarely been isolated or detected in children with primary HHV-6 infection, and is not clearly associated with any infectious illness in healthy populations. HHV-6 persists in latent or replicative states throughout life, and actively replicates in salivary glands (variant B), which probably accounts for the route of transmission. Secondary infection by HHV-6 is usually silent except in immuno-compromised patients (Campadelli-Fiume et al., 1999; Dockrell, 2003). Antibodies against HHV-6 are found in most of the general population, and steadily persist through life before declining in older subjects (Levy, 1997).

The CD46 cellular HHV-6 receptor (Santoro et al., 1999) is expressed ubiquitously including in CNS, but only in humans and certain higher mammals and primates, which explains the narrow range of species that can be infected with this virus. CD46 binds to the C3b and C4b proteins and inactivates the complement system, thus one of its presumed functions is to protect the cells from self-lysis by complement. HHV-6 is capable of infecting $CD4^+$, $CD8^+$, NK, and $\gamma\delta$ T cells, B cells, macrophages, dendritic cells, fibroblasts, epithelial cells, and a variety of lymphoid or CNS-derived cell lines (Lusso et al., 1987; Ablashi et al., 1991; Levy, 1997; Campadelli-Fiume et al., 1999; Dockrell, 2003). Both variants infect primary fetal astrocytes, but HHV-6A appears to have a greater neurotropism *in vivo* (Hall et al., 1998). Infection *in vitro* by HHV-6 is monophasic and generally followed by decreased cell proliferation, immune suppression and/or cell death (Grivel et al., 2003; Opsahl and Kennedy, 2005; Smith et al., 2005). The CD46 receptor is shared by a number of pathogens including measles virus, and signaling through this molecule is one of the most potent mechanisms of T-cell stimulation and activation. Several isoforms of CD46 that differ by their cytoplasmic domains are expressed in humans, and engagement of these two classes of CD46 receptors appears to have opposite consequences on the polarization of the immune response toward Th1 or Th2 phenotypes (Marie et al., 2002; Riley-Vargas et al., 2004; Russell, 2004).

In vivo, HHV-6 induces CD4 T-cell depletion, as shown in a severe common immune deficiency (SCID) mouse model implanted with human fetal thymus and liver (Gobbi et al., 1999, 2000), and may contribute to human immunodeficiency virus (HIV)-associated immunosuppression (Lusso and Gallo, 1995b). It is also clear that HHV-6 infection interferes with other viruses, including EBV, cytomegalovirus (CMV), and HIV for which either enhancing or suppressing effects have been described (Ablashi et al., 1988; Levy, 1997). A number of attempts have been made to create models of infection with HHV-6 in primates (macaques and chimpanzees), which as in the SCID mouse model primarily support the concept that HHV-6 acts as a cofactor in the simian acquired immunodeficiency syndrome (Lusso et al., 1990, 1994).

HHV-6 and MS

An association between HHV-6 and MS was recently suggested by findings of HHV-6B DNA sequences in diseased oligodendrocytes within MS plaques (Challoner et al., 1995; Opsahl and Kennedy, 2005). This observation, however, has not been confirmed by subsequent attempts (Coates and Bell, 1998), and could not be formally confirmed by immunohistochemistry. HHV-6 DNA has also been found in the brain of normal subjects and in Alzheimer's disease (Luppi et al., 1995; Lin et al., 2002). Serologic studies have reported elevated titers of anti-HHV-6 antibodies in patients with relapsing remitting MS compared to controls (Sola et al., 1993; Soldan et al., 1997; Ablashi et al., 1998). However, a large number of subsequent studies that examined IgG/IgM reactivity in serum and/or CSF, the presence of HHV-6 DNA or viral transcripts in serum, CSF, and brain, peripheral T-cell proliferative responses to HHV-6, or virus recovery in culture have not unequivocally confirmed these results (Ablashi et al., 1988, 1998; Krueger and

Sander, 1989; Enbom, 2001; Moore and Wolfson, 2002; Krueger and Ablashi, 2003; DeRanieri et al., 2004; Dewhurst, 2004; Fotheringham and Jacobson, 2005)

More compelling for an association between HHV-6 and certain forms of CNS demyelination which possibly represent extremes of the spectrum of MS presentations are numerous case reports of encephalomyelitis, and acute and chronic myelitis where a clear relationship between the infection and CNS disease was strongly suggested (Mackenzie et al., 1995; McCullers et al., 1995; Carrigan et al., 1996; Gilden et al., 1996; Novoa et al., 1997; Campadelli-Fiume et al., 1999; Singh and Paterson, 2000; Kleinschmidt-DeMasters and Gilden, 2001; Portolani et al., 2001, 2002; Moore and Wolfson, 2002; Dockrell, 2003; Ward, 2005). Of relevance to MS are recent observations that both HHV-6 variants show capability to modulate T-cell inflammatory responses toward Th1 (pro-inflammatory) phenotypes (Mayne et al., 2001), and a small study showing the presence of intrathecal oligoclonal bands directed against HHV-6 antigens in a significant percentage of patients with MS (Derfuss et al., 2005). In addition, a consequence of infection of endothelial cells by HHV-6 appears to be an increase in vascular endothelium permeability (Caruso et al., 2002). Thus, in keeping with the concept of heterogeneity in MS pathophysiology, it is possible that an association between MS and HHV-6 exists for certain clinical or neuropathological subtypes that are yet to be identified.

In addition to MS and encephalomyelitis, and association with HHV-6 exposure and HHV-6 reactivity has also been claimed for chronic fatigue syndrome (CFS) and narcoplepsy. CFS is an incapacitating disease of adult of all ages, which shares certain clinical features with MS (the fatigue) and is also likely immune-mediated. Similar to MS, studies of antibody reactivity have been inconsistent in proving a relationship between CFS and HHV-6 (Wallace et al., 1999; Ablashi et al., 2000; Enbom, 2001; Nicolson et al., 2003).

We describe here, a new experimental system created to understand causal and time-dependent relationships between HHV-6 infection and the occurrence of CNS inflammatory or neurodegenerative conditions that mimic human MS in a controlled fashion.

The common marmoset model of EAE

Common marmosets (white ear-tuffed marmoset, *Callithrix jacchus jacchus*) are small New World non-human primates (˜400 gm at adult age) that have been extensively used for pharmacological studies, models of Parkinsonism, and aging due to their ease of breeding in captivity (Brack and Rothe, 1981; Gore et al., 2001; Abbott et al., 2003; Mansfield, 2003; Zuhlke and Weinbauer, 2003). These relatives of ours, and the related tamarins are differentially susceptible to a number of autoimmune diseases, and spontaneously develop colitis, thyroiditis, and a wasting syndrome with kidney failure of unclear pathophysiology (Levy et al., 1972; Clapp et al., 1985). Marmosets have a very restricted class I due to a large evolutionary deletion (Watkins et al., 1990; Antunes et al., 1998), which may explain their susceptibility to a number of viruses. They have close phylogeny to humans, with a number of immune and nervous system genes highly conserved (Uccelli et al., 1997, 2001; von Büdingen et al., 2001, 2002; Mesleh et al., 2002).

C. jacchus marmosets have been the subject of intense investigations of EAE in the last decade. Active immunization with whole human white matter, or myelin/oligodendrocyte glycoprotein (MOG) in adjuvant, produce chronic, relapsing-remitting disorders of mild to moderate clinical severity, reminiscent of human MS. The neuropathology of *C. jacchus* EAE consists of large concentric areas of primary demyelination, macrophage infiltration, astrogliosis, and death of oligodendrocytes (Massacesi et al., 1995; Brok et al., 2001; Genain and Hauser, 2001). Ultrastructural features of myelin breakdown are similar in marmoset EAE and human MS (Raine et al., 1999; Genain and Hauser, 2001). The causal mechanisms underlying the marmoset, MS-like EAE lesion have been elucidated and are a complex interplay of myelin-directed autoaggressive response and pathogenic autoantibody responses (Genain and Hauser, 2001) (Fig. 1).

Marmoset model of HHV-6 infection

Infection of marmoset cells in vitro

Common marmosets are susceptible to infection by herpesviruses (Falk et al., 1976; Ablashi et al., 1978; de Thè et al., 1980; Johnson and Jondal, 1981; Wedderburn et al., 1984; Provost et al., 1987; Cox et al., 1996; Farrell et al., 1997; Ramer et al., 2000; Jenson et al., 2002), and express a CD46 molecule that is highly homologous to the HHV-6 receptor (Murakami et al., 1998). Using trans-well co-cultures with HHV-6-infected human T-cell lines, we have demonstrated that marmoset peripheral blood mononuclear cells (PBMC) can be infected *in vitro* with both HHV-6 variants A and B. *In vitro* infection is facilitated by stimulation of PBMC with phytohemagglutinin and occurs within 5–10 days after exposure. The lymphocyte subtypes that are infected in the marmoset are under investigation (Fig. 2).

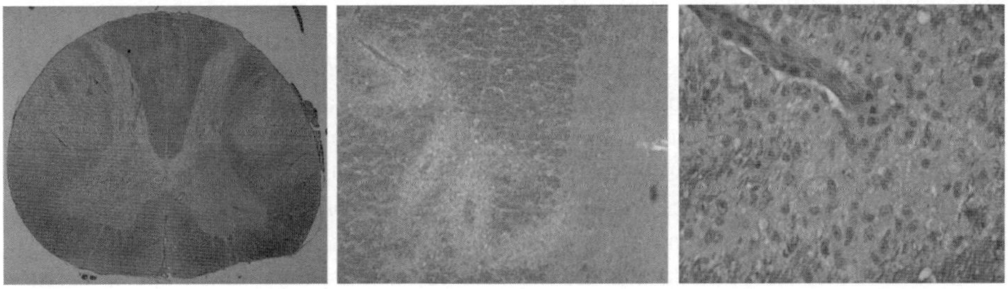

Fig. 1 Neuropathology of acute *C. jacchus* EAE. Perivascular inflammatory demyelinating infiltrates in spinal cord and brain periventricular white matter (left, middle). High-power view to show monocyte/macrophage infiltration (LFB/PAS). (for colour version: see colour section on page 359).

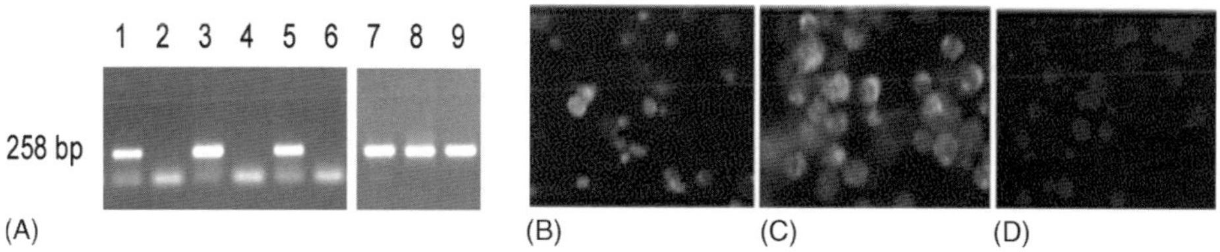

Fig. 2 *In vitro* infection of marmoset PBMC with HHV6 variants. A. Detection of HHV-6 DNA by nested PCR (expected fragment size 258 bp). Lane 1: marmoset PBMC infected with HHV-6A, 10 days after infection; 2: uninfected T-cell line HSB_2; 3 and 8: HSB_2 infected with HHV-6A; 4: uninfected T-cell line MOLT3; 5 and 9: $MOLT_3$ infected with HHV-6B; 6: no DNA template; 7: Marmoset PBMC infected with HHV-6B. B–D. Staining for HHV-6 nuclear antigen p41 by immunofluorescence (IFA): HHV-6A-infected marmoset PBMC (B), HHV-6A-infected HSB_2 cells (C), and uninfected marmoset PBMC (D).

Infection of marmosets **in vivo**

We have infected seven adult marmosets with HHV-6 *in vivo* using various protocols:

- Intravenous (i.v.) administration of the animal's own PBMC infected *in vitro* with HHV-6A or B (as verified by IFA and PCR), followed by intravenous injection of a cell lysate containing identical live virus variant 6–7 weeks later (HHV-6A: animal ♯190-94 and U031-00, closed symbols in Fig. 3; HHV-6B: U054-01, open symbols).
- Two i.v. injections of lysates from MOLT3 HHV-6B-infected cells at 5-weeks interval (125-, open symbols).
- One inoculation of HSB2 cells infected with HHV-6A (HHV-6A$^+$ HSB2), followed by injection of either HHV-6A-infected HSB2 cells (550-99, closed symbols), or uninfected cells (367-94, open symbols).

Initial infection was near asymptomatic. However, animals re-exposed to live HHV-6A rapidly developed weight loss and hypotonic paralysis with sensory deficits following the second inoculation. The most severe syndromes were observed in animals 190-94 and U076-03 (Fig. 3).

Imaging studies and neuropathology

Neuropathology obtained for all animals showed evidence of CNS blood brain barrier breakdown and inflammation in the animals that received repeated inoculations of replicating HHV-6A. Demyelination undistinguishable from that seen in marmoset EAE was clearly present. In one animal that was sacrificed early after the second inoculation (190-94), a corresponding MRI-visible, T2-weighted hyperintense brain stem lesion was seen (Fig. 4A and B).

The presence of HHV-6 virus could be demonstrated by immunohistochemistry in the vicinity of inflammatory infiltrates (Fig. 4C). We have not so far been able to

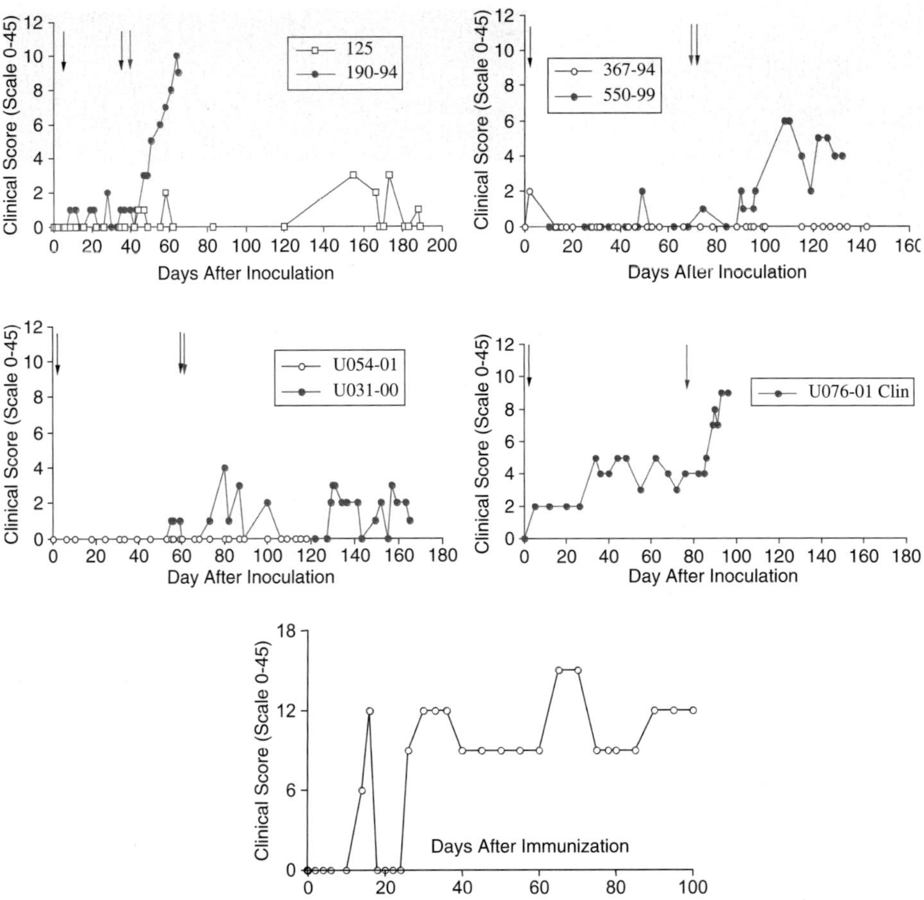

Fig. 3 Clinical course (neurological signs) in the seven animals initially studied. Marmoset EAE-grading scale (0–45) (Villoslada et al., 2000). Animals developed weight loss, paraparesis, and sensory deficits as in the MS-like EAE model. Initial and later injections are indicated by arrows. For comparison, a typical course of clinical signs for marmoset white-matter-induced EAE is shown at the bottom left.

induce significant clinical disease or neuropathology using HHV-6B lysates or a single injection of HHV-6A^{+} cells.

Mechanisms of HHV-6-Induced Inflammatory CNS Demyelination

Reactivity to viral antigens

Monitoring of serum antibody responses to viral antigens was performed by flow cytometry of serum on cell lines infected with HHV-6A and B with a fluorescein isothiocynate (FITC)-labeled anti-monkey IgG as secondary antibody. These

Fig. 4 A, Coronal MRI section showing T2 hyper-intensity lateral to median CSF space in pons (arrow). B, Corresponding demyelinating inflammatory infiltrate (animal 190-94; LFB/PAS, also see Fig. 1). C, Staining for early nuclear antigen p41/p38 (Advanced Biotechnologies, Inc.), demonstrating viral persistence/replication within lesions. Positively stained cells (arrows) have not yet been formally identified but appear to be oligodendrocytes. (for colour version: see colour section on page 359).

experiments indicated that marmosets bred in captivity in our colony are naïve to HHV-6. Antibody reactivity appeared after the first exposure and titers increase after the second inoculation. Reactivity in each subject was observed against the infecting variant, and not against the other HHV-6 variant or against uninfected cell lines, indicating specificity of the staining (Fig. 5).

In vivo *persistence and replication of HHV-6*

The presence of HHV-6 DNA can be monitored serially by nested PCR. Consistent with the known tropism of HHV-6 variants, we have been able to detect HHV-6B, and not A, in the blood of infected animals (not shown). In contrast to blood (HHV-6B) and CNS (HHV-6A detected by IHC; see Fig. 4C), evidence of viral persistence or replication so far was not detected in other organs sampled at euthanasia.

Reactivity to CNS myelin antigens

Viral infections can result in molecular mimicry, a phenomenon by which the host's immune system recognizes a viral peptide that resembles a myelin protein peptide (Fujinami and Oldstone, 1985; Oldstone, 1998), which triggers an immune attack. Such homology to an immunodominant peptide of MBP was recently described within the HHV-6 U24 protein (Cirone et al., 2002; Tejada-Simon et al., 2003). Molecular mimicry that leads to cross-activation of MBP-reactive T-cell clones has been demonstrated for other viruses as a possible mechanism for triggering MS attacks, or perpetrating disease (Wucherpfennig and Strominger, 1995; Talbot et al., 2001; Lang et al., 2002). Two marmosets inoculated with HHV-6A, and one inoculated with HHV-6B (as control) are currently being followed chronically, as shown in Fig. 6 (top panel). Monitoring of T-cell immune reactivity (PBMC) in serial blood samples obtained from these animals indicates that the second HHV-6

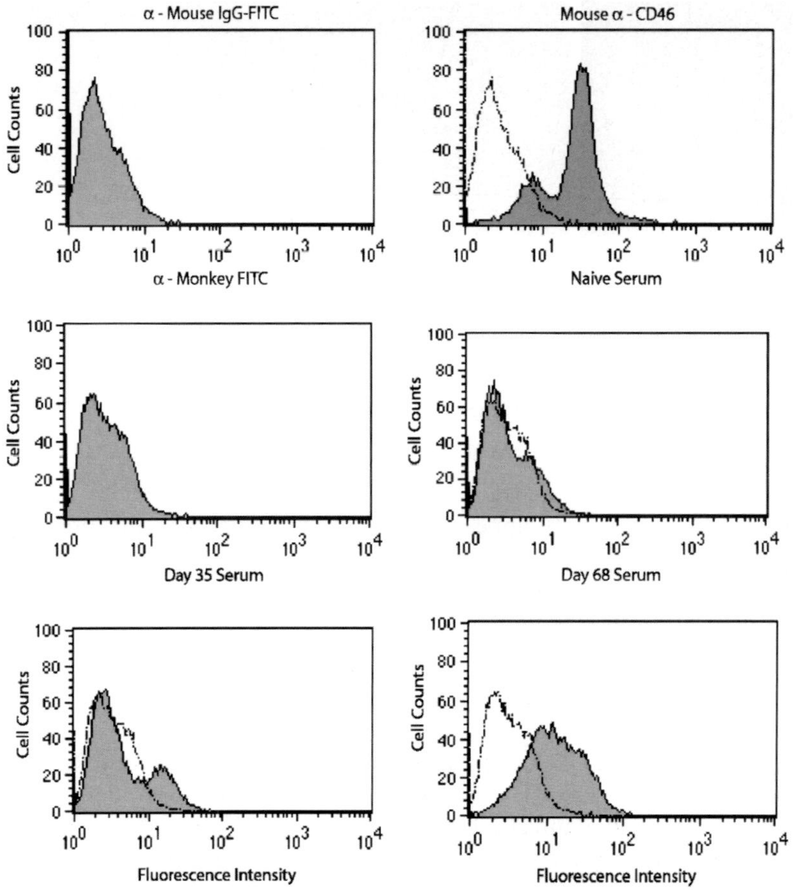

Fig. 5 Serum reactivity to HHV-6A$^+$-HSB$_2$ cells in animal 190-94. Top panels, isotype control (left) and CD46 (right, isotype control overlapped as an open trace). Middle panels, control anti-monkey IgG antibody (left) and naïve serum (right). Bottom, serum after the first inoculation (day 35), and at euthanasia after the second inoculation. Strong positivity is detected at euthanasia with a trend toward positivity at day 35. The open trace represents the negative signal obtained with anti-monkey IgG-FITC.

inoculation is followed by a transient state of immunosuppression (decreased reactivity to phytohemagglutinin (PHA)), and later by appearance of reactivity to MOG (Fig. 6, middle and bottom panels). The appearance of clinical signs and several clinical relapses before T-cell reactivity is detected in the circulation suggests that in this model, T-cell responses are not the primary determinant of disease and pathology but rather, appear secondary to damage created by viral persistence and replication within the CNS (see Fig. 4C).

We have not to date, detected significant antibody (IgG) responses to myelin proteins, in contrast to the typical humoral responses that occur at the early stages of marmoset EAE after active immunization with myelin antigens (Genain et al., 1996).

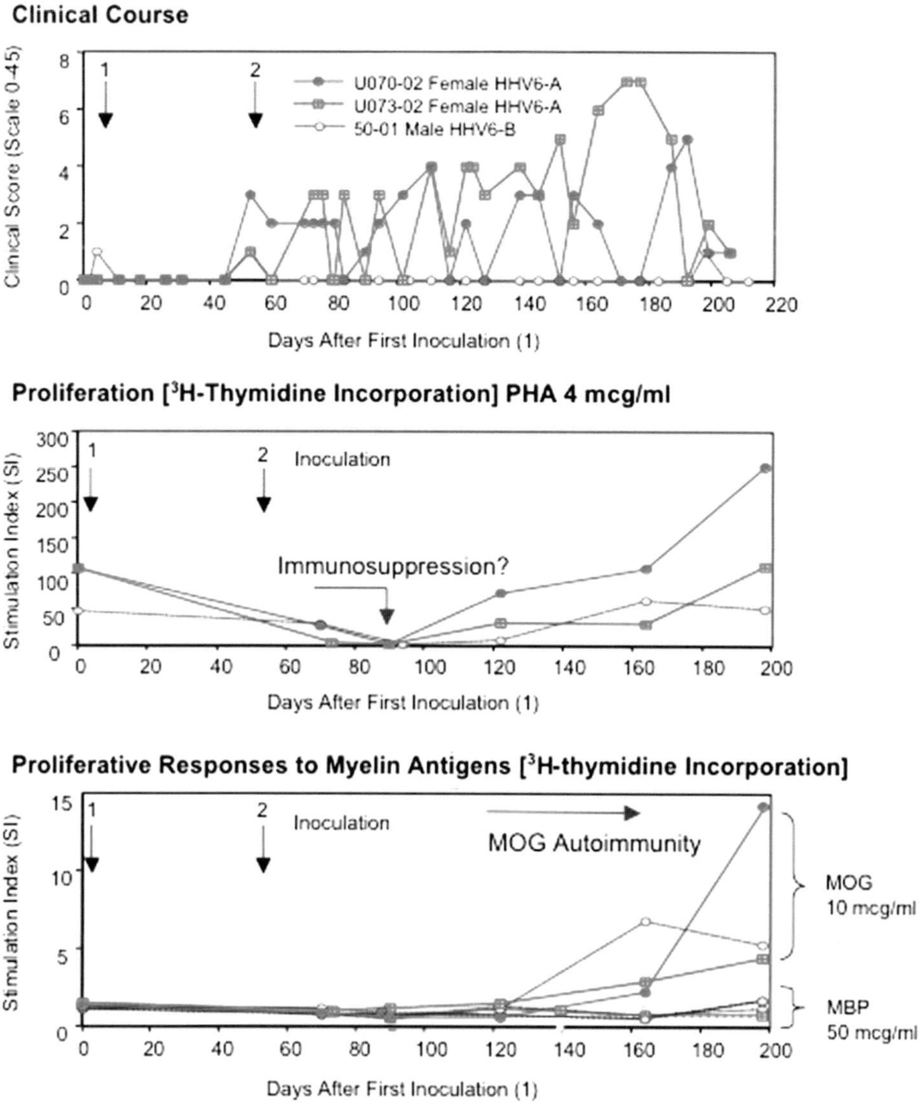

Fig. 6 Chronic HHV-6-mediated CNS disease and immune responses to PHA and myelin antigens. Note the relapsing remitting course of mild signs (upper panel, closed symbols), and the absence of disease in the animal inoculated with HHV-6B (open symbols). Inoculations are denoted by the black arrows (1 and 2). Middle panel, proliferative responses in PBMC after stimulation with PHA. Bottom panel, T-cell proliferative responses measured in PBMC against two myelin antigens, MOG and MBP.

HHV-6 has lytic effects on CNS glial cells

Apoptosis, or death of oligodendrocytes and neurons have been suggested to participate in the pathogenesis of the lesions MS and EAE (Raine, 1997b; Lucchinetti et al., 2000). There are several studies suggesting that the HHV-6 variants may be toxic to glial cells such as astrocytes in CNS, although potential protective effects have been also reported (Kong et al., 2003; De Bolle et al., 2005; Donati et al., 2005). Preliminary studies were conducted in the marmoset model of HHV-6 infection using the TUNEL reaction and staining for caspase 3 on sections of brain from animals infected with HHV-6A that developed clinical signs and inflammatory infiltrates. These experiments clearly demonstrated the presence of numerous apoptotic cells in the vicinity of the lesions. Further immunohistochemical characterization of the affected cell types is in progress, as well as *in vitro* experiments to investigate whether apoptosis can occur in human glial and neuronal cell lines after exposure to the HHV-6 variants.

Conclusions

C. jacchus marmosets are naïve to HHV-6A and B, and can be infected by these viruses. Repeated infection of adult animals with HHV-6A produces a mild, chronic relapsing CNS disease with pathologically, perivascular inflammatory demyelination similar to MS. This model is the first to causally link a ubiquitous human virus to a chronic disorder mimicking MS, and affords to model interactions between such microbes and complex neuroimmune responses in outbred species. HHV-6 infection by both variants A and B, which are capable of persistence and replication in marmosets as in humans, may cause transient immunosuppression. However, only HHV-6A infestation results in MS-like CNS inflammatory demyelination. Mimicry with myelin antigens does not appear to be a primary or causal mechanism for inflammatory CNS damage in this model.

As outlined earlier, there is considerable circumstantial and experimental evidence to support the hypothesis that microbial infections—and especially those with ubiquitous human viruses—may be linked to MS and other autoimmune disorders. Epidemeiological studies suggest that children with active EBV infection may be at increased risk to develop MS (Alotaibi et al., 2004). More strikingly, several studies indicate that the postponement of natural infections from infant age to a later stage of childhood, tend to be associated with an increased risk of developing MS (Alter and Cendrowski, 1976; Alter et al., 1987; Bachmann and Kesselring, 1998; Granieri et al., 2001; Hernan et al., 2001; Bach, 2002; Ponsonby et al., 2005). It remains that in the absence of a practical laboratory model to study the temporal and mechanistic relationships between HHV-6 infection and MS, the exact risk for a given individual to develop this disease after HHV-6 exposure is purely conjectural.

Such an experimental system has now been created, providing strong evidence that HHV-6 variant A may be causally linked to MS in humans. Potential implications are far-reaching, beginning with the opportunity to identify and validate biomarkers suitable to diagnose underlying infectious causes of MS, and predict an "MS risk" in young adults including perhaps at a pre-clinical stage. With that regard, data obtained from the marmoset model may enhance the ability to model these interactions by means of bioinformatics (Krueger et al., 2004). Finally, these continued studies will most likely help identify novel therapeutic targets for curative and preventative intervention for subgroups of MS patients where disease is driven by HHV-6 infection.

Acknowledgments

This work was supported by grants from the DANA Foundation, the National Multiple Sclerosis Society, the HHV-6 Foundation, and the Cure MS Now and Lunardi Foundations.

References

Abbott DH, Barnett DK, Colman RJ, Yamamoto ME, Schultz Darken NJ. Comp Med 2003; 53: 339–350.

Ablashi DV, Eastman HB, Owen CB, Roman MM, Friedman J, Zabriskie JB, Peterson DL, Pearson GR, Whitman JE. J Clin Virol 2000; 16: 179–191.

Ablashi DV, Josephs SF, Buchbinder A, Hellman K, Nakamura S, Llana T, Lusso P, Kaplan M, Dahlberg J, Memon S, Imam F, Ablashi KL, Markham PD, Kramarsky B, Krueger GRF, Biberfeld P, Wong-Staal F, Salahuddin SZ, Gallo RC. J Virol Methods 1988; 21: 29–48.

Ablashi DV, Lapps W, Kaplan M, Whitman JE, Richert JR, Pearson GR. Mult Scler 1998; 4: 490–496.

Ablashi DV, Pearson G, Rabin H, Armstrong G, Easton J, Valerio M, Cicmanec J. Biomedicine 1978; 29: 7–10.

Ablashi DV, Salahuddin SZ, Josephs SF, Balachandran N, Krueger GR, Gallo RC. In Vivo 1991; 5: 193–199.

Alotaibi S, Kennedy J, Tellier R, Stephens D, Banwell B. JAMA 2004; 291: 1875–1879.

Alter M, Cendrowski W. Neurology 1976; 26: 201–204.

Alter M, Zhen-xin Z, Davanipour Z, Sobel E, Min Lai S, LaRue L. Ital J Neurol Sci 1987; 8: 23–28.

Antunes S, de Groot N, Brok H, Doxiadis G, Menezes A, Otting N, Bontrop R. Proc Natl Acad Sci USA 1998; 95: 11745–11750.

Bach JF. N Engl J Med 2002; 347: 911–920.

Bachmann S, Kesselring J. Neuroepidemiology 1998; 17: 154–160.

Brack M, Rothe H. Vet Pathol 1981; 18: 45–54.

Brok HP, Bauer J, Jonker M, Blezer E, Amor S, Bontrop RE, Laman JD, Hart BA. Immunol Rev 2001; 183: 173–185.

Brosnan CF, Raine CS. Brain Pathol 1996; 6: 243–257.

Campadelli-Fiume G, Mirandola P, Menotti L. Emerg Infect Dis 1999; 5: 353–366.

Carrigan D, Harrington D, Knox K. Neurology 1996; 47: 145–148.

Caruso A, Rotola A, Comar M, Favilli F, Galvan M, Tosetti M, Campello C, Caselli E, Alessandri G, Grassi M, Garrafa E, Cassai E, Di Luca D. J Med Virol 2002; 67: 528–533.
Caserta MT, McDermott MP, Dewhurst S, Schnabel K, Carnahan JA, Gilbert L, Lathan G, Lofthus GK, Hall CB. J Pediatr 2004; 145: 478–484.
Cepok S, Zhou D, Srivastava R, Nessler S, Stei S, Bussow K, Sommer N, Hemmer B. J Clin Invest 2005; 115: 1352–1360.
Challoner PB, Smith KT, Parker JD, MacLeod D, Coulter SN, Rose TM, Schultz ER, Bennett L, Garber RL, Chang M, Schad PA, Stewart PM, Nowinski RC, Brown JP, Burmer GC. Proc Natl Acad Sci 1995; 92: 7440–7444.
Cirone M, Cuomo L, Zompetta C, Ruggieri S, Frati L, Faggioni A, Ragona G. J Med Virol 2002; 68: 268–272.
Clapp N, Lushbaugh C, Humason G, Gangaware B, Henke M, McArthur A.. In: Carcinoma of the Large Bowel and its Precursors: Progress in Clinical and Biological Research (Ingalls M, Mastromarino J, editors). New York: Allen R. Liss; 1985; pp. 247–261.
Coates AR, Bell J. Nat Med 1998; 4: 537–538.
Cox C, Chang S, Karran L, Griffin B, Wedderburn N. J Gen Virol 1996; 77(Pt 6): 1173–1180.
Dal Canto M, Melvold R, Kim B, Miller S. Microsc Res Tech 1995; 32: 215–229.
De Bolle L, Naesens L, De Clercq E. Clin Microbiol Rev 2005; 18: 217–245.
DeRanieri JT, Clements PT, Clark K, Wolcik Kuhn D, Manno MS. J Sch Nurs 2004; 20: 69–75.
Derfuss T, Hohlfeld R, Meinl E. J Neurol 2005; 252: 968–971.
de Thè G, Dubouch P, Fontaine C, Wedderburn N, Carter RL, Edwards MB, Cohen B. Intervirology 1980; 14: 284–291.
Dewhurst S. Herpes 2004; 11(Suppl 2): 105A–111A.
Dockrell DH. J Med Microbiol 2003; 52: 5–18.
Donati D, Martinelli E, Cassiani-Ingoni R, Ahlqvist J, Hou J, Major EO, Jacobson S. J Virol 2005; 79: 9439–9448.
Enbom M. Apmis 2001; 109: 401–411.
Evans C, Horwitz M, Hobbs M, Oldstone M. J Exp Med 1996; 184: 2371–2384.
Falk L, Deinhardt F, Wolfe L, Johnson D, Hilgers J, de Thè G. Int J Cancer 1976; 17: 785–788.
Farrell PJ, Hollyoake M, Niedobitek G, Agathanggelou A, Morgan A, Wedderburn N. J Gen Virol 1997; 78(Pt 6): 1417–1424.
Fotheringham J, Jacobson S. Herpes 2005; 12: 4–9.
Fujinami R, Oldstone M. Science 1985; 230: 1043–1045.
Genain C, Hauser S. Immunol Rev 2001; 183: 159–172.
Genain CP, Abel K, Belmar N, Villinger F, Rosenberg DP, Linington C, Raine CS, Hauser SL. Science 1996; 274: 2054–2057.
Gilden DH, Devlin ME, Burgoon MP, Owens GP. Mult Scler 1996; 2: 179–183.
Gobbi A, Stoddart CA, Locatelli G, Santoro F, Bare C, Linquist-Stepps V, Moreno ME, Abbey NW, Herndier BG, Malnati MS, McCune JM, Lusso P. J Virol 2000; 74: 8726–8731.
Gobbi A, Stoddart CA, Malnati MS, Locatelli G, Santoro F, Abbey NW, Bare C, Linquist-Stepps V, Moreno MB, Herndier BG, Lusso P, McCune JM. J Exp Med 1999; 189: 1953–1960.
Gore MA, Brandes F, Kaup FJ, Lenzner R, Mothes T, Osman AA. J Med Primatol 2001; 30: 179–184.
Granieri E, Casetta I, Tola MR, Ferrante P. Neurol Sci 2001; 22: 179–185.

Grivel JC, Santoro F, Chen S, Faga G, Malnati MS, Ito Y, Margolis L, Lusso P. J Virol 2003; 77: 8280–8289.
Hall CB, Caserta MT, Schnabel KC, Long C, Epstein LG, Insel RA, Dewhurst S. Clin Infect Dis 1998; 26: 132–137.
Hernan MA, Zhang SM, Lipworth L, Olek MJ, Ascherio A. Epidemiology 2001; 12: 301–306.
Hohlfeld R. Brain 1997; 120: 865–916.
Jenson HB, Ench Y, Zhang Y, Gao SJ, Arrand JR, Mackett M. J Gen Virol 2002; 83: 1621–1633.
Johnson DR, Jondal M. Proc Natl Acad Sci USA 1981; 78: 6391–6395.
Johnson R. Ann Neurol 1994; 36: S54–S60.
Johnson RT, Griffin DE, Hirsh RL, Wolinski JS, Roedenbeck S, Soriano ILd, Vaisberg A. N Engl J Med 1984; 310: 137–141.
Kleinschmidt-DeMasters BK, Gilden DH. Brain Pathol 2001; 11: 440–451.
Kong H, Baerbig Q, Duncan L, Shepel N, Mayne M. J Neurovirol 2003; 9: 539–550.
Krueger GR, Ablashi DV. Intervirology 2003; 46: 257–269.
Krueger GR, Brandt ME, Wang G, Buja LM. Anticancer Res 2004; 24: 187–197.
Krueger GR, Sander C. Pathol Res Pract 1989; 185: 915–929.
Kurtzke JF. Clin Microbiol Rev 1993; 6: 382–427.
Kurtzke JF, Hyllested K. Neurology 1986; 36: 307–328.
Lang HL, Jacobsen H, Ikemizu S, Andersson C, Harlos K, Madsen L, Hjorth P, Sondergaard L, Svejgaard A, Wucherpfennig K, Stuart DI, Bell JI, Jones EY, Fugger L. Nat Immunol 2002; 3: 940–943.
Lassmann H. Mult Scler 1998; 4: 93–98.
Levy BM, Hampton S, Dreizen S, Hampton JK. J Comp Pathol 1972; 82: 99–103.
Levy JA. Lancet 1997; 349: 558–563.
Levy JA, Ferro F, Greenspan D, Lennette ET. Lancet 1990a; 335: 1047–1050.
Levy JA, Ferro F, Lennette ET, Oshiro L, Poulin L. Virology 1990b; 178: 113–121.
Lin WR, Wozniak MA, Cooper RJ, Wilcock GK, Itzhaki RF. J Pathol 2002; 197: 395–402.
Lucchinetti C, Brück W, Parisi J, Scheithauer B, Rodriguez M, Lassmann H. Ann Neurol 2000; 47: 707–717.
Luppi M, Barozzi P, Maiorana A, Marasca R, Trovato R, Fano R, Ceccherini-Nelli L, Torelli G. J Med Virol 1995; 47: 105–111.
Lusso P, Gallo RC. Baillieres Clin Haematol 1995a; 8: 201–223.
Lusso P, Gallo RC. Immunol Today 1995b; 16: 67–71.
Lusso P, Markham PD, DeRocco SE, Gallo RC. J Virol 1990; 64: 2751–2758.
Lusso P, Salahuddin SZ, Ablashi DV, Gallo RC, Di Marzo Veronese F, Markham PD. Lancet 1987; 2: 743.
Lusso P, Secchiero P, Crowley RW. AIDS Res Hum Retroviruses 1994; 10: 181–187.
Mackenzie IR, Carrigan DR, Wiley CA. Neurology 1995; 45: 2015 2017.
Mansfield K. Comp Med 2003; 53: 383–392.
Marie JC, Astier AL, Rivailler P, Rabourdin-Combe C, Wild TF, Horvat B. Natl Immunol 2002; 3: 659–666.
Massacesi L, Genain CP, Lee-Parritz D, Letvin NL, Canfield D, Hauser SL. Ann Neurol 1995; 37: 519–530.
Mayne M, Cheadle C, Soldan SS, Cermelli C, Yamano Y, Akhyani N, Nagel JE, Taub DD, Becker KG, Jacobson S. J Virol 2001; 75: 11641–11650.

McCullers JA, Lakeman FD, Whitley RJ. Clin Infect Dis 1995; 21: 571–576.
Mesleh MF, Belmar N, Lu CW, Krishnan VV, Maxwell RS, Genain CP, Cosman M. Neurobiol Dis 2002; 9: 160–172.
Moore FG, Wolfson C. Acta Neurol Scand 2002; 106: 63–83.
Murakami Y, Seya T, Kurita M, Fukui A, Ueda S, Nagasawa S. Biochem J 1998; 330: 1351–1359.
Nicolson GL, Gan R, Haier J. Apmis 2003; 111: 557–566.
Novoa LJ, Nagra RM, Nakawatase T, Edwards-Lee T, Tourtellotte WW, Cornford ME. J Med Virol 1997; 52: 301–308.
Oksenberg JR, Hauser SL. In: Multiple Sclerosis: Advances in Clinical Trial Design, Treatment and Future Perspectives (Goodkin DE, Rudick RA, editors). London: Springer; 1996; pp. 17–46.
Oldstone M. Faseb J 1998; 12: 1255–1265.
Opsahl ML, Kennedy PG. Brain 2005; 128: 516–527.
Ponsonby AL, van der Mei I, Dwyer T, Blizzard L, Taylor B, Kemp A, Simmons R, Kilpatrick T. JAMA 2005; 293: 463–469.
Portolani M, Pecorari M, Gennari W, Sabbatini AM, Meacci M, Beretti F, Frigieri G, Bergonzini P. Minerva Pediatr 2002; 54: 459–464.
Portolani M, Pecorari M, Tamassia MG, Gennari W, Beretti F, Guaraldi G. J Med Virol 2001; 65: 133–137.
Provost PJ, Keller PM, Banker FS, Keech BJ, Klein HJ, Lowe RS, Morton DH, Phelps AH, McAleer WJ, Ellis RW. J Virol 1987; 61: 2951–2955.
Raine C. Demyelinating diseases. In: Textbook of Neuropathology (Davis R, Robertson D, editors). Baltimore, MD: Williams & Wilkins; 1997a; pp. 627–714.
Raine C, Cannella B, Hauser S, Genain C. Ann Neurol 1999; 46: 144–160.
Raine CS. Multiple sclerosis and chronic relapsing EAE: Comparative ultrastructural neuropathology. In: Multiple Sclerosis: Pathology, Diagnosis and Management (Hallpike JF, Adams GWM, Tourtelotte WW, editors). Baltimore, MD: William & Wilkins; 1983; pp. 413–460.
Raine CS. J Neuroimmunol 1997b; 77: 135–152.
Ramer JC, Garber RL, Steele KE, Boyson JF, O'Rourke C, Thomson JA. Comp Med 2000; 50: 59–68.
Riley-Vargas RC, Gill DB, Kemper C, Liszewski MK, Atkinson JP. Trends Immunol 2004; 25: 496–503.
Russell S. Tissue Antigens 2004; 64: 111–118.
Santoro F, Kennedy PE, Locatelli G, Malnati MS, Berger EA, Lusso P. Cell 1999; 99: 817–827.
Singh N, Paterson DL. Transplantation 2000; 69: 2474–2479.
Smith AP, Paolucci C, Di Lullo G, Burastero SE, Santoro F, Lusso P. J Virol 2005; 79: 2807–2813.
Sola P, Merelli E, Marasca R, Poggi M, Luppi M, Montorsi M, Torelli G. J Neurol Neurosurg Psychiatry 1993; 56: 917–919.
Soldan S, Berti R, Salem N, Secchiero P, Flamand L, Calabresi P, Brennan M, Maloni H, McFarland H, Lin H, Patnaik M, Jacobson S. Nat Med 1997; 3: 1394–1397.
Stohlman SA, Hinton DR. Brain Pathol 2001; 11: 92–106.
Storch MK, Piddlesden S, Haltia M, Iivanainen M, Morgan P, Lassmann H. Ann Neurol 1998; 43: 465–471.

Talbot PJ, Arnold D, Antel JP. Curr Top Microbiol Immunol 2001; 253: 247–271.
Tejada-Simon MV, Zang YC, Hong J, Rivera VM, Zhang JZ. Ann Neurol 2003; 53: 189–197.
Tourtelotte WW, Potvin AR, Fleming JO, Murthy KN, Levy J, Syndulco K, Potvin JH. Neurology 1980; 30: 240–244.
Trapp B, Peterson J, Ransohoff R, Rudick R, Mork S, Bo L. N Engl J Med 1998; 338: 278–285.
Uccelli A, Giunti D, Caroli F, Fiorone M, Seri M, Mancardi G, Hauser S, Genain C. Eur J Immunol 2001; 31: 474–479.
Uccelli A, Oksenberg JR, Jeong MC, Genain CP, Rombos T, Jaeger EEM, Giunti D, Lanchbury JS, Hauser SL. J Immunol 1997; 158: 1201–1207.
Villoslada P, Hauser S, Bartke I, Unger J, Heald N, Rosenberg D, Cheung S, Mobley W, Fisher S, Genain C. J Exp Med 2000; 191: 1799–1806.
von Büdingen H-C, Hauser S, Fuhrmann A, Nabavi C, Lee J, Genain C. Proc Natl Acad Sci USA 2002; 99: 8207–8212.
von Büdingen H-C, Hauser S, Nabavi C, Genain C. Immunogenetics 2001; 53: 557–563.
Wallace 2nd HL, Natelson B, Gause W, Hay J. Clin Diagn Lab Immunol 1999; 6: 216–223.
Ward KN. Curr Opin Infect Dis 2005; 18: 247–252.
Watkins D, Chen Z, Hughes A, Hodi F, Letvin N. J Immuno 1990; 144: 3726–3735.
Wedderburn N, Edwards JM, Desgranges C, Fontaine C, Cohen B, de Thè G. J Infect Dis 1984; 150: 878–882.
Wucherpfennig KW, Strominger JL. Cell 1995; 80: 695–705.
Zerr DM, Meier AS, Selke SS, Frenkel LM, Huang ML, Wald A, Rhoads MP, Nguy L, Bornemann R, Morrow RA, Corey L. N Engl J Med 2005; 352: 768–776.
Zuhlke U, Weinbauer G. Toxicol Pathol 2003; 31(Suppl): 1230–1237.

Human Herpesvirus-6: General Virology, Epidemiology and Clinical Pathology
Perspectives in Medical Virology
Gerhard Krueger and Dharam Ablashi (Editors)

DOI 10.1016/S0168-7069(06)12025-X

Computational Simulation of HHV-6 Infection

Guanyu Wang, Gerhard R.F. Krueger
Department of Pathology and Laboratory Medicine, The University of Texas — Houston Medical School, Houston, TX 77030, USA

Introduction

Human Herpesvirus-6 (HHV-6) is a ubiquitous virus that infects almost all children by age two, causing mild flu-like symptoms in some with occasional rash, high fever, encephalitis, and seizures. In most cases, the virus infection remains clinically inapparent resulting in persistent latent infection. In patients with impaired immune function, the virus remains in active state at low levels for years once reactivated. It may then become associated with a variety of disorders such as infectious mononucleosis (IM), Kikuchi's lymphadenitis, chronic fatigue syndrome (CFS) (Ablashi et al., 1991, 2000; Josephs et al., 1991; Yamanishi, 1992; Wagner et al., 1996; Ablashi and Krueger, 2003; Krueger and Ablashi, 2003; Glaser et al., 2005), allergic reactions (Descamps et al., 1997, 2001; Tohyama et al., 1998), demyelinating encephalopathies, and multiple sclerosis (MS) (Challoner et al., 1995; Ablashi et al., 1998; Berti et al., 2000, 2002; Cermelli et al., 2003; Goodman et al., 2003; Clark, 2004; Rotola et al., 2004). More detailed data can be found in other chapters of this book. The quite variable dynamics of a cellular immune response following acute or chronic HHV-6 infection can be well characterized by a mathematical model. Such a computational model has the advantage to reducing intractable real-world complexities to tractable mathematical problems, while upholding the realistic biological dynamics.

During past years we have developed several mathematical models to simulate the cellular immune response toward viral infection. A basic model has first been published to simulate T-cell population changes following various viral infections (Krueger et al., 2002, 2003, 2004). This conceptual model was further developed to simulate specifically HHV-6 infections (Wang et al., 2003) by using well-realistic laboratory data: After binding to the CD4 cells, the virus becomes internalized into susceptible T cells within several hours and replicates. After viral infection, the CD4 cells harboring viral components become antigenic, stimulating the proliferation of CD8 T cells. These activated CD8 T cells later differentiate into immune effector cells that will kill the infected CD4 T cells. Infected CD4 T cells will also die secondary to virus replication *per se*. CD4 cells proliferate after viral antigenic stimulation of the peptide–MHC II complexes on the surface of some antigen-presenting cells (APCs). Viruses are partially cleared by immune elimination, resulting in a decrease in the intensity of antigenic stimulation. Our modeling approach was as follows: first we constructed the mathematical model according to the above biological mechanisms, and then the model parameters were identified/optimized by fitting simulation runs with respective clinical data (Bertram et al., 1991; Horwitz and Krueger, 1992). Clinical data from patients with acute or chronic infections are shown in Fig. 1 and in Tables 1 and 2. The virus load (v^*) in terms of DNA copies ($\log_{10}/5\,\mu l$) represent total concentrations of virus in whole blood. The data for the peripheral T-cell counts (z^*) are given as x-times deviations

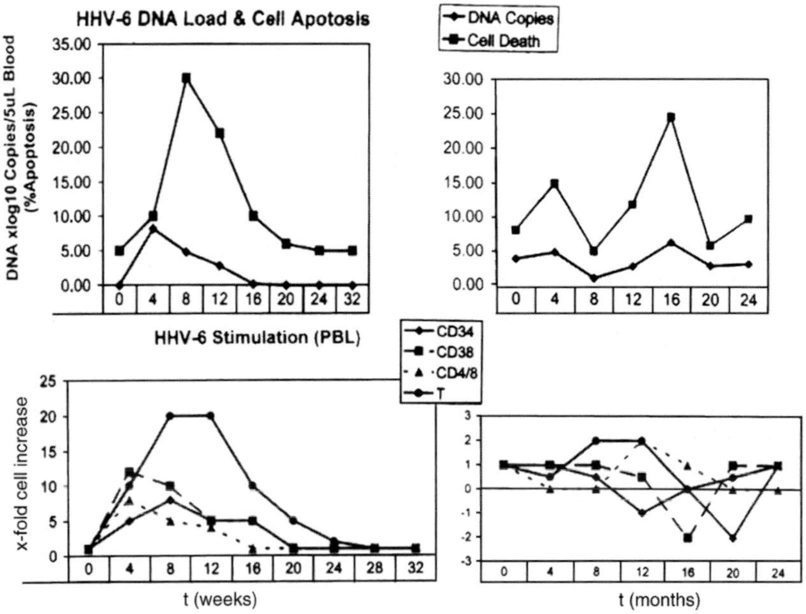

Fig. 1 DNA load and cellular changes in acute and chronic HHV-6 infection. Each ten patients with acute and chronic HHV-6 infections were studied at every date. Mean values are shown.

Table 1

Clinical data of HHV-6 virus load, T-cell count, during an 8-month course of −progressive infection

t_i ($i = 0, 1, \ldots, N = 8$)	0	1	2	3	4	5	6	7	8
Virus copies	–	8.2	4.8	2.8	0.2	0.0	0.0	0.0	0.0
T-cell count	2.0	11	21	21	11	6.0	3.0	2.0	2.0

Table 2

Clinical data of HHV-6 virus load, T-cell count, during an 24-month course of chronic infection

t_i ($i = 0, 1, \ldots, N = 6$)	0	4	8	12	16	20	24
Virus copies	3.8	4.8	1.0	2.8	6.4	3.0	3.3
T-cell count	2.0	1.5	3.0	3.0	1.0	1.5	2.0

from mean values of age-matched healthy individuals. The latter represented laboratory standards from healthy blood donors at Immunopathology Laboratory, The University of Cologne, Germany. For example, $z^*(0) = 2$ means that T-cell values of patients at the time of admission were twice those of age-matched healthy controls (i.e. increased by onefold).

The validity of the model has been verified by the realistic simulation runs generated by the model (Wang et al., 2003). However, the model was too complex because it considered many immunological mechanisms, some of which may be unimportant. Furthermore, only the acute infection was simulated, leaving the applicability of the model to the chronic infection an undetermined problem. In this chapter the model is drastically simplified and both acute and chronic infections are simulated.

Model development

Mathematical modeling is accomplished in several steps: design of a basic model of T-cell pools and their interaction for a healthy person, design of model variation for simulating viral stimulation, and optimization of the model by parameter adjustment through search algorithms. In corresponence with the clinical data, the simulated T-cell count (z) is given as relative value with respect to the age-matched healthy individuals. In addition, the T-cell population is divided into CD4 helper cells (x) and CD8 cytotoxic cells (y). For healthy people one has $x + y = z = 1$.

Basic model: healthy persons without virus infection

T-cell counts remain fairly constant throughout the life of a healthy individual reflecting the internal balance between cell death and cell regeneration. This feature of the immune system is called homeostasis and can be modeled as

$$\dot{x} = S_1 - ax \tag{1}$$

$$\dot{y} = S_2 - by \tag{2}$$

where x is the CD4 cell count, S_1 represents the source rate of the CD4 cells, and a the gain in the CD4 cell death rate. y, S_2, and b are defined for CD8 cell populations, respectively. In a state of health (no viral infection), cell counts are in homeostasis,

$$\dot{x} = \dot{y} = 0 \tag{3}$$

and thus possess constant values

$$x = \frac{\rho_0}{1+\rho_0} = x_0 \tag{4}$$

$$y = \frac{1}{1+\rho_0} = y_0 \tag{5}$$

where ρ_0 is the CD4/CD8 ratio in healthy persons. ρ_0 is assumed to be 2 in this paper. It follows that

$$x + y = 1, \quad x : y = \rho_0 \tag{6}$$

Substitution of Eqs. (1)–(3) by Eqs. (4) and (5), resulting in

$$S_1 = ax_0 \text{ and } S_2 = by_0 \tag{7}$$

Therefore, our basic model for T-cell populations in a healthy person has the following simplified form:

$$\dot{x} = a(x_0 - x) \tag{8}$$

$$\dot{y} = b(y_0 - y) \tag{9}$$

Extended model for viral infection

$$\dot{x}_u = a(x_0 - x_u) + cvx_u - dvx_u \tag{10}$$

$$\dot{x}_i = dvx_u - eyx_i \tag{11}$$

$$\dot{y} = b(y_0 - y) + fvy \tag{12}$$

$$\dot{v} = gx_i - hv \tag{13}$$

where x_u is the uninfected CD4 cell population, x_i the infected CD4 cells, y the CD8 cells, and v the virus. The parameters are as follows: cvx_u is the clonal expansion of CD4 cells when stimulated by the viral-peptide-MHC II complex, as presented on the surface of some APCs. dvx_u is the source rate of the infected CD4 cells. eyx_i is the death rate of the infected CD4 cells. It correlates with y because it is the CD8 cells attack the infected CD4 cells. fvy is the viral stimulation of CD8 cells to proliferate. gx_i identifies virus replication in infected CD4 cells. hv is virion clearance, which is the result of immune elimination, or nonspecific removal by the reticuloendothelial system.

The present model has only eight parameters (a, b, ..., h) and is much simplified, as compared with the one in (Wang et al., 2003), which has 17 parameters. The key is that some less important factors affecting the viral dynamics are omitted, such as the inverse proportional relation between the virus load and the passive cell death, the time delay between infection and cell death. Such factors do not qualitatively affect the viral dynamics, as suggested by the good data fitting without these factors (see below).

The final step in making the mathematical model useful for simulating cellular changes as they actually occur in human patients consists of adjusting and optimizing the values of individual parameters in given equations. The task of this procedure is to define parameter values of equations in a way that the simulation results for specific viral effects resemble closely those determined in human patients. In our present model, resulting T-cell changes should imitate actual levels of different T-cell populations in acute/chronic infections with HHV-6.

Parameter optimization

To identify the eight parameters $\boldsymbol{p} = [p_1, p_2, \ldots, p_8] = [a, b, c, \ldots, h]$, the conventional Gauss–Newton method (Stortelder, 1996; Wang, 2004) is used that involves the minimization of an objective, defined here as

$$J(\boldsymbol{p}) = \sum_i (x_u(\boldsymbol{p}, t_i) + x_u(\boldsymbol{p}, t_i) + y(\boldsymbol{p}, t_i) - z^*(t_i))^2 + (v(\boldsymbol{p}, t_i) - v^*(t_i))^2 \tag{14}$$

where $\boldsymbol{p} = [p_1, p_2, \ldots, p_8] = [a, b, c, \ldots, h]$ is the set of parameters to be determined. $x_u(\boldsymbol{p}, t)$, $x_i(\boldsymbol{p}, t)$, $y(\boldsymbol{p}, t)$, and $v(\boldsymbol{p}, t)$ constitute one simulation run generated by Eqs. (10)–(13) under the parameter set $\boldsymbol{p}$. The clinical data z^* (t) and v^* (t) have been presented in Tables 1 and 2.

The detailed Gauss–Newton optimization method can be summarized as the following algorithm:

Step 0: Arbitrarily choose a set of initial values $\boldsymbol{p}^{(0)}$. Set $\boldsymbol{p} = \boldsymbol{p}^{(0)}$.

Step 1: Use the current $\boldsymbol{p}$ to integrate Eqs. (10)–(13), which obtains $z(t) = x_u(t) + x_i(t) + y(t)$ and $v(t)$. Then, $z(t)$ and $v(t)$ are sampled at time t_i. Their differences with the clinical data z^* (t_i) and v^* (t_i) are denoted by $\boldsymbol{w}$, namely

$$w = [z(t_1) - z^*(t_1) \cdots z(t_N) - z^*(t_N), v(t_1) - v^*(t_1) \cdots v(t_N) - v^*(t_N)]^{\mathrm{T}}$$

Step 2: Use the current $\boldsymbol{p}$ to integrate the following variation equations:

$$\begin{aligned}
\frac{\mathrm{d}}{\mathrm{d}t}\left(\frac{\partial x_{\mathrm{u}}}{\partial a}\right) &= x_0 - x_{\mathrm{u}} - a\frac{\partial x_{\mathrm{u}}}{\partial a} + (c-d)\frac{\partial v}{\partial a}x_{\mathrm{u}} + (c-d)v\frac{\partial x_{\mathrm{u}}}{\partial a}\\
\frac{\mathrm{d}}{\mathrm{d}t}\left(\frac{\partial x_{\mathrm{i}}}{\partial a}\right) &= d\frac{\partial v}{\partial a}x_{\mathrm{u}} + dv\frac{\partial x_{\mathrm{u}}}{\partial a} - e\frac{\partial y}{\partial a}x_{\mathrm{i}} - ey\frac{\partial x_{\mathrm{i}}}{\partial a}\\
\frac{\mathrm{d}}{\mathrm{d}t}\left(\frac{\partial y}{\partial a}\right) &= -b\frac{\partial y}{\partial a} + f\frac{\partial v}{\partial a}y + fv\frac{\partial y}{\partial a}\\
\frac{\mathrm{d}}{\mathrm{d}t}\left(\frac{\partial v}{\partial a}\right) &= g\frac{\partial x_{\mathrm{i}}}{\partial a} - h\frac{\partial v}{\partial a}
\end{aligned} \tag{15}$$

$$\begin{aligned}
\frac{\mathrm{d}}{\mathrm{d}t}\left(\frac{\partial x_{\mathrm{u}}}{\partial b}\right) &= -a\frac{\partial x_{\mathrm{u}}}{\partial b} + (c-d)\frac{\partial v}{\partial b}x_{\mathrm{u}} + (c-d)v\frac{\partial x_{\mathrm{u}}}{\partial b}\\
\frac{\mathrm{d}}{\mathrm{d}t}\left(\frac{\partial x_{\mathrm{i}}}{\partial b}\right) &= d\frac{\partial v}{\partial b}x_{\mathrm{u}} + dv\frac{\partial x_{\mathrm{u}}}{\partial b} - e\frac{\partial y}{\partial b}x_{\mathrm{i}} - ey\frac{\partial x_{\mathrm{i}}}{\partial b}\\
\frac{\mathrm{d}}{\mathrm{d}t}\left(\frac{\partial y}{\partial b}\right) &= y_0 - y - b\frac{\partial y}{\partial b} + f\frac{\partial v}{\partial b}y + fv\frac{\partial y}{\partial b}\\
\frac{\mathrm{d}}{\mathrm{d}t}\left(\frac{\partial v}{\partial b}\right) &= g\frac{\partial x_{\mathrm{i}}}{\partial b} - h\frac{\partial v}{\partial b}
\end{aligned} \tag{16}$$

$$\begin{aligned}
\frac{\mathrm{d}}{\mathrm{d}t}\left(\frac{\partial x_{\mathrm{u}}}{\partial c}\right) &= -a\frac{\partial x_{\mathrm{u}}}{\partial c} + vx_{\mathrm{u}} + (c-d)\frac{\partial v}{\partial c}x_{\mathrm{u}} + (c-d)v\frac{\partial x_{\mathrm{u}}}{\partial c}\\
\frac{\mathrm{d}}{\mathrm{d}t}\left(\frac{\partial x_{\mathrm{i}}}{\partial c}\right) &= d\frac{\partial v}{\partial c}x_{\mathrm{u}} + dv\frac{\partial x_{\mathrm{u}}}{\partial c} - e\frac{\partial y}{\partial c}x_{\mathrm{i}} - ey\frac{\partial x_{\mathrm{i}}}{\partial c}\\
\frac{\mathrm{d}}{\mathrm{d}t}\left(\frac{\partial y}{\partial c}\right) &= -b\frac{\partial y}{\partial c} + f\frac{\partial v}{\partial c}y + fv\frac{\partial y}{\partial c}\\
\frac{\mathrm{d}}{\mathrm{d}t}\left(\frac{\partial v}{\partial c}\right) &= g\frac{\partial x_{\mathrm{i}}}{\partial c} - h\frac{\partial v}{\partial c}
\end{aligned} \tag{17}$$

$$\begin{aligned}
\frac{\mathrm{d}}{\mathrm{d}t}\left(\frac{\partial x_{\mathrm{u}}}{\partial d}\right) &= -a\frac{\partial x_{\mathrm{u}}}{\partial d} - vx_{\mathrm{u}} + (c-d)\frac{\partial v}{\partial d}x_{\mathrm{u}} + (c-d)v\frac{\partial x_{\mathrm{u}}}{\partial d}\\
\frac{\mathrm{d}}{\mathrm{d}t}\left(\frac{\partial x_{\mathrm{i}}}{\partial d}\right) &= vx_{\mathrm{u}} + d\frac{\partial v}{\partial d}x_{\mathrm{u}} + dv\frac{\partial x_{\mathrm{u}}}{\partial d} - e\frac{\partial y}{\partial d}x_{\mathrm{i}} - ey\frac{\partial x_{\mathrm{i}}}{\partial d}\\
\frac{\mathrm{d}}{\mathrm{d}t}\left(\frac{\partial y}{\partial d}\right) &= -b\frac{\partial y}{\partial d} + f\frac{\partial v}{\partial d}y + fv\frac{\partial y}{\partial d}\\
\frac{\mathrm{d}}{\mathrm{d}t}\left(\frac{\partial v}{\partial d}\right) &= g\frac{\partial x_{\mathrm{i}}}{\partial d} - h\frac{\partial v}{\partial d}
\end{aligned} \tag{18}$$

$$
\begin{aligned}
\frac{\mathrm{d}}{\mathrm{d}t}\left(\frac{\partial x_{\mathrm{u}}}{\partial e}\right) &= -a\frac{\partial x_{\mathrm{u}}}{\partial e} + (c-d)\frac{\partial v}{\partial e}x_{\mathrm{u}} + (c-d)v\frac{\partial x_{\mathrm{u}}}{\partial e} \\
\frac{\mathrm{d}}{\mathrm{d}t}\left(\frac{\partial x_{\mathrm{i}}}{\partial e}\right) &= d\frac{\partial v}{\partial e}x_{\mathrm{u}} + dv\frac{\partial x_{\mathrm{u}}}{\partial e} - yx_{\mathrm{i}} - e\frac{\partial y}{\partial e}x_{\mathrm{i}} - ey\frac{\partial x_{\mathrm{i}}}{\partial e} \\
\frac{\mathrm{d}}{\mathrm{d}t}\left(\frac{\partial x_{\mathrm{i}}}{\partial e}\right) &= d\frac{\partial v}{\partial e}x_{\mathrm{u}} + dv\frac{\partial x_{\mathrm{u}}}{\partial e} - yx_{\mathrm{i}} - e\frac{\partial y}{\partial e}x_{\mathrm{i}} - ey\frac{\partial x_{\mathrm{i}}}{\partial e} \\
\frac{\mathrm{d}}{\mathrm{d}t}\left(\frac{\partial v}{\partial e}\right) &= g\frac{\partial x_{\mathrm{i}}}{\partial e} - h\frac{\partial v}{\partial e}
\end{aligned} \tag{19}
$$

$$
\begin{aligned}
\frac{\mathrm{d}}{\mathrm{d}t}\left(\frac{\partial x_{\mathrm{u}}}{\partial f}\right) &= -a\frac{\partial x_{\mathrm{u}}}{\partial f} + (c-d)\frac{\partial v}{\partial f}x_{\mathrm{u}} + (c-d)v\frac{\partial x_{\mathrm{u}}}{\partial f} \\
\frac{\mathrm{d}}{\mathrm{d}t}\left(\frac{\partial x_{\mathrm{i}}}{\partial f}\right) &= d\frac{\partial v}{\partial f}x_{\mathrm{u}} + dv\frac{\partial x_{\mathrm{u}}}{\partial f} - yx_{\mathrm{i}} - e\frac{\partial y}{\partial f}x_{\mathrm{i}} - ey\frac{\partial x_{\mathrm{i}}}{\partial f} \\
\frac{\mathrm{d}}{\mathrm{d}t}\left(\frac{\partial y}{\partial f}\right) &= -b\frac{\partial y}{\partial f} + vy + f\frac{\partial v}{\partial f}y + fv\frac{\partial y}{\partial f} \\
\frac{\mathrm{d}}{\mathrm{d}t}\left(\frac{\partial v}{\partial f}\right) &= g\frac{\partial x_{\mathrm{i}}}{\partial f} - h\frac{\partial v}{\partial f}
\end{aligned} \tag{20}
$$

$$
\begin{aligned}
\frac{\mathrm{d}}{\mathrm{d}t}\left(\frac{\partial x_{\mathrm{u}}}{\partial g}\right) &= -a\frac{\partial x_{\mathrm{u}}}{\partial g} + (c-d)\frac{\partial v}{\partial g}x_{\mathrm{u}} + (c-d)v\frac{\partial x_{\mathrm{u}}}{\partial g} \\
\frac{\mathrm{d}}{\mathrm{d}t}\left(\frac{\partial x_{\mathrm{i}}}{\partial g}\right) &= d\frac{\partial v}{\partial g}x_{\mathrm{u}} + dv\frac{\partial x_{\mathrm{u}}}{\partial g} - yx_{\mathrm{i}} - e\frac{\partial y}{\partial g}x_{\mathrm{i}} - ey\frac{\partial x_{\mathrm{i}}}{\partial g} \\
\frac{\mathrm{d}}{\mathrm{d}t}\left(\frac{\partial y}{\partial g}\right) &= -b\frac{\partial y}{\partial g} + f\frac{\partial v}{\partial g}y + fv\frac{\partial y}{\partial g} \\
\frac{\mathrm{d}}{\mathrm{d}t}\left(\frac{\partial v}{\partial g}\right) &= x_{\mathrm{i}} + g\frac{\partial x_{\mathrm{i}}}{\partial g} - h\frac{\partial v}{\partial g}
\end{aligned} \tag{21}
$$

$$
\begin{aligned}
\frac{\mathrm{d}}{\mathrm{d}t}\left(\frac{\partial x_{\mathrm{u}}}{\partial h}\right) &= -a\frac{\partial x_{\mathrm{u}}}{\partial h} + (c-d)\frac{\partial v}{\partial h}x_{\mathrm{u}} + (c-d)v\frac{\partial x_{\mathrm{u}}}{\partial h} \\
\frac{\mathrm{d}}{\mathrm{d}t}\left(\frac{\partial x_{\mathrm{i}}}{\partial h}\right) &= d\frac{\partial v}{\partial h}x_{\mathrm{u}} + dv\frac{\partial x_{\mathrm{u}}}{\partial h} - yx_{\mathrm{i}} - e\frac{\partial y}{\partial h}x_{\mathrm{i}} - ey\frac{\partial x_{\mathrm{i}}}{\partial h} \\
\frac{\mathrm{d}}{\mathrm{d}t}\left(\frac{\partial y}{\partial h}\right) &= -b\frac{\partial y}{\partial h} + f\frac{\partial v}{\partial h}y + fv\frac{\partial y}{\partial h} \\
\frac{\mathrm{d}}{\mathrm{d}t}\left(\frac{\partial v}{\partial h}\right) &= g\frac{\partial x_{\mathrm{i}}}{\partial h} - v - h\frac{\partial v}{\partial h}
\end{aligned} \tag{22}
$$

which obtains

$$\frac{\partial z}{\partial p_k}(t) = \frac{\partial x_{\mathrm{u}}}{\partial p_k}(t) + \frac{\partial x_{\mathrm{i}}}{\partial p_k}(t) + \frac{\partial y}{\partial p_k}(t), \text{ and } \frac{\partial v}{\partial p_k}(t) \quad \text{for } k = 1, 2, \ldots, 8$$

They are sampled at time and form the following Jacobian matrix

$$\Phi = \begin{bmatrix} U \\ V \end{bmatrix}$$

where

$$U = \begin{bmatrix} \frac{\partial z}{\partial p_1}(t_1) & \frac{\partial z}{\partial p_2}(t_1) & \ldots & \frac{\partial z}{\partial p_8}(t_1) \\ \frac{\partial z}{\partial p_1}(t_2) & \frac{\partial z}{\partial p_2}(t_2) & \ldots & \frac{\partial z}{\partial p_8}(t_2) \\ \ldots & \ldots & \ldots & \ldots \\ \frac{\partial z}{\partial p_1}(t_N) & \frac{\partial z}{\partial p_2}(t_N) & \ldots & \frac{\partial z}{\partial p_8}(t_N) \end{bmatrix}$$

and

$$V = \begin{bmatrix} \frac{\partial v}{\partial p_1}(t_1) & \frac{\partial v}{\partial p_2}(t_1) & \ldots & \frac{\partial v}{\partial p_8}(t_1) \\ \frac{\partial v}{\partial p_1}(t_2) & \frac{\partial v}{\partial p_2}(t_2) & \ldots & \frac{\partial v}{\partial p_8}(t_2) \\ \ldots & \ldots & \ldots & \ldots \\ \frac{\partial v}{\partial p_1}(t_N) & \frac{\partial v}{\partial p_2}(t_N) & \ldots & \frac{\partial v}{\partial p_8}(t_N) \end{bmatrix}$$

Step 3: Calculate the parameter increment $\delta \boldsymbol{p}$, according to $\delta \boldsymbol{p} = -(\boldsymbol{\Phi}^{\mathrm{T}}\boldsymbol{\Phi} + \lambda I)^{-1} \boldsymbol{\Phi}^{\mathrm{T}}\boldsymbol{w}$, where I is the identity matrix, λ a constant used to compensate for the possible singularity of the matrix $\boldsymbol{\Phi}^{\mathrm{T}}\boldsymbol{\Phi}$.

Step 4: $\boldsymbol{p} \rightarrow \boldsymbol{p} + \delta \boldsymbol{p}$. Calculate the objective J. If it converges, then exit. Otherwise go to Step 1.

Steps 1–4 constitute only one iteration. There may be many iterations before the algorithm converges and exists. If by convergence a satisfactory solution cannot be found, the whole process should be repeated with a new set of initial values.

Simulation results

Acute HHV-6 infection

The identified parameter values are as follows: $a = 6.469$; $b = 0.625$; $c = 0.819$; $d = 0.022$; $e = 5.066$; $f = 0.211$; $g = 40.81$; $h = 0.472$. By using these parameter values, Eqs. (10)–(13) generate simulation runs fitting well with the clinical data (see Fig. 2). As Fig. 2 shows, simulation runs of acute HHV-6 infection by the extended

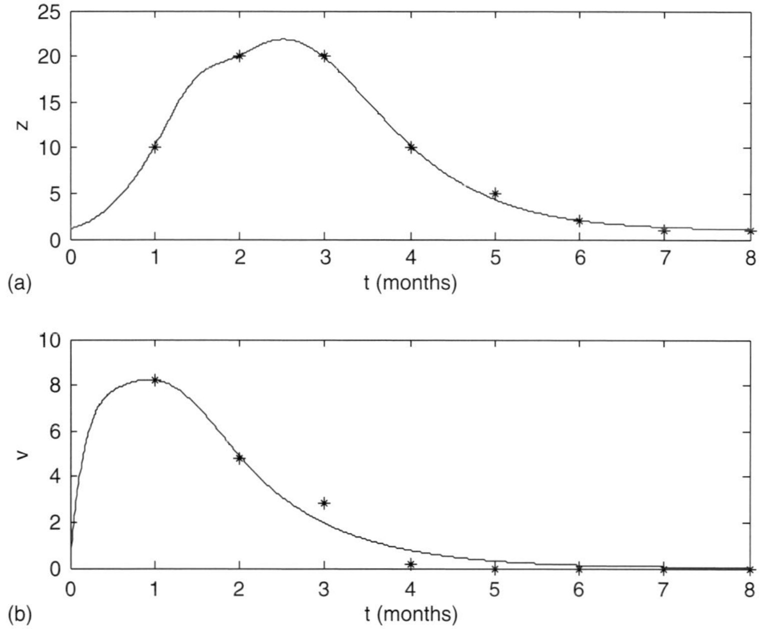

Fig. 2 Eight-month simulation run under parameter condition shown in the text compared with the clinical data. z shows changes of the total T-cell count, v shows changes of the virus load. Solid lines depict the simulation run; stars depict the actual patients' data.

computational model with parameter optimization is able to realistically simulate cell changes in the infected patient. This pertains equally well to all other cell populations such as T-helper and suppressor cells, thymic, and prethymic immature cell populations.

Chronic HHV-6 infection

The identified parameter values are as follows: $a = 3.089$; $b = 0.842$; $c = 0.165$; $d = 0.045$; $e = 17.9$; $f = 0.347$; $g = 2099.4$; $h = 8.65$. By using these parameter values, Eqs. (10)–(13) generate simulation runs fitting well with the clinical data (see Fig. 3). Simulation runs of chronic HHV-6 infection as shown in Fig. 3 depict a "clean" basic pattern of viral loads and respective cell changes. Although it matches the few available patients' data, its regularity appears rather unrealistic in an actual patient's situation. In chronic persistent HHV-6 infection, various extraneous influences can be expected during such a long 25 months period (e.g. immunosuppressive or immunostimulatory). These will influence viral load and the host's response to it, so that the actual curve of cell populations and viral load will become more irregular. This must be accounted for by adjusting mathematical procedures and validating these by more patient data from longitudinal studies.

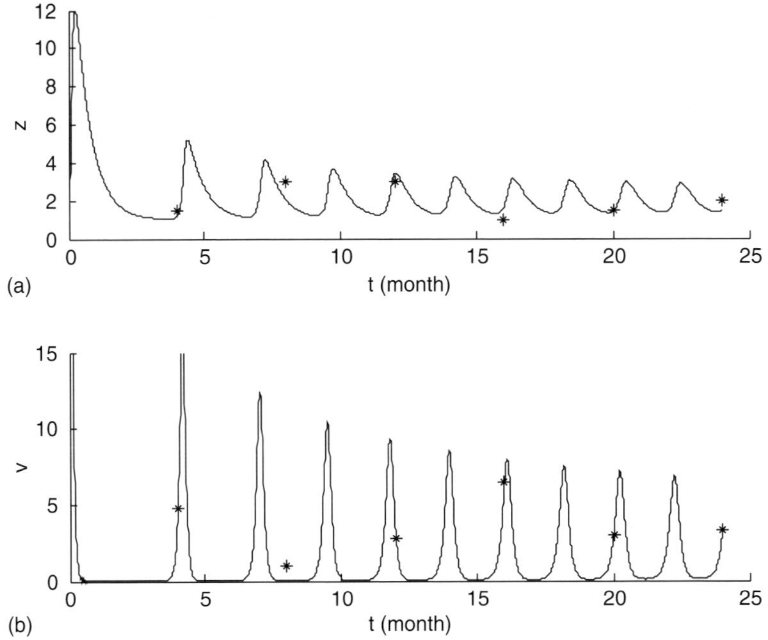

Fig. 3 Twenty four-month simulation run under parameter condition shown in the text compared with the clinical data. z shows changes of the total T-cell count, v shows changes of the virus load. Solid lines depict the simulation run; stars depict the data.

Essentially unchanged will stay, however, the gross pattern of simulation runs indicating that there exists a steady fluctuation of viral load and cell pools in a chronic infection. This is important to note, and it clearly decreases the value of the usual cross-sectional clinical studies in a chronically infected patient.

Discussion

The mechanism of HHV-6 infection is simple and well characterized (Krueger et al., 2001a,b; Ablashi and Krueger, 2003; Krueger and Ablashi, 2003; Wang et al., 2003). One wonders, though, why infections by the same virus may cause different disease outcomes, such as CFS, IM, MS, and lymphoproliferative disorders (LPD). IM is caused to a certain percentage by the accute HHV-6 infection and is self-limited (Purtilo et al., 1985; Krueger and Ferrer, 1994; Krueger et al., 2001a, 2003), because everything returns to normal after the first round of infection. Dynamically, this represents a stable steady state, or a fixed point. The chronic HHV-6 infection is prolonged and induces undulant cell reaction patterns (Ablashi et al., 2000; Krueger et al., 2001b, 2003). Both T-cell counts and the viral concentration change consistently throughout the patient's life, manifesting the intense combat between the virus and the immune system. They will never return to their base level. Dynamically, this signals a chaotic state (states other than fixed points, periodic, quasi-periodic are all classified as chaotic).

The power of mathematical modeling is that diverse phenomena and dynamics can be studied using the same model. Namely, different diseases can be studied under the same paradigm. The described model contributes to explaining such variable disease courses. The model is a nonlinear one, which generates different dynamical behaviors of immunocompetent T-cell populations under different parameter conditions. Clinically this implies that diverse disease outcomes can be attributed to different sets of model parameters of patients such as rate of infection, duration of exposure to the virus, intensity of immune response etc.

By comparing model parameters of acute infection and chronic infection, we find that the values of c, g, and h change drastically. As we know $cvx_u = c_1vx_u - c_2vx_u$, where c_1vx_u represents virus stimulating CD4 cells and c_2vx_u the death of CD4 cells because of viral replication. In the chronic infection, the virus appears to be more 'vicious'; it causes CD4 cells to die faster. This parellels to the large value of g in the chronic infection. Indeed, a larger virus production consumes the cells faster. Therefore, parameters c and g can be used as the bifurcation parameters in studying and controling the abrupt dynamic changes. In brief, we can find mathematically a law perturbing the two parameters, which effectively attracts the chronic infection dynamics to the benigh dynamics, letting the dynamics returns to normal very naturally. Therefore, the control strategе may serve as the theoretical basis for further therapeutic interventions.

The present model can even be used to study HIV infection, because the immune responses following HIV infection are similar to those of HHV-6 infection. The differences consist in that in HIV infection, the model uses another set of parameters, which characterize a more vicious virus.

References

Ablashi DV, Eastman HB, Owen CB, Roman MM, Friedman J, Zabriskie JB, Peterson DL, Pearson GR, Whitman JE. Frequent HHV-6 reactivation in multiple sclerosis (MS) and chronic fatigue syndrome (CFS) patients. J Clin Virol 2000; 16: 179–191.

Ablashi DV, Krueger GR. The Human Herpesviruses HHV-6, HHV-7, and HHV-8. In: Viral Infections and Treatment (Rubsamen-Waigmann H, Deres K, Hewlett G, Welker R, editors). Wuppertal, Germany: Marcel-Dekker; 2003; pp. 659–705.

Ablashi DV, Lapps W, Kaplan M, Whitman JE, Richert JR, Pearson GR. Human herpesvirus-6 (HHV-6) infection in multiple sclerosis: a preliminary report. Multiple Sclerosis 1998; 4: 490–496.

Ablashi DV, Zompetta C, Lease C, Josephs SF, Balachandra N, Komaroff AL, Krueger GR, Henry B, Lukau J, Salahuddin SZ. Human herpesvirus 6 (HHV6) and chronic fatigue syndrome (CFS). Can Dis Wkly Rep 1991; 17(Suppl 1E): 33–40.

Berti R, Brennan MB, Soldan SS, Ohayon JM, Casareto L, McFarland HF, Jacobson S. Increased detection of serum HHV-6 DNA sequences during multiple sclerosis (MS) exacerbations and correlation with parameters of MS disease progression. J Neurovirol 2002; 8: 250–256.

Berti R, Soldan SS, Akhyani N, McFarland HF, Jacobson S. Extended observations on the association of HHV-6 and multiple sclerosis. J Neurovirol 2000; 6(Suppl 2): S85–S87.

Bertram G, Dreiner N, Krueger GR, Ramon A, Ablashi DV, Salahuddin SZ, Balachandram N. Frequent double infection with Epstein–Barr virus and human herpesvirus-6 in patients with acute infectious mononucleosis. In Vivo 1991; 5: 271–279.

Cermelli C, Berti R, Soldan SS, Mayne M, D'Ambrosia JM, Ludwin SK, Jacobson S. High frequency of human herpesvirus 6 DNA in multiple sclerosis plaques isolated by laser microdissection. J Infect Dis 2003; 187: 1377–1387.

Challoner PB, Smith KT, Parker JD, MacLeod DL, Coulter SN, Rose TM, Schultz ER, Bennett JL, Garber RL, Chang M, Schad PA, Stewart PM, Nowinski RC, Brown JP, Burmer GC. Plaque-associated expression of human herpesvirus 6 in multiple sclerosis. Proc Natl Acad Sci USA 1995; 92: 7440–7444.

Clark D. Human herpesvirus type 6 and multiple sclerosis. Herpes 2004; 11(Suppl 2): 112A–119A.

Descamps V, Bouscarat F, Laglenne S, Aslangul E, Veber B, Descamps D, Saraux JL, Grange MJ, Grossin M, Navratil E, Crickx B, Belaich S. Human herpesvirus 6 infection associated with anticonvulsant hypersensitivity syndrome and reactive haemophagocytic syndrome. Br J Dermatol 1997; 137: 605–608.

Descamps V, Valance A, Edlinger C, Fillet AM, Grossin M, Lebrun-Vignes B, Belaich S, Crickx B. Association of human herpesvirus 6 infection with drug reaction with eosinophilia and systemic symptoms. Arch Dermatol 2001; 137: 301–304.

Glaser R, Padgett DA, Litsky ML, Baiocchi RA, Yang EV, Chen M, Yeh PE, Klimas NG, Marshall GD, Whiteside T, Herberman R, Kiecolt-Glaser J, Williams MV. Stress-associated changes in the steady-state expression of latent Epstein–Barr virus: implications for chronic fatigue syndrome and cancer. Brain Behav Immun 2005; 19: 91–103.

Goodman AD, Mock DJ, Powers JM, Baker JV, Blumberg BM. Human herpesvirus 6 genome and antigen in acute multiple sclerosis lesions. J Infect Dis 2003; 187: 1365–1376.

Horwitz CA, Krueger GRF. HHV-6 Induced Mononucleosis-like Illnesses. In: Human Herpesvirus-6 (Steeper TA, Bertram G, editors). Amsterdam: Elsevier Sci Publ; 1992; pp. 159–174.

Josephs SF, Henry B, Balachandran N, Strayer D, Peterson D, Komaroff AL, Ablashi DV. HHV-6 reactivation in chronic fatigue syndrome. Lancet 1991; 337: 1346–1347.

Krueger GR, Ablashi DV. Human Herpesvirus-6: a short review of its biological behavior. Intervirology 2003; 46: 257–269.

Krueger GRF, Bertram G, Ramon A, Koch B, Ablashi DV, Brandt ME, Wang G, Buja LM. Dynamics of infection with human herpesvirus-6 in EBV-negative infectious mononucleosis: data acquisition for computer modeling. In Vivo 2001a; 15: 373–380.

Krueger GRF, Brandt ME, Wang G, Berthold F, Buja LM. A computational analysis of Canale–Smith syndrome: chronic lymphadenopathy simulating malignant lymphoma. Anticancer Res 2002; 22: 2365–2372.

Krueger GRF, Brandt ME, Wang G, Buja LM. TCM-1: a nonlinear dynamical computational model to simulate cellular changes in the T-cell system: conceptional design and validation. Anticancer Res 2003; 23: 123–136.

Krueger GR, Brandt ME, Wang G, Buja LM. Computational simulation of chronic persistent virus infection: factors determining differences in clinical outcome of HHV-6, HIV-1 and HTLV-1 infections including aplastic, hyperplastic and neoplastic responses. Anticancer Res 2004; 24: 187–197.

Krueger GRF, Ferrer AV. A unifying concept of viral immunopathogeneis of proliferative and aproliferative diseases (working hypothesis). In Vivo 1994; 8: 493–500.

Krueger GRF, Koch B, Hoffmann A, Rojo J, Brandt ME, Wang G, Buja LM. Dynamics of chronic active herpesvirus-6 infection in patients with chronic fatigue syndrome: data acquisition for computer modeling. In Vivo 2001b; 15: 461–466.

Purtilo D, Tatsumi E, Manolov G, Manolova Y, Harada S, Lipscomb H, Krueger G. Epstein–Barr virus as an etiological agent in the pathogenesis of lymphoproliferative and aproliferative diseases in immune deficient patients. Int Rev Exp Path 1985; 27: 113–183.

Rotola A, Merlotti I, Caniatti L, Caselli E, Granieri E, Tola MR, Di Luca D, Cassai E. Human herpesvirus 6 infects the central nervous system of multiple sclerosis patients in the early stages of the disease. Multiple Sclerosis 2004; 10: 348–354.

Stortelder WJH. Parameter estimation in dynamic system. Math Comput Simul 1996; 42: 135–142.

Tohyama M, Yahata Y, Yasukawa M, Inagi R, Urano Y, Yamanishi K, Hashimoto K. Severe hypersensitivity syndrome due to sulfasalazine associated with reactivation of human herpesvirus 6. Arch Dermatol 1998; 134: 1113–1117.

Wagner M, Krueger GR, Ablashi D, Whitman JE. Chronic fatigue syndrome (CFS): a critical evaluation of testing for active human herpesvirus-6 (HHV-6) infection: review of data from 107 cases. J Chron Fatigue Syndr 1996; 2: 3–16.

Wang G. Parameter optimization in large-scale dynamical systems: a method of contractive mapping. Math Comput Simul 2004; 66: 565–576.

Wang G, Krueger GRF, Buja LM. A mathematical model to simulate the cellular dynamics of infection with human herpesvirus-6 in EBV negative infectious mononucleosis. J Med Virol 2003; 71: 569–577.

Yamanishi K. Chronic fatigue syndrome and virus infection: human herpesvirus 6 (HHV-6) infection. Nippon Rinsho 1992; 50: 2612–2616.

Human Herpesvirus-6: General Virology, Epidemiology and Clinical Pathology
Perspectives in Medical Virology
Gerhard Krueger and Dharam Ablashi (Editors)

DOI 10.1016/S0168-7069(06)12026-1

Experimental Therapeutic Approaches

Joseph H. Brewer
Section of Infectious Disease - St. Luke's Hospital and Plaza Infectious Disease, 4320 Wornall Road, Suite 440, Kansas City, MO 64111, USA

Introduction

The potential therapeutic approaches for human herpesvirus-6 (HHV-6) infections include antiviral therapy and immune therapies. Several experimental therapy approaches have possible benefits for such infection based on theoretical grounds as well as clinical studies. These therapies may alter viral infection via immune-mediated mechanisms, which relate to either humoral immunity or cell-mediated immune function. There may also be direct or indirect antiviral effects. Herein, these therapy alternatives are reviewed in terms of experimental and clinical data. Most of the clinical studies regarding these therapies that have implications for patients with HHV-6 infection have generally been done in groups of patients with disease states or syndromes that are possibly associated with HHV-6 infection. The main examples of such clinical syndromes that are addressed in this section are chronic fatigue syndrome (CFS) and multiple sclerosis (MS).

Immunoglobulin

Several commercially available intravenous immunoglobulin (IVIG) preparations have been studied (Krause et al., 2002) to evaluate antimicrobial activity against various microbial pathogens. High antibody titers were found for several of the herpes viruses including HHV-6. The relative importance of humoral immunity (immunoglobulin) for the control of herpes infections, including HHV-6, is not entirely clear. If indeed immunoglobulin represents an important aspect of controlling HHV-6, then it is intuitive that IVIG may be a potentially effective therapy for HHV-6-associated infection. On the other hand, if cell-mediated immunity is the dominant mechanism for immune control for HHV-6, then immunoglobulin may be of limited value.

Several clinical studies have been done in patients with CFS utilizing IVIG compared to placebo. One double-blind, placebo-controlled study of 30 CFS patients (Peterson et al., 1990) did not demonstrate the benefit of IVIG in terms of symptom amelioration or improvement in functional status. A similar study of 99 CFS patients from Australia (Vollmer-Conna, 1997) also showed lack of statistically significant benefit of IVIG as compared to albumen. However, another randomized, double, blind study (Lloyd et al., 1990) comparing monthly high-dose IVIG (2 g/kg) versus placebo, demonstrated improvement in several parameters including decrease in symptoms, increased functional status and improved immunologic measures. The reason for the difference in outcomes in these studies remains unclear.

A study of relapsing remitting MS patients (Fazekas et al., 1997) compared IVIG at smaller doses (0.15–0.2 mg/kg) to placebo. In this study of 150 patients, benefits were noted in the IVIG group in terms of disability status (Kurtzke expanded disability status score).

The potential role of IVIG for the treatment of HHV-6-related infection remains conjectural. IVIG is quite expensive and has a variety of adverse effects that are not uncommonly associated with the infusions.

Interferon

HHV-6 infection induces production of various cytokines from infected macrophages including interferon (Inoue et al., 1993). These cytokines, including interferon, play a role in controlling and containing the viral infection. Since interferon has broad antiviral properties, recombinant interferon may be potentially useful in the treatment of infection with HHV-6. With regard to interferon therapy, there is very little information regarding the treatment of HHV-6 with any of the three types of interferon. Alpha interferon has been studied in CFS patients showing improvement in the quality of life scores (See & Tilles, 1996). Interferon beta has been used in the treatment of MS for over 20 years. Interferon beta has been shown to be effective in decreasing progression of MS and reducing disability (Fillipini, 2003), particularly with relapsing and remitting MS. Albeit the mechanism is widely held to be immune modulation, the antiviral effects of interferon beta may be involved, as well.

It has been shown (Hong et al., 2002) *in vitro* that interferon beta at concentrations of 0.5 μg/ml reduced the replication of HHV-6 in a susceptible line of T cells. Additionally, this group examined the sera of MS patients treated with interferon beta compared to control patients not on treatment for cell-free detection of HHV-6 DNA and IgM antibody reactivity to HHV-6. The sera of patients treated with interferon beta had reduced levels of HHV-6 DNA detected and lower levels of IgM antibody reactivity. They also looked at paired sera of seven patients before and after treatment with interferon beta. The sera obtained after treatment showed decreased levels of HHV-6 DNA as compared to the pretreatment sera. Another study (Alvarez-Lafuente, 2004) evaluated relapsing remitting MS patients treated with interferon beta ($n = 105$) compared to similar patients who were not being treated ($n = 84$). Utilizing a quantitative real time PCR for HHV-6, the HHV-6 viral load was twice as high in the untreated patients versus the treated cases. These effects were only seen during relapse. No differences were seen when patients were in remission. All of these cases, interestingly, were HHV-6 variant A. Thus, interferon beta may exert at least some of the effects in the treatment of MS related to the antiviral properties of these agents for HHV-6.

Interferon therapy is associated with numerous adverse effects including fever, fatigue, myalgia, nausea and headache, among others (Fillipini, 2003).

Ampligen

Ampligen is a chain of nucleic acids, which is mismatched double-stranded RNA, poly(I): poly(C12U), that has been shown to have antiviral activity. *In vitro* studies (Ablashi et al., 1994) have demonstrated activity against HHV-6. In a study of HSB-2 cells (a B cell line) infected with the HHV-6 variant A (GS strain), ampligen was effective at blocking viral infection at concentrations of 100 and 200 pg/ml (Ablashi et al., 1994). In this study, when cells were pretreated with ampligen, there was a 98% inhibition of infection. If the cells were infected first and then treated with ampligen, there was still a 95% inhibition of infection. When the ampligen was removed from the virus-infected cell culture, HHV-6 infection reappeared slowly but never reached the same level as before. No toxicity to uninfected HSB-2 cells was noted. Ampligen inhibited HHV-6 DNA polymerase activity.

Studies that address the activity of ampligen in HHV-6-associated infection have been done in patients with CFS. Ampligen has been evaluated in a randomized, double-blind, placebo-controlled study (Strayer et al., 1995) of CFS patients. Efficacy was assessed by Karnofsky performance score (KPS), cognition scale (SCL-90-R), activity of daily living (ADL) scale and exercise treadmill performance. After 24 weeks of treatment, patients receiving ampligen had higher scores for global performance and perceived cognition than the placebo group. Patients on ampligen had improved KPS, increased capacity for ADL, reduced cognitive impairment and improved work on the treadmill. Patients receiving ampligen also required less medication to control their symptoms. In another study (Strayer et al., 1994), 15 patients with CFS were studied for performance (KPS), neuropsychological testing, exercise capacity and HHV-6 antigen positivity in cell culture. After treatment with ampligen for 12–48 weeks, there was an improvement from baseline for KPS, cognition and exercise tolerance. In the viral analysis, there was a reduction in the expression of HHV-6. An evaluation of levels of 2,5-oligoadenylate synthetase (2-5A) and RNAase L (Suhadolnik, 1994) in 15 patients with CFS before and after treatment with ampligen was compared to healthy controls. Patients had lower levels of 2-5A and increased levels of RNAase L activity. Also, the levels of HHV-6 replication activity in peripheral blood mononuclear cells (PBMC) were studied before and after treatment. Therapy with ampligen resulted in significant downregulation of the 2-5A/RNAase L pathway, as well as significant decrease in HHV-6 replication in PBMC.

Thus, ampligen has demonstrated *in vitro* activity for HHV-6 in infected cells, decreased HHV-6 replication in CFS patients and improved clinical parameters (performance, cognitive skills and symptom reduction) in patients treated with ampligen compared to placebo. Presently, it remains an investigational therapy.

Transfer factor

Transfer factor (TF) was originally described as dialyzable leukocyte extract over 40 years ago. This small dialyzable molecule can transfer cell-mediated immunity from an immune donor to a non-immune recipient (Fudenberg and Pizza, 1989). The molecular structure of TF is uncertain, however, it has been postulated to be a small ribonucleopeptide (Dwyer, 1983). TF appears to have the properties of a cytokine that can induce an immune response in the recipient, probably via the effects on T cells (Fudenberg and Pizza, 1989). TF is antigen specific but the degree of epitope specificity has not been clarified. It is not species specific and thus can be transferred from one species to another without an allergic reaction in the recipient (Fudenberg and Pizza, 1989). There are several potential sources for TF. One very accessible source for substantial amounts of TF is bovine colostrum (Jones, 2003).

TF has been used as a therapeutic modality in a variety of infections (Fudenberg and Pizza, 1989). Numerous studies (Steele et al., 1980; Jones, 2003; Viza et al., 1987) have investigated the treatment of herpes virus infections with TF including herpes simplex, varicella zoster virus, cytomegalovirus and Epstein–Barr virus. One study utilizing an oral formulation of TF derived from bovine colostrum was highly effective in controlling infection with cytomegalovirus and Epstein–Barr virus in children (Jones, 2003).

A report was published (Ablashi et al., 1996) on the treatment of two patients with CFS and active HHV-6 infection treated with a TF that was specific for HHV-6. One of these patients improved rapidly and resumed normal activities, whereas the other patient did not improve. Another study (Brewer and Wilson, 2003) was reported in CFS patients with active HHV-6 infection (HHV-6 viremia demonstrated by positive rapid viral culture) and low natural killer cell function who were treated with TF. One group of 28 patients was given a TF derived from bovine colostrum that had specific activity for HHV-6 compared to 10 patients who were given TF from colostrum devoid of activity for HHV-6. The group given the HHV-6-specific TF showed a significant improvement in symptom profile scores as well as improved NK immune function compared to the placebo group. This study, among others, has shown TF to be remarkably free of adverse effects (Fudenberg and Pizza, 1989; Brewer and Wilson, 2003).

Although, the data for TF in the treatment of HHV-6-related infection is limited, it may represent an attractive alternative that merits further investigation. It has proven to be a potentially effective therapy for various herpes virus infections including HHV-6 and has a good record of safety.

Isoprinosine

Isoprinosine (inosine pranobex) is a synthetic purine derivative licensed in 1971. It has been demonstrated to have immune modulating and antiviral properties. This compound has been used to treat immune deficiencies and various viral infections. Isoprinosine modulates T cell and NK function (Diaz-Mitoma et al., 2003), thus, the major effects relate to cell-mediated immunity. Safety studies in clinical trials and post-marketing experience have shown it to be quite free of adverse effects (Diaz-Mitoma et al., 2003). It has not been specifically studied *in vitro* for HHV-6 infection. However, isoprinosine has been recently studied in patients with CFS and low NK function (Diaz-Mitoma et al., 2003). In this single-blinded study, clinical improvement was noted in six out of ten patients treated with isoprinosine compared to a group receiving placebo. The clinical improvement was directly associated with improved NK function from baseline. This is the only controlled study to date evaluating this compound in any of the HHV-6-associated infections. This agent is not currently available in the US, but is sold in many other countries as an antiviral drug.

References

Ablashi DV, Berneman Z, Strayer DR, Suhadolnik RJ, Reichenbach NL, Hitzges P, Komaroff A. Clin Infect Dis 1994; 18: S113.

Ablashi DV, Berneman ZN, Williams M, Strayer DR, Kramarsky B, Suhadolnik RJ, Reichenbach N, Hiltzges P, Komaroff AL. In Vivo 1994; 8(4): 587.

Ablashi DV, Levine PH, DeVinci C, Whiteman JE, Pizza G, Viza D. Biotherapy 1996; 9: 81.

Alvarez-Lafuente R, De Las Heras V, Bartolome M, Picazo JJ, Arroyo R. Eur Neurol 2004; 52: 87.

Brewer JH, Wilson G. Poster Presentation, AACFS 6th International Conference. Chantilly, VA; 2003.

Diaz-Mitoma F, Turgonyi E, Kumar A, Lim W, Larocque L, Hyde BM. J Chronic Fatigue Syndrome 2003; 11: 71.

Dwyer JM. In: Immunobiology of Transfer Factor (Kirkpatrick CH, Burger DR, Lawrence HS, editors), New York: Academic Press; 1983; p. 233.

Fazekas F., Deisenhammer F., Strasser-Fuchs S., Nahler G., Mamoli B. Lancet 1997; 349: 589

Fillipini G, Munari L, Incorvaia B, Ebers GC, Polman C, D'Amico R, Rice GPA. Lancet 2003; 361: 545.

Fudenberg H, Pizza G. Ann Rev Pharmacol Toxicol 1989; 29: 309.

Hong J, Tejada-Simon MV, Rivera VM, Zang YC, Zhang JZ. Multiple Sclerosis 2002; 8: 237.

Inoue N, Dambaugh TR, Pellet PE. Infect Agents Dis 1993; 2: 343.
Jones JF, Minnich LL, Jeter WS, Pritchett RF, Fulginiti VA, Wedgewood RJ. Lancet 2003; 2: 122.
Krause I, Wu R, Sherer Y, Patanik M, Peter JB, Schoenfeld Y. Transfusion Med 2002; 12: 133.
Lloyd A, Hickie I, Wakefield D, Boughton C, Dwyer J. Am J Med 1990; 89: 561.
Peterson PK, Shepard J, Macres M, Schenck C, Crosson J, Renchtman D, Lurie N. Am J Med 1990; 89: 554.
See DM, Tilles JG. Immunol Invest 1996; 25: 153.
Steele RW, Myers MG, Monroe VM. N Engl J Med 1980; 303: 355.
Strayer DR, Carter WA, Brodsky I, Cheney P, Peterson D, Salvato P, Thompson C, Loveless M, Shapiro DE, Elsasser W, Gillespie DH. Clin Infect Dis 1994; 18: S88.
Strayer DR, Carter W, Strauss KI, Brodsky I, Suhadolnik RJ, Ablashi D, Henry B, Mitchell WM, Bastein S, Peterson D. J Chronic Fatigue Syndrome 1995; 1(1): 35.
Suhadolnik RJ, Reichenbach NL, Hitzges P, Adelson ME, Peterson DL, Cheney P, Salvato P, Thompson C, Loveless M, Muller WE. In Vivo 1994; 8: 599.
Viza D, Vich JM, Phillips J, Davies DA. J Exp Pathol 1987; 3: 407.
Vollmer-Conna U, Hickie I, Hadzi-Pavlovic D, Tymms K, Wakefield D, Dwyer J, Lloyd A. Am J Med 1997; 103: 38.

Colour Section

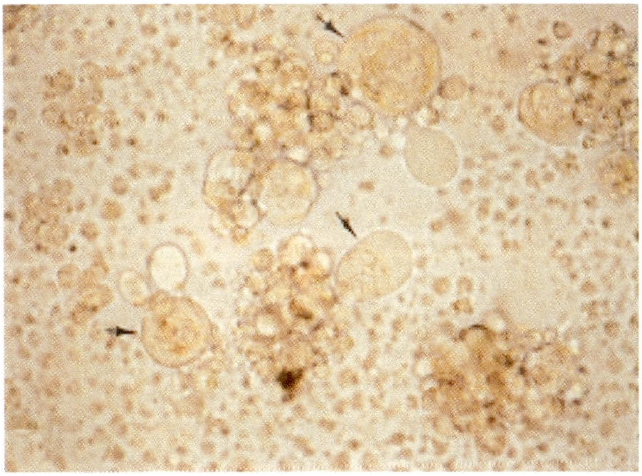

Plate 1 Peripheral blood mononuclear cells (PHA stimulated), cultured from AIDS patient with B-cell lymphoma, showing large refractile cells. (see page 4).

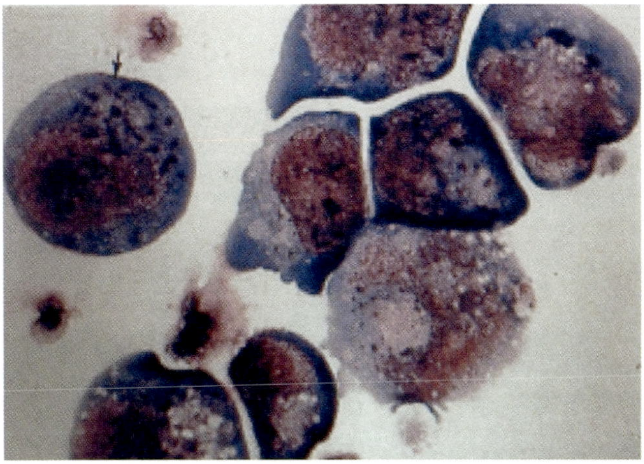

Plate 2 Giemsa-stained peripheral blood mononuclear cells, containing refractile cells, showing multinucleated giant cells. (see page 4).

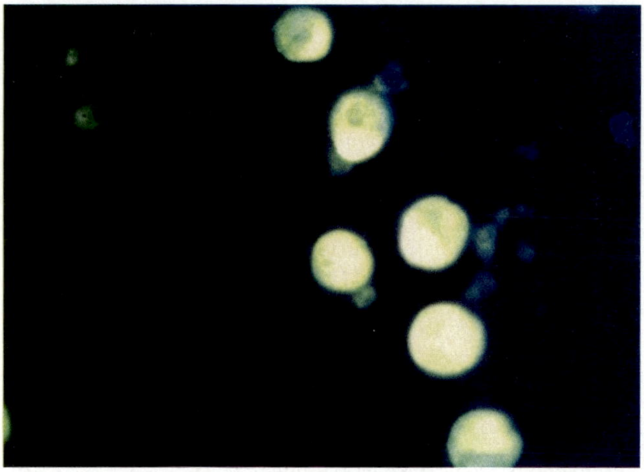

Plate 3 Immunofluorescent staining of HHV-6-infected human cordblood mononuclear cells with patient GS serum. (see page 6).

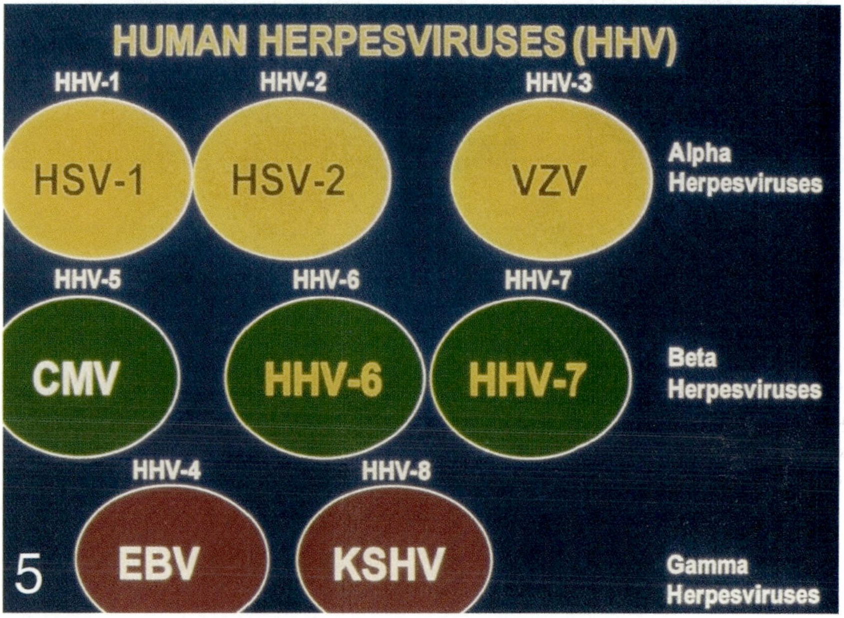

Plate 4 Sub-families of Human Herpesviruses. (see page 8).

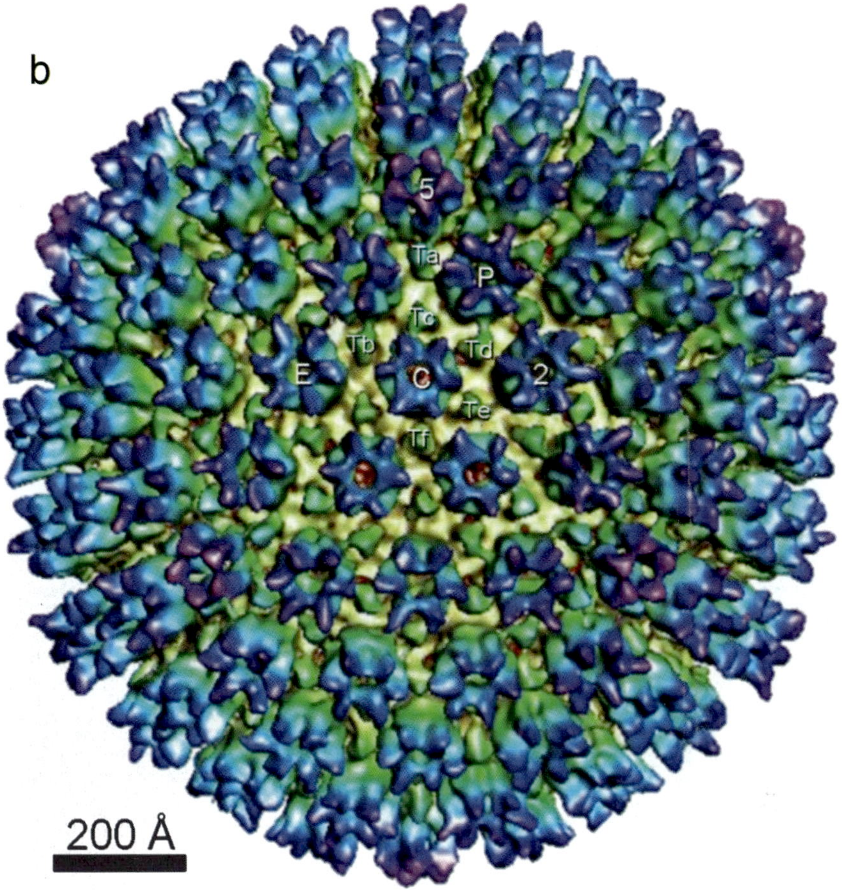

Plate 5 CryoEM imaging and 3D reconstruction HSV-6 capsid. (a) A gallery of cryoEM particle images of HHV-6 capsids. (b) Shaded surface representation of HHV-6 capsid reconstruction at 30 Å resolution. The structure is color coded according to capsid radius so that the capsid shell is in yellow, the triplexes are in green, and the upper domains of the pentons and hexons are in purple. (see page 17).

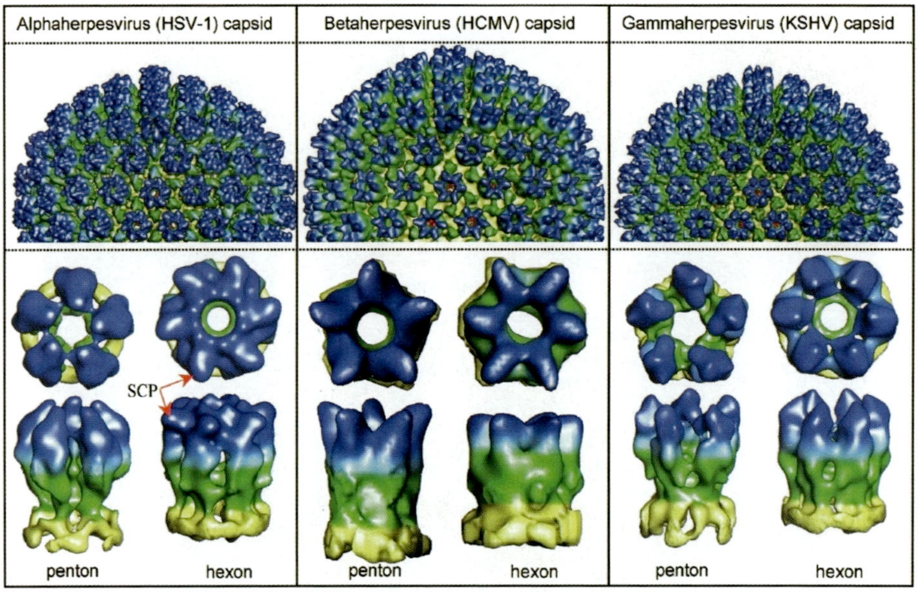

Plate 6 Comparison of HSV-1, HCMV, and KSHV capsids. A penton and a hexon are extracted from the cryoEM structures for detail comparison. Overall structure is similar; minor differences in hexon morphology may be a result of different structures of the distally located SCPs. (Modified from a figure by P. Lo). (see page 18).

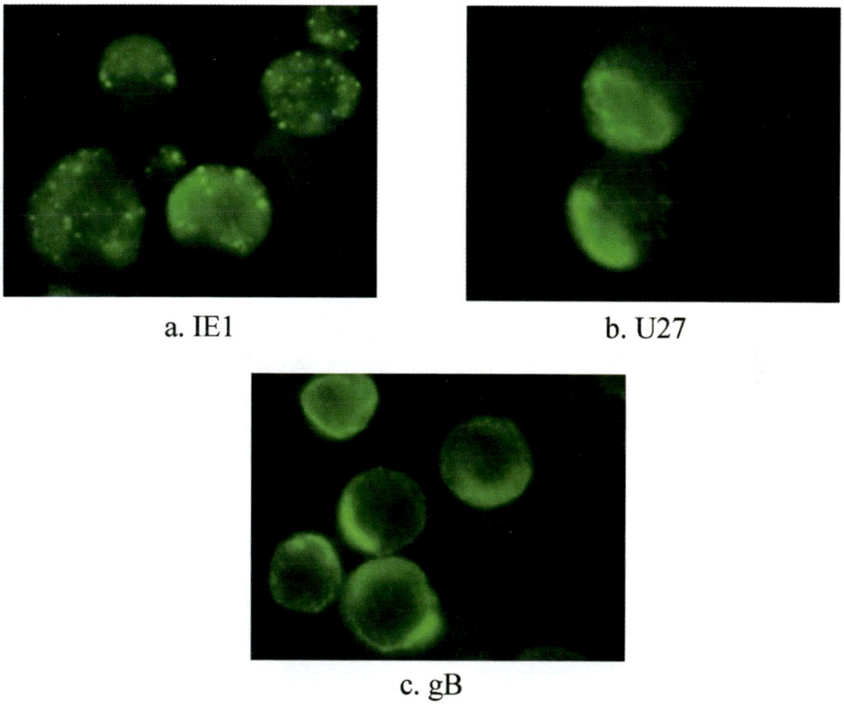

Plate 7 Indirect immunofluorescence assay of HHV-6-infected cells. The HSB-2 cells infected with HHV-6A (strain GS) were stained with monoclonal antibody for IE1 (a) U27 (b) or gB (c) at 86 h p.i. (a) IE1 locates in nucleus with punctuated pattern. (b) U27 locates in nucleus like forming replication compartment. (c) gB locates in the cytoplasm. (see page 49).

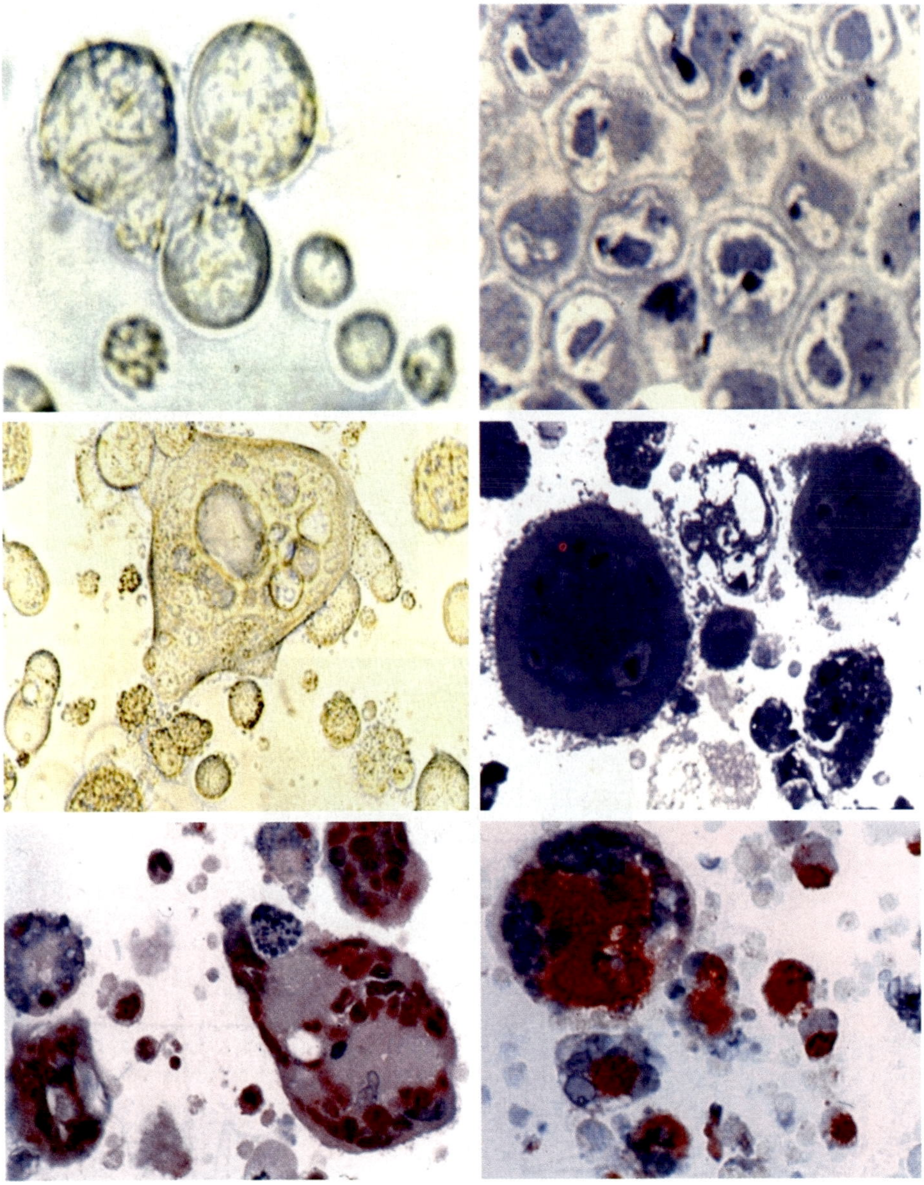

Plate 8 Cellular reactions in HHV-6-infected cells. *Top Row*: HHV-6A-infected HSB2 cells in culture: blastic transformation of infected cells (left) and nuclear inclusions in semithin section (right). *Center Row:* Prominent giant cell formation after infection with HHV-6A: SupT1 cells in culture (left) and L-428 Hodgkin cell line (semithin section of culture; right). *Bottom Row:* Antigen expression in HHV-6A-infected SupT1 cells: p41 early antigen (left) and HAR3 (i.e. mixture of late gp antigens; right). All APAAP immunohistochemistry. (see page 136).

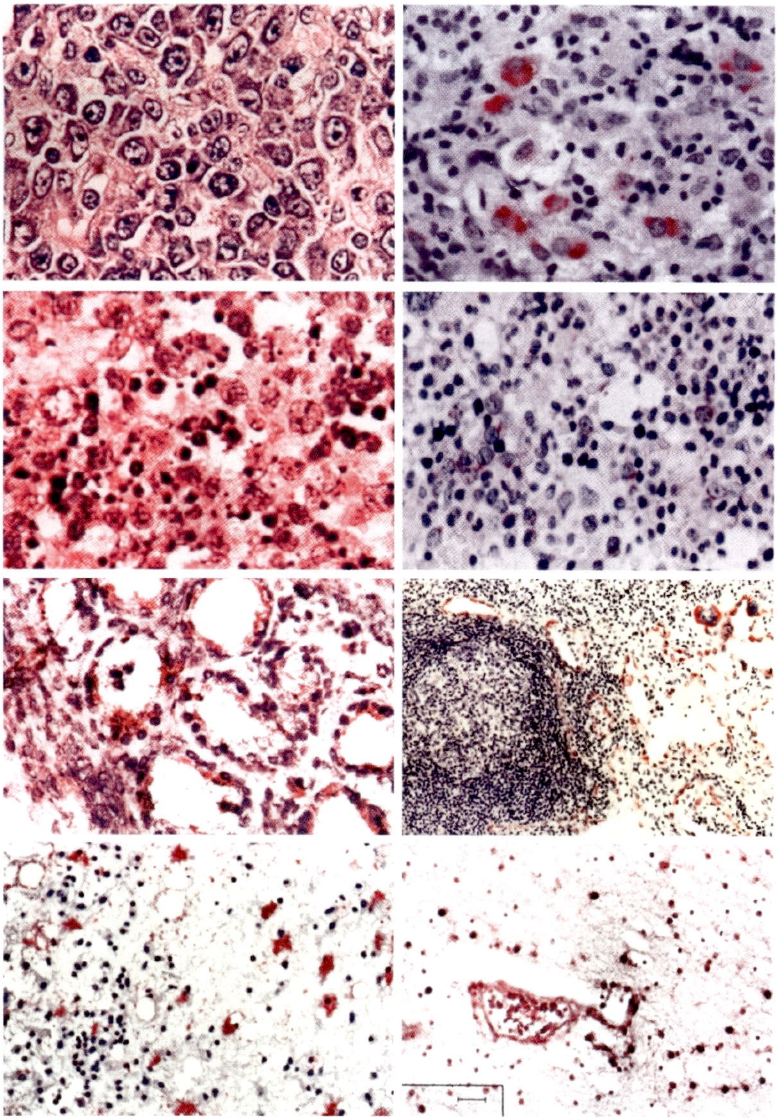

Plate 9 Some characteristic tissue reactions in HHV-6 infections. *Top Row:* Blastic transformation of cells in tonsils with prominent nucleoli (left) with expression of HHV-6 p41 early antigen (red cells in APAAP reaction; right). *Upper Center Row:* Prominent apoptosis of HHV-6 infected cells in Kikuchi's lymphadenitis (left); hematopoietic stem cells in bone marrow expressing HHV-6 p41 antigen (red dots in APAAP reaction; right). *Lower Center Row:* Expression of HHV-6 gp135 in epithelial cells of the salivary gland (lip biopsy; left); HHV-6A-associated lymphoid interstitial pneumonitis, LIP, in an HIV infected patient (red cells carry p41 HHV-6 antigen; APAAP reaction). *Bottom Row:* Acute necrotizing encephalitis with HHV-6 p41 and DNA expression in numerous astroglial cells (red cells by APAAP left, black dots for HHV-6 DNA by *in situ* hybridization (right; Wagner et al., 1997). (see page 140).

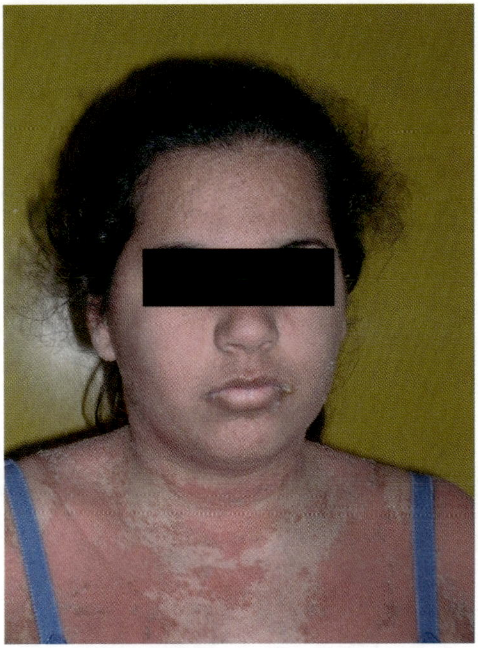

Plate 10 Facial edema and an exfoliative dermatitis in DRESS. (see page 152).

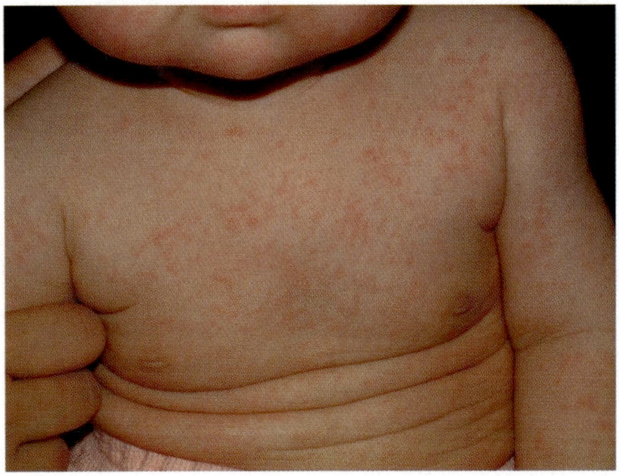

Plate 11 Typical skin eruption observed in patient with primary HHV-6 infection. Maculopapular skin rash is observed on the trunk and extremity. (see page 164).

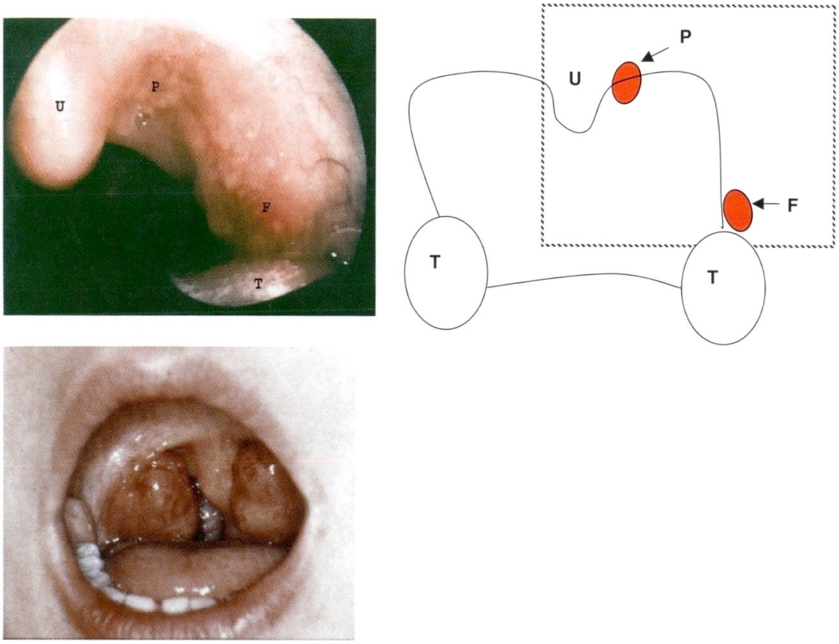

Plate 12 Enanthema with Nagayama's papules at the parauvular region (P) and fossa supratonsillaris (F), (U) Uvula, (T) tonsils; on the right a scheme of the throat with the photo section. Tonsillar hyperplasia in primary HHV-6 infection (lower photo with courtesy of G. Bertram, University ENT Clinic Cologne, Germany). (see page 177).

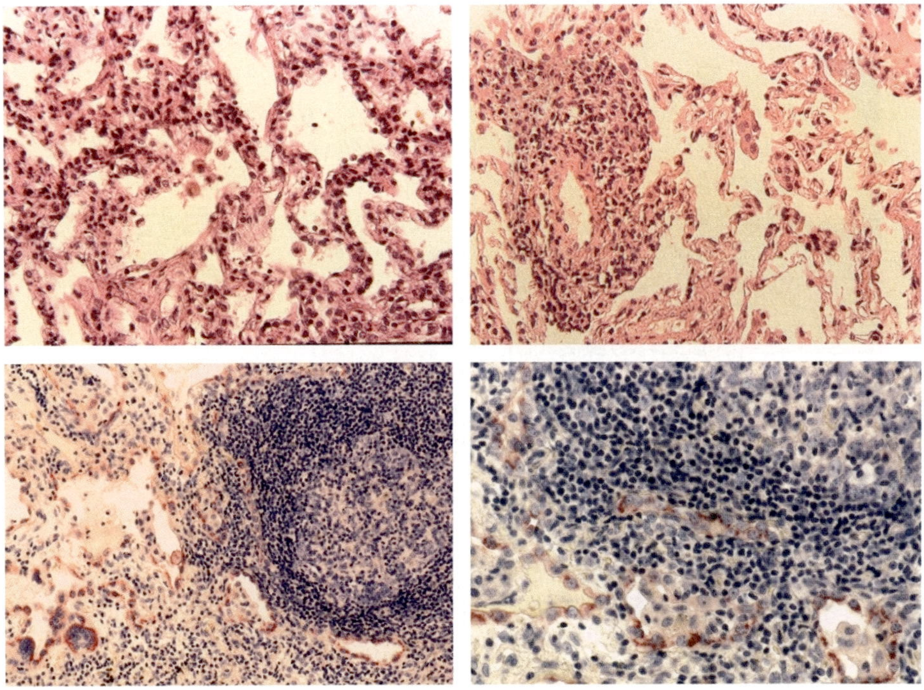

Plate 13 Top row: non-specific interstitial pneumonitis (NIP) in patient with acute necrotizing encephalitis following primary HHV-6 infection (Wagner et al., 1997). Bottom row: lymphoid interstitial pneumonitis (LIP) in patient with HIV infection and HHV-6 reactivation red cells immunohistochemical APAAP reaction for HHV-6 p41 antigen (courtesy of G. Krueger, Immunopathology Laboratory, University of Cologne, Germany). (see page 180).

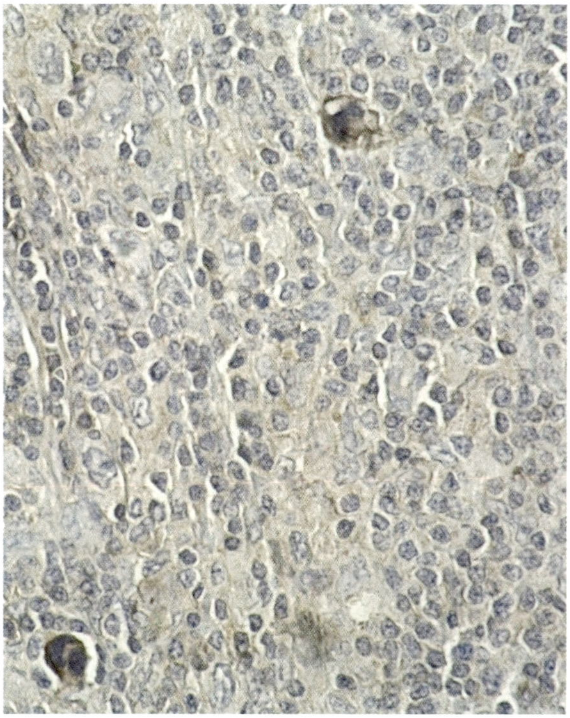

Plate 14 Expression of DR7 protein in tissue samples from patients suffering from HD, previously detected positive by HHV-6 structural monoclonal antibodies. DR7 was strongly positive and was principally found in RS cells, and to a lesser extent in other lymphoid cells. (see page 196).

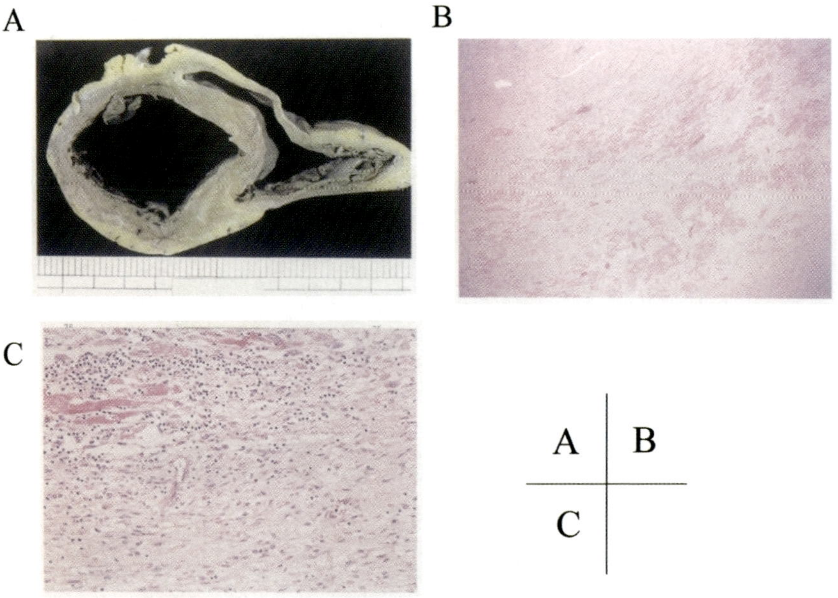

Plate 15 (A) Postmortem specimen showing a dilatation of the left ventricular cavity and thinning of the left ventricle and intraventricular septum; (B) A loss of myofibers with interstitial fibrosis (HE stain, × 40); and (C) Inflammatory cellular infiltration (mainly rounded cell) (HE stain, × 1 0 0) was observed in the left ventricle. (see page 229).

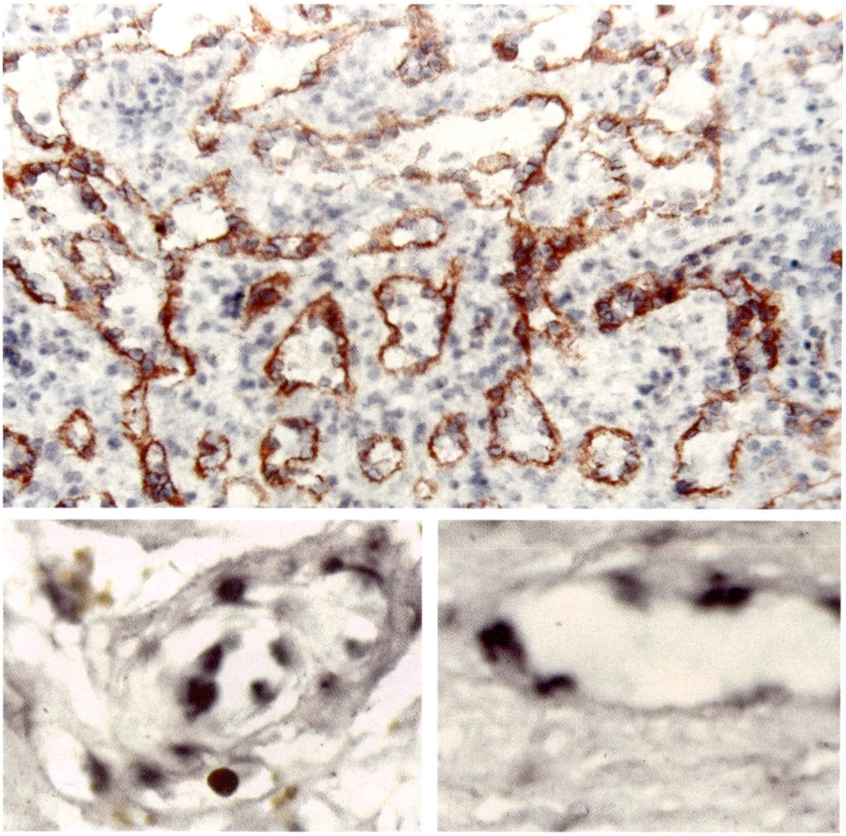

Plate 16 HHV-6 infection of vascular endothelial cells. *Top*: Splenic sinusoidal endothelial cells containing HHV-6 late antigens (red-stained cells; APAAP reaction using HAR 1-3 antibody). *Bottom*: HHV-6 DNA in endothelial cells of cardiac arteriole in an AIDS patient (left) and of a brain venule (right) in a case of necrotizing encephalitis in a child with active HHV-6 infection (black cells; *in situ* hybridization with pZVH14 probe). (see page 235).

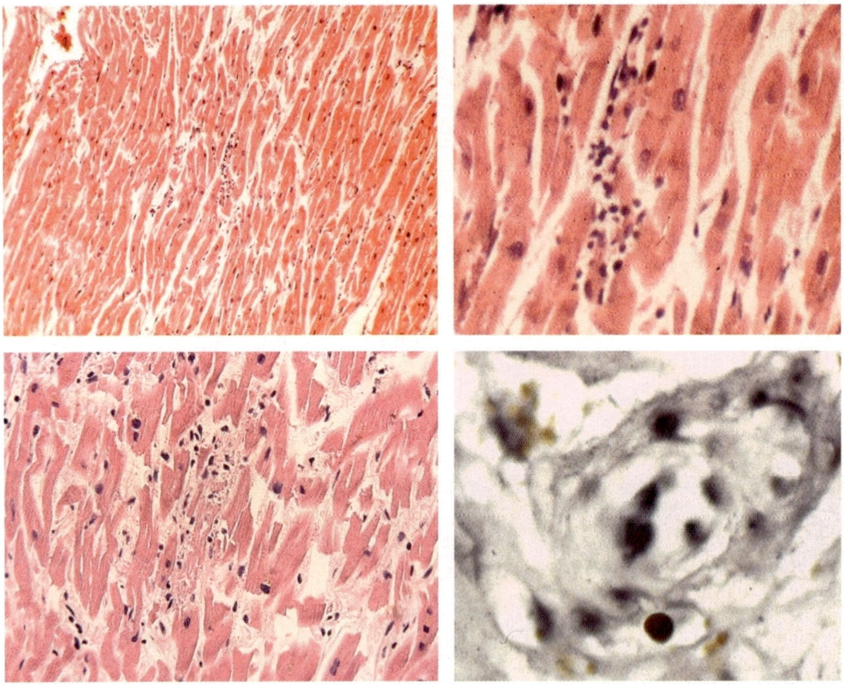

Plate 17 Various forms of myocarditis accompanying HHV-6 reactivation in AIDS patients (hematoxylin and eosin stain of autopsy specimens). *Lower right* shows an interstitial cardiac arteriole from such cases containing HHV-6 DNA (*in situ* hybridization with pZVH14 probe). (see page 237).

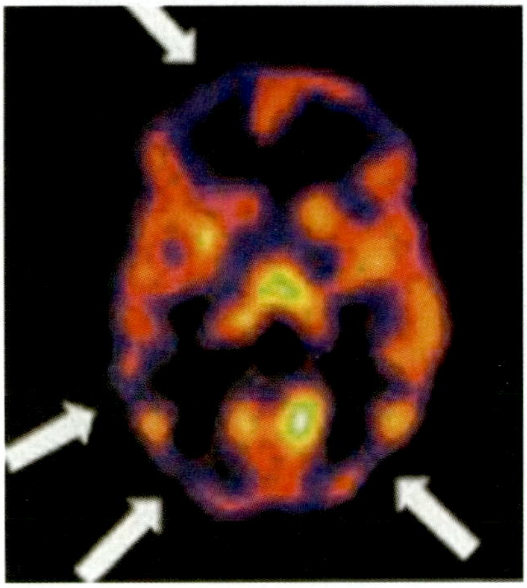

Plate 18 Axial SPECT image showing multiple foci of decreased perfusion (arrows) in the brain. (see page 257).

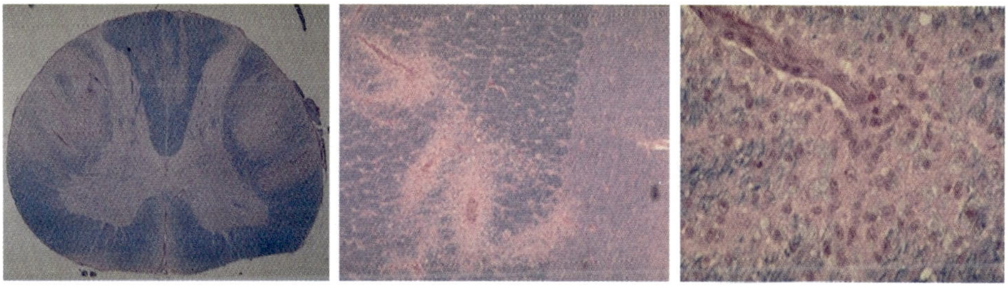

Plate 19 Neuropathology of acute *C. jacchus* EAE. Perivascular inflammatory demyelinating infiltrates in spinal cord and brain periventricular white matter (left, middle). High-power view to show monocyte/macrophage infiltration (LFB/PAS). (see page 310).

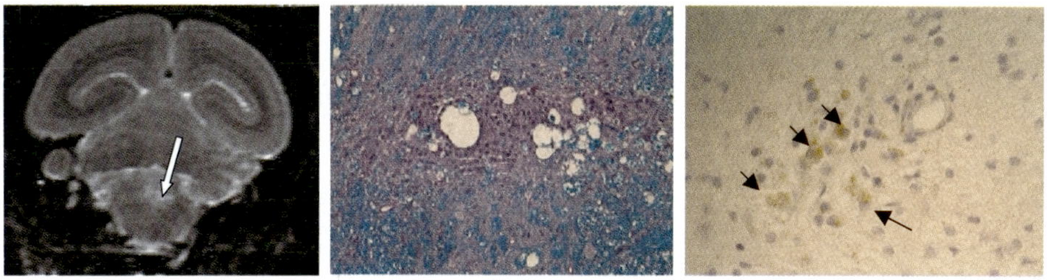

(A) (B) (C)

Plate 20 A, Coronal MRI section showing T2 hyper-intensity lateral to median CSF space in pons (arrow). B, Corresponding demyelinating inflammatory infiltrate (animal 190-94; LFB/PAS, also see Fig. 1). C, Staining for early nuclear antigen p41/p38 (Advanced Biotechnologies, Inc.), demonstrating viral persistence/replication within lesions. Positively stained cells (arrows) have not yet been formally identified but appear to be oligodendrocytes. (see page 313).

List of Contributors

D.V. Ablashi
HHV-6 Foundation
285 San Ysidro Road
Santa Barbara, CA 93108, USA

H. Agut
Department of Virology
EA 2387, Pierre et Marie Curie Paris 6 University
CERVI, Pitie-Salpetriere Hospital
83 boulevard de l'Hopital
75013 Paris, France

N. Ashizawa
Department of Cardiovascular Medicine
Graduate School of Biomedical Sciences
Nagasaki University, Sakamoto 1-12-4
Nagasaki 852-8523, Japan

L. Atwell
Sierra Internal Medicine
865 Tahoe Blvd., #306, Incline Village
NV 89451, USA

D. Bordessoule
Service d'Hematologie et de Therapie Cellulaire
CHRU de Limoges
87042 Limoges cedex, France

B. Bradel-Tretheway
Department of Microbiology and Immunology
The University of Rochester Medical Center
575 Elmwood Avenue
Rochester, NY 14642, USA

J.H. Brewer
Section of Infectious Disease – St. Luke's Hospital and Plaza Infectious Disease
4320 Wornall Road, Suite 440
Kansas City, MO 64111, USA

J. Britz
Cylex Inc.
8980-I Old Annapolis Road
Columbia, MD 21045, USA

L.M. Buja
Health Science Center at Houston
The University of Texas
7000 Fannin St., UCT Suite 1715
Houston, TX 77030, USA

L. De Bolle
Pharmaceutical Care Unit
Faculty of Pharmaceutical Sciences
Ghent University, Harelbekestraat 72
B-9000 Gent, Belgium

L. Dagna
Unit of Human Virology
Department of Biological and Technological Research (DIBIT)
Universita' Vita-Salute San Raffaele School of Medicine
San Raffaele Scientific Institute
20132 Milan, Italy

E. De Clercq
Rega Institute
Katholieke Universiteit Leuven
Minderbroedersstraat 10
B-3000 Leuven, Belgium

F. Denis
Department of Virology
Equipe d'accueil EA 4021
Dupuytren University Teaching Hospital
87042 Limoges, France

V. Descamps
Department of Dermatology
Bichat Claude Bernard Hospital
Assistrance Publique—Hopitaux de Paris
46 rue Henri Huchard
F-75018 Paris, France

S. Dewhurst
Department of Microbiology and
Immunology
University of Rochester Medical School
575 Elmwood Avenue
Rm 3-9609 (box 672)
Rochester, NY 16642, USA

S. Fukae
Department of Internal Medicine
Nagasaki University School of Medicine
Sakamoto 1-12-4, Nagasaki 852-8523, Japan

R.C. Gallo
Institute of Human Virology
University of Maryland Baltimore
725 West Lombard Street
Baltimore, MD 21201, USA

A. Gautheret-Dejean
Department of Virology
EA 2387
Pierre et Marie Curie Paris 6 University
CERVI, Pitie-Salpetriere Hospital
83 boulevard de l'Hopital
75013 Paris, France

C.P. Genain
Department of Neurology
University of California at San Francisco
513 Parnassus Avenue
San Francisco
CA 94143-0435, USA

R. Glaser
Institute for Behavioral Medicine Research
Kathryn & Gilbert Mitchell Chair in Medicine, Molecular Virology, Immunology and Medical Genetics
The Ohio State University Medical Center
2175 Graves Hall, 333 W. 10th Avenue
Columbus, OH 43210, USA

U.A. Gompels
Department of Infectious and Tropical Diseases
London School of Hygiene and Tropical Medicine
University of London Keppel Street
London WC1E 7HT, UK

A.D. Goodman
Multiple Sclerosis Center
Chief, Neuroimmunology Unit
Department of Neurology
University of Rochester Medical Center
601 Elmwood Avenue, Rm. 6-8521, Box 605
Rochester, NY 14642, USA

C.B. Hall
Department of Pediatrics and Medicine, Infectious Diseases
University of Rochester School of Medicine & Dentistry
601 Elmwood Avenue, Box 689
Rochester, NY 14642, USA

S. Jacobson
Chief, Viral Immunology Section, Neuroimmunology Branch
Bld. 10, Rm 5N214
National Institute of Neurological Disorders and Stroke
NIH, Bethesda, MD 20892, USA

F.C. Kasolo
University Teaching Hospital
Lusaka, Zambia; and
AFRO VPD
World Health Organisation
Harare, Zimbabwe

G.R.F. Krueger
Department of Pathology & Laboratory Medicine
The University of Texas—Houston Medical School
6431 Fannin St., MSB 2.246
Houston, TX 77030, USA

A. Lacroix
Department of Virology
Equipe d'accueil EA 4021
Dupuytren University Teaching Hospital
87042 Limoges, France

P. Ljungman
Department of Hematology
Karolinska University Hospital
Stockholm 14186, Sweden

P. Lusso
Unit of Human Virology
Department of Biological and Technical Research
San Raffaele Scientific Institute
Via Olgettina 58, 20132 Milan, Italy

E. Mahe
Department of Dermatology
Bichat Claude Bernard Hospital
Assistance Publique – Hôpitaux de Paris 46 rue Henri Huchard
75018 Paris, France

Yasuko Mori
Laboratory of Virology and Vaccinology
National Institute of Biomedical Innovation
7-6-8, Saito-Asagi, Ibaraki
Osaka 567-0085, Japan

L. Naesens
Rega Institute
Katholieke Universiteit Leuven
Minderbroedersstraat 10
B-3000 Leuven, Belgium

D.L. Peterson
Sierra Internal Medicine
865 Tahoe Blvd., #306, Incline Village
NV 89451, USA

S. Ranger-Rogez
Department of Virology
Equipe d'accueil EA 4021
Dupuytren University Teaching Hospital
87042 Limoges, France

F. Santoro
Unit of Human Virology
Department of Biological and Technological Research (DIBIT)
San Raffaele Scientific Institute
Via Olgettina 58, 20132 Milan, Italy

S.M. Schmidt
Abteilung Allgemeine Paediatrie
Zentrum fuer Kinder—und Jugendmedizin
Ernst-Moritz-Arndt Universitaet
Soldmannstrasse 15, 17475 Greifswald, Germany

B. Schneider
Immunonephrology Laboratory (Lab 414)
Department of Research
University Hospital Basel
Hebelstrasse 20, Basel 4031, Switzerland

N. Singh
Infectious Disease Section
VA Medical Center
University Dr. C, Pittsburgh
PA 15240, USA

S.S. Soldan
Department of Neurology
University of Pennsylvania
415 Curie Boulevard, CRB Room 255
Philadelphia, PA 19104, USA

J.K. Stoops
Department of Pathology & Laboratory Medicine
The University of Texas—Houston Medical School
6431 Fannin St., MSB 2.216
Houston, TX 77030, USA

G. Wang
Department of Bioengineering
Rice University, 6100 Main
Houston, TX 77005, USA

S.K.W. Wiersbitzky
Klinik und Poliklinik fuer Kinder—und Jugendmedizin
Ernst-Moritz-Arndt Universitaet
Soldmannstrasse 15, 17475 Greifwald, Germany

H. Wiersbitzky
Funktionsbereich Kinderradiologie
Zentrum fuer Radiologie
Ernst-Moritz-Arndt Universitaet
Soldmannstrasse 15, 17475 Greifswald, Germany

M.V. Williams
Department of Molecular Virology, Immunology and Medical Genetics
College of Medicine and Public Health
Ohio State University
2094 Graves Hall, 333 West Tenth Avenue
Columbus, OH 43210, USA

T. Yoshikawa
Department of Pediatrics
Fujita Health University School of Medicine
Toyoake, Aichi 470 1192, Japan

Z.H. Zhou
Department of Pathology & Laboratory Medicine
The University of Texas—Houston Medical School
6431 Fannin St., MSB 2.280
Houston, TX 77030, USA

Index